Peter A. Rinck

Magnetic Resonance in Medicine

The Basic Textbook of the European Magnetic Resonance Forum

4th, Completely Revised Edition

Including MR Image Expert, version 2.5 and Dynalize, version 1.0 Demo

With contributions by

Atle Bjørnerud, Patricia de Francisco, Jürgen Hennig, Richard A. Jones, Jørn Kværness, Robert N. Muller, Timothy E. Southon, and Geir Torheim

290 figures, 45 tables

Blackwell Wissenschafts-Verlag Berlin · Vienna 2001
Boston · Copenhagen · Edinburgh · London · Melbourne · Oxford · Tokyo

Magnetic Resonance in Medicine: The Basic Textbook of the European Magnetic Resonance Forum. ed. by Peter A. Rinck. 4th completely revised edition.

© 2001 by EMRF Foundation, Minusio, Switzerland.

The third English edition was published as: Rinck PA (ed.) Magnetic Resonance in Medicine. Oxford: Blackwell Scientific Publications. 1993.

First German edition:
Rinck PA, Petersen SB, Muller RN (eds): Magnetresonanz-Imaging und -Spektroskopie in der Medizin. Stuttgart, New York: Georg Thieme Verlag. 1986.

First Spanish edition:
Rinck PA, Petersen SB, Muller RN: Introducción a la Resonancia Magnética Biomédica. Buenos Aires: Anejo Producciones 1986.

First Japanese edition:
Rinck PA, Petersen SB, Muller RN (eds): An Introduction to Biomedical Nuclear Magnetic Resonance. Niigata: Nishimura Corp. 1991.

First Italian edition:
Rinck PA (ed): Risonanza Magnetica in Medicina. III edizione. Edizione italiana a cura di A. Giovagnoni. Naples: Guido Gnocchi Editore. 1994.

First Russian edition:
Ринкк ПА: Магнитный Резонанс в Медицине. Основной учебник Европейского Форума по магнитному резонанс. Oxford: Blackwell Scientific. 1995.

A catalogue record for this title is available from the British Library.

The names *European Workshop on Magnetic Resonance in Medicine* and *European Magnetic Resonance Forum* and their respective logos are the property of the European Magnetic Resonance Forum Foundation, Swiss Office, P.O. Box 1235, CH-6648 Minusio-Locarno, Switzerland.

Please visit the EMRF Foundation's home page at www.emrf.org.

Important note:

Medicine is an ever-changing science. Research and clinical experience are continually broadening our knowledge, in particular the knowledge of proper treatment and drug therapy. Insofar as this book mentions any dosage or application, or proposes certain imaging or spectroscopy procedures, readers may rest assured that the editor, the co-authors, and the publisher have made every effort to ensure that such references are strictly in accordance with the state of knowledge at the time of production of this book.

Some of the product names, patents, and registered designs referred to in this book are in fact registered trademarks or proprietary names even though specific reference to this fact is not always made in the text. Therefore, the appearance of a name without designation as proprietary is not to be construed as a representation by the editor or publisher that it is in the public domain.

Printed in Germany.
ISBN 0-632-05986-9

Peter A. Rinck

Magnetic Resonance in Medicine

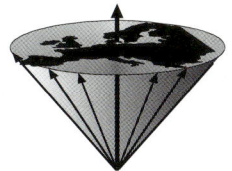

Foundation
Fondation
Fondazione
Stiftung

EUROPEAN MAGNETIC RESONANCE FORUM

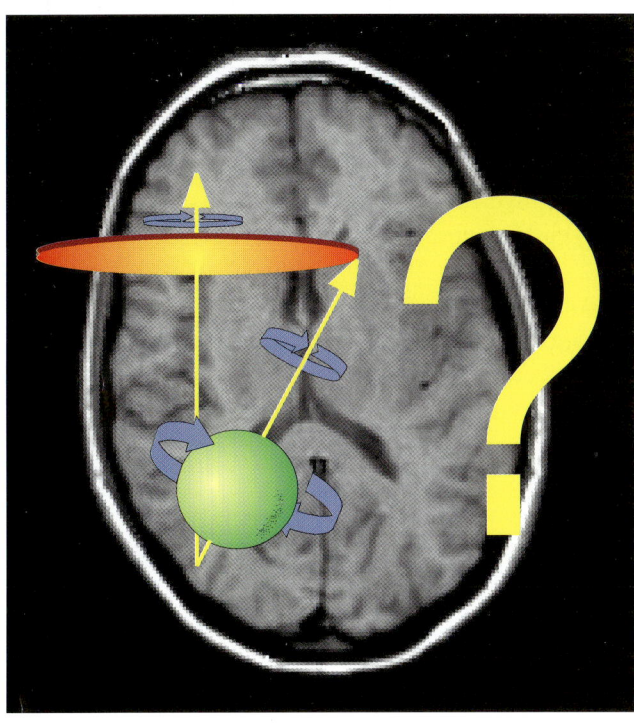

Foreword

»Why, sometimes I've believed as many as six impossible things before breakfast.«

The White Queen
in Lewis Caroll's *Alice Through the Looking Glass.*

In magnetic resonance (MR) imaging beginners and even the more knowledgeable often have to believe what they read or hear about the topic — and often things seem to be impossible. Sometimes it seems like a fairy tale. This book tries to tell the tale and lead its readers from believing to understanding.

Readers should be able to acquire a basic knowledge which enables them to pursue studies of their own and to cope with some of the most common problems, such as image contrast and artifacts or questions concerning possible hazards to patients. The editor and contributors have not attempted to cover the field completely nor to be exhaustive in the topics discussed, as the field of magnetic resonance is in a permanent stage of development and therefore changing day by day.

The intention of this book is rather to lead the readers to a fundamental understanding to be able to explore the details and new developments by themselves.

The first version of this book was written in 1982 and published by the University of Mons Press. It was intended to be used as the primer for the first European Workshop on Magnetic Resonance in Medicine and later the European Magnetic Resonance Forum to teach the basics and some of the applications of MR imaging and MR spectroscopy. Many organizers of other teaching courses and universities the world over have found it valuable for their students.

After Paul C. Lauterbur, the father of MR imaging, saw the first edition of this book, he wrote back to us:

»It looks like a fine book, especially for residents, nurses, and technicians.«

Initially we thought this statement was not very encouraging, but in hindsight this was exactly what we had intended to write: an introduction for beginners.

The public reaction to and the reviews of the last edition were extremely positive. The book is very popular with radiologists, cardiologists, medical students and radiographers, who encouraged us to revise it again. The text and all the figures have been completely redone and new chapters have been added.

The text has been written in such a way that newcomers to the field will be provided with the concepts, explanations and illustrations necessary to master the basics of magnetic resonance in medicine.

As with everything in life, MR imaging does not only require knowledge of facts but also background information for decision making. Therefore in this edition we have included some subjective, critical, and opinion-oriented sections, intended to offset the technical nature of the book and provide some insights into more practical questions faced by MR users.

A new addition is **MR Image Expert**, an interactive learning and teaching tool for MR imaging, in particular for facilitating the understanding of MR image contrast. It simulates the most important aspects of MR imaging on a personal computer. Teaching tools such as textbooks and slide presentations are used to introduce the newcomer to the field, but the best teacher would be an MR imaging machine itself. However, learning by experience is time-consuming, expensive, and in cases with human pathology, impossible. Thus, the next best tool is an MR imaging simulator, a computer program which simulates an MRI machine without performing real examinations. **MR Image Expert** is such a simulator.

Dynalize is the second software included. It allows the presentation and analysis of dynamic MR imaging studies.

One problem has increased during recent years: differences between the equipment of manufacturers. Some manufacturers have even developed their own specialized terminology which is only applicable to their machines. No standardization seems to be envisaged. We have tried to explain the most common techniques which have been implemented in most machines and have focused upon the basics. We have left the bulk of applications in medicine to others who have written outstanding books on applications in many medical disciplines.

We are grateful to Knut Nordli for his comments. Karen Janssen and Marc Kouwenhoven of Philips Medical Systems have kindly supplied a number of images for the angiography chapter.

All previous editions have been published in English, some of them having been translated into German, French, Italian, Polish, Russian, Spanish, and Japanese. We intend to have this new edition also translated into other languages and hope these translations will be available in the near future.

If you find any mistakes in this book, rest assured that they were left intentionally so as not to provoke the gods with something which is perfect.

Anyway, we appreciate any comments concerning this book. Please forward them to the

European Magnetic Resonance Forum Foundation
Belgium Office
University of Mons-Hainaut
Faculty of Medicine and Pharmacy
NMR Laboratory
B-7000 Mons
Belgium
or e-mail them to: office@emrf.org

March 2001
Peter A. Rinck and the contributors

Contents

Prelude

How it all began

The *Finale* after Chapter 18 tells more about the history of MR imaging.

In late 1972, a prospective contributor to the British scientific journal, *Nature*, received a letter from the editor of the journal that read as follows:

> »With regret I am returning your manuscript which we feel is not of sufficiently wide significance for inclusion in *Nature*. This action should not in any way be regarded as an adverse criticism of your work, nor even an indication of editorial policies on studies in this field. A choice must inevitably be made from the many contributions received; it is not even possible to accommodate all those manuscripts which are recommended for publication by the referees.«

The paper submitted was very short and described a new imaging technique dubbed *zeugmatography*. For those who did not study Greek at school, *zeugma* is the yoke, or as the author put it: »That what is used for joining.«

The author did not mean that two horses were to be joined with a yoke; rather, he meant two magnetic fields were to be joined. His method was derived from an analytical technique that had been used in chemistry since the late 1940s, called *nuclear magnetic resonance*, or, for short, *NMR*.

The author of the paper was Paul C. Lauterbur, Professor of Chemistry at the State University of New York at Stony Brook. Lauterbur wanted this paper to be published in *Nature* and wrote back to the editor proposing to change the style of the paper:

> »Several of my colleagues have suggested that the style of the manuscript was too dry and spare, and that the more exuberant prose style of the grant application would have been more appropriate. If you should agree, after reconsideration, that the substance meets your standards, ... I would be willing to incorporate some of the material below in a revised manuscript ...«

The answer from the editor was short and positive:

> »Would it be possible to modify the manuscript so as to make the applications more clear ?« [1]

Finally, the paper was accepted and published in the 16 March 1973 issue of *Nature* under the title:

Image Formation by Induced Local Interaction: Examples Employing Magnetic Resonance.

From reading this title, one would not think that a revolutionary idea in medical imaging was hidden behind it. However, this idea was the foundation of MR imaging, which has developed into one of the most outstanding medical innovations of the twentieth century, only comparable with Wilhelm Conrad Roentgen's invention of the medical application of x-rays.

Magnetic resonance, or nuclear magnetic resonance (NMR) as natural scientists still call it, is a phenomenon that was first mentioned in the scientific literature more than 50 years ago. In 1946, independently of each other, two scientists in the United States described a physico-chemical phenomenon that was based upon the magnetic properties of certain nuclei in the periodic system [2,3]. They found that when these nuclei were placed in a magnetic field, they absorbed energy in the radiofrequency range and re-emitted this energy during the transition to their original orientation. Because the strength of the magnetic field and the radiofrequency must match each other, the phenomenon was called *nuclear magnetic resonance: nuclear* because it is only the nuclei of the atoms that react; *magnetic* because it happens in a magnetic field; and *resonance* because of the direct dependence of field strength and frequency.

The two scientists, Felix Bloch working at Stanford University and Edward M. Purcell working at Harvard, received the Nobel Prize for Physics in 1952. In 1991, the Nobel Prize in Chemistry was awarded to Richard R. Ernst of Zurich for his contributions to the field of NMR spectroscopy.

With the introduction of NMR to clinical imaging, the adjective *nuclear* was dropped by marketing people and radiologists because it sounded like *nuclear warfare* or *nuclear power plant*, with which NMR has nothing in common at all. It was thought that the general public would be unable to distinguish between one *nuclear* and the other. Thus, today we talk about MR imaging and MR spectroscopy.

However, it should always be kept in mind that it is the nucleus we talk about because there is another kind of resonance that also can be used for imaging: *electron spin resonance (ESR)*. ESR involves the electrons of an atom.

NMR signals carry encoded information about the physical and chemical environment of the nuclei. Originally, NMR was used as an analytical method to study the composition of chemical compounds. Today, there are applications in a wide range of areas in chemistry, physics, biology, medicine, and food science.

However, nobody could determine from where within a sample the NMR signal stems. It could originate at the left or right end, at the top or at the bottom. Lauterbur's zeugmatography changed this. He joined the strong magnetic field and a second weaker field, the gradient field. Because the strength of the magnetic field is proportional to the radiofrequency, the frequency of, for instance, a hydrogen nucleus of a water molecule at one end of a sample differs from the signal of another hydrogen nucleus at the other end of the sample. Thus, the location of these nuclei can be calculated. Once their location is known, an image can be created of a slice through a human body, for example. Basically, therefore, MR imaging requires a strong static magnetic field produced by a large magnet, a second weaker magnetic field that varies across the sample, a radio transmitter and receiver, and a powerful computer to calculate an image.

Compared to x-ray and radioisotope methods, MR imaging uses energy on the opposite end of the electromagnetic spectrum, and to date, no permanent harmful side effects of MR imaging have been reported. The energy of MR imaging is nine orders of magnitude lower than that of x-rays and radioisotope techniques.

Although Lauterbur did not suggest distinct applications of the new technique in his paper, he did refer to the fact that it had been shown that cancer tissue had different signal properties compared to normal tissue, and he believed that zeugmatography could be used for medical imaging. Thus, he urged his university to file a patent application, but because neither the university patent lawyer nor the university administration itself believed in his idea, no patent application was filed and Lauterbur never obtained a patent on his invention.

Despite the nonbelievers within the university, it only took eight years for the first whole-body MR machines to appear in clinical settings, although these machines were crude prototypes compared to today's equipment. In 1982, approximately a dozen research groups worked with whole-body imagers. Today, approximately 20,000 machines are operated worldwide, the majority of them in the United States and Japan, a quarter in Europe.

The hope that MR imaging, or other adaptations of MR in medicine, would be able to characterize cancerous cells in the body has not come true, but many other important applications of MR imaging have been found during the last decade. Today, MR imaging influences decisions in most areas of medicine, from neurology to orthopedics, from pediatrics to radiation therapy. Because of its high soft-tissue contrast, its lack of side effects, its three-dimensional capabilities, and its high patient acceptability, more and more patients are referred for MR examinations. MR imaging is even more interdisciplinary than roentgenology, although it is also complex and sophisticated.

References

1. Hollis DP. Abusing cancer science. Chehalis, WA. The Strawberry Fields Press 1987.
2. Bloch F, Hanson WW, Packard M. Nuclear induction. Phys Rev 1946; 69: 127.
3. Purcell EM, Torrey HC, Pound RV. Resonance absorption by nuclear magnetic moments in a solid. Phys Rev 1946; 69: 37-38.

Chapter One — Introductory Fundamentals

Magnetism and Electricity

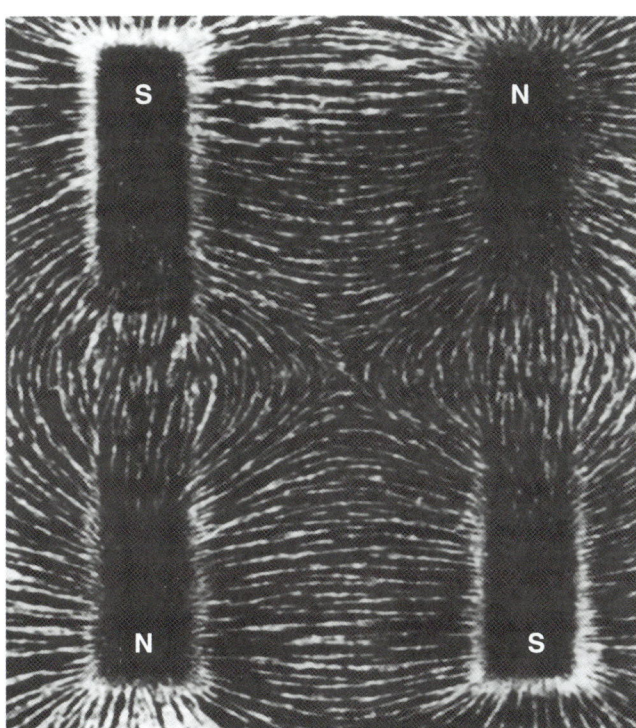

Figure 1-1:
Making magnetic field lines visible: iron filings sprinkled on a pane of glass resting on two bar magnets are oriented by lines of magnetic force marking their direction and locations. They are concentrated at the ends of the magnet where the field is most intense (N = magnetic north pole, S = magnetic south pole).

Magnetic Units	
Quantity	Magnetic Flux Density (magnetic induction)
Symbol	B
SI name	Tesla
SI symbol	T
prior name	Gauss
prior symbol	G
Quantity	Magnetic Field Strength
Symbol	H
Quantity	Magnetization
Symbol	M or H_1
SI name	ampere/meter
SI symbol	A/m

Table 1-1:
Fundamental magnetic units.

This book aims to explain MR imaging, a technology which is based upon magnetism and electricity. To comprehend the theory and practice of MR imaging, one has to understand the fundamentals of magnetism and electricity.

Thus, at the beginning of this book, we will review and remind you of some of them.

The word 'magnetism' is derived from Magnesia, a town in the Western part of Asia Minor, close to the Turkish city of Izmir. According to Pliny the Elder, the shepherd Magnes was walking in the mountains at around 1000 BC and was drawn to the earth by the tacks in his sandals. When he investigated the cause he discovered that lodestone, a magnetic oxide of iron, was responsible for this attraction. He also could magnetize metal by rubbing lodestone on it.

This happened 3000 years ago and it has taken all this time to try to explain the phenomenon. Still today, magnetism has a touch of witchcraft about it because it is created by something which is not directly visible (Figure 1-1).

Magnetic phenomena are fundamentally electrical in nature. Since the first half of the nineteenth century it had been known that electric currents create magnetic fields. However, only 100 years later Werner Heisenberg was able to show that ferromagnets owed their properties to cooperative electrical actions of the atoms.

For physicists, magnetic field strength is measured in ampere per meter (A/m), whereas the magnetic flux density is measured in Gauss (G), the ancient unit, or Tesla (T), the modern SI unit. 1 T equals 10 kG or 10,000 G. Often the unit Tesla is loosely used for magnetic field strength, as we do in this book too. Table 1-1 gives an overview of the units.

The earth's magnetic field, at the equator, is approximately 0.5 G or 0.00005 T. The field of an electric can opener is approximately 0.2 G and of a computer monitor 0.1 G (both at 30 cm distance).

Most clinical MR imaging machines operate between 1,000 and 15,000 G or 0.1 and 1.5 T. At low fields one finds permanent magnets, at medium and high fields the equipment is resistive or superconducting — both of them different kinds of electromagnets.

Interactions between Magnetism and Electricity

The magnetic field of a permanent magnet is easily perceivable; the field created by electric current is slightly more difficult. A fundamental discovery which opened the way to understanding magnetic phenomena was made in 1820 by the Danish physicist Hans Christian Oersted, who demonstrated that a compass needle was deflected from its normal north-south orientation when a current-carrying wire was placed parallel to the needle.

The deflection follows the direction of the current and the resulting magnetic field surrounds the wire (Figure 1-2).

If one passes electric current in the same direction through two wires parallel to each other, they will attract each other; if the current is passed in opposite directions, the wires will move away from each other.

In other words, the electric current in the wire creates a magnetic field depending on the direction of the current (Figure 1-3). If the direction of the current is indicated by the thumb of the right hand, the direction of the magnetic field corresponds to the direction of the fingers.

But the reverse is also true: a magnetic field creates movement of electrons. If we have an oscillating magnetic field around a wire, a voltage is induced and there will be electric current in the wire — which is the principle of an electric generator. In MR imaging such a current is induced and gives rise to the MR signal.

As in a generator, the greater the magnetic field, the greater the voltage will be.

The magnetic field around a wire or a wire loop diminishes non-linearly with distance.

However, when one places two wire loops with opposite currents flowing in the distance of their diameters parallel to each other, one can create a linearly changing magnetic field between them. This field in the middle between the two loops is also described as a field with a constant field gradient (Figure 1-4).

Both static and gradient magnetic fields are essential for MR imaging.

The third ingredient of MR imaging consists of radiofrequency (RF) waves.

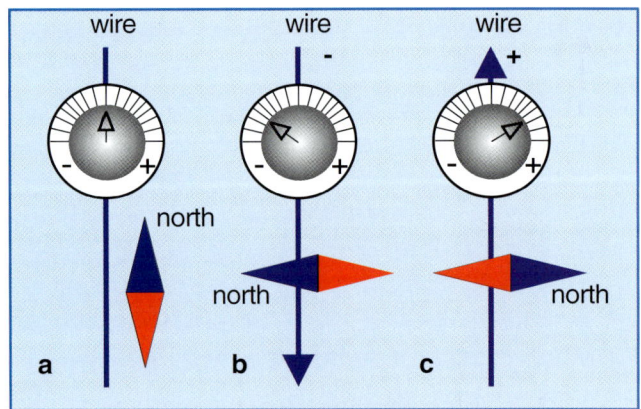

Figure 1-2:
Deflection of a compass needle placed parallel to an electric current-carrying wire. (a) = no current; (b) = current flowing to the bottom; (c) = current flowing to the top.

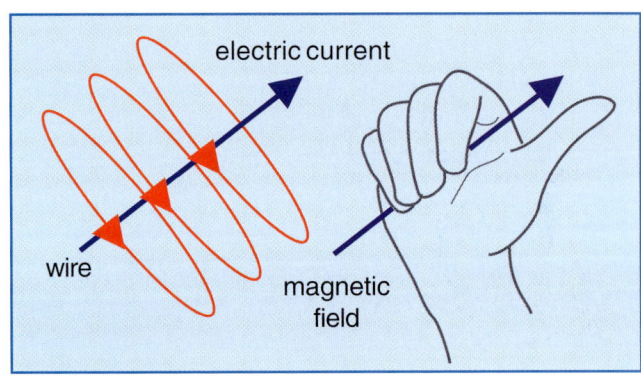

Figure 1-3:
The right-hand rule: the electric current moves in the direction of the thumb of the right hand, the magnetic field spreads perpendicularly in the direction of the fingers.

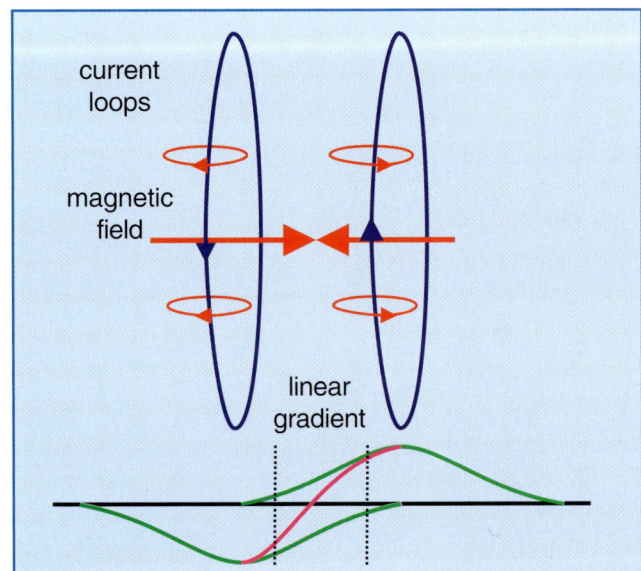

Figure 1-4:
Overlapping magnetic fields (green) with linear gradient (magenta).

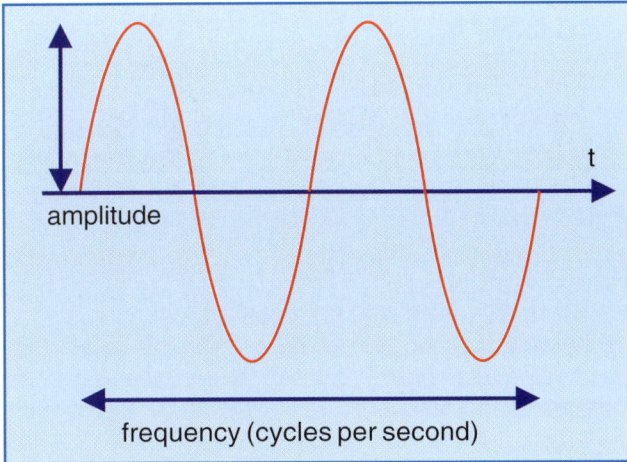

Figure 1-5:
Amplitude, frequency, and phase (t = time).

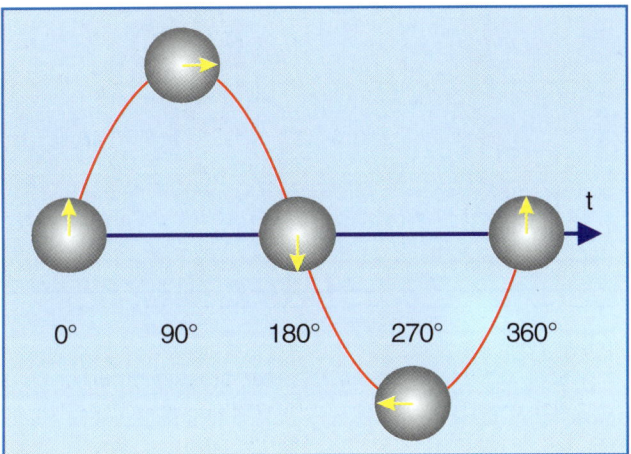

Figure 1-6:
Phase (t = time).

The Signal and its Components

Magnetic resonance signals are time-varying electric currents or voltages. They come in waves which are induced by an oscillating magnetic field. The form of the wave reflects the information in the signal.

Basically, magnetic resonance signals are sine and cosine waves which are described and defined by the three factors: *amplitude*, *frequency*, and *phase*.

Amplitude is also called *signal strength*. It reflects the final brightness of a picture element of a magnetic resonance image. Frequency and phase determine shape and spatial detail in an MR image. *Amplitude* is the difference between the peak of the curve and zero (Figure 1-5). Its usual unit is volt (V), in our context usually mV.

Frequency is the number of complete cycles per second. It is measured in Hertz (Hz; 1 Hz = 1 cycle per second). In magnetic resonance imaging we usually deal with kilohertz (1 kHz = 1,000 Hz) or megahertz (MHz = 1,000,000 Hz).

Phase specifies the initial amplitude of a wave. Phase can only be compared for waves with the same frequency. Phase differences are expressed in degrees. One complete cycle equals 360° (Figure 1-6).

If there are two waves with the same frequency which are shifted from each other, there is a *phase shift* or *phase offset* (Figures 1-7 and 1-8).

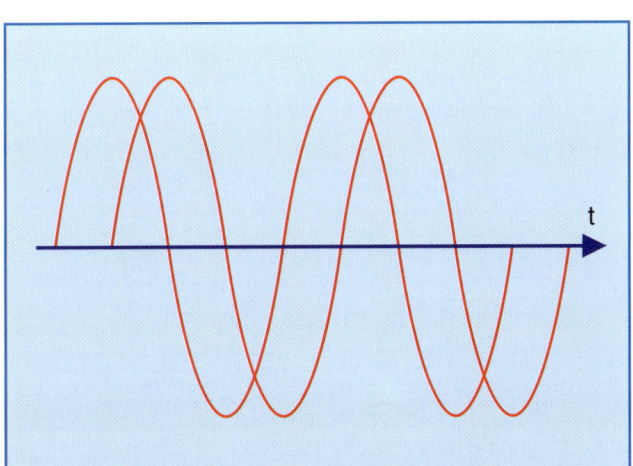

Figure 1-7:
Phase shift. Two sine waves are shifted 90° from each other.

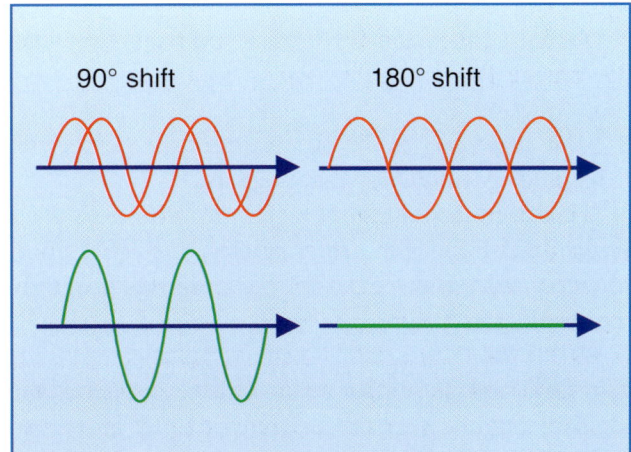

Figure 1-8:
Phase shift and resultant voltage changes (green).

Pulse, Bandwidth, and Fourier Transform

Transmission of signals without wire is done not by changing electric currents but by changing electromagnetic radiation. In magnetic resonance, radiation is not transmitted continuously but chopped into *pulses*. These pulses can be arranged in different pulse sequences.

The form of the pulses is pivotal for its desired purpose. There are *selective* pulses which are used to create a single slice through an object to be examined, and *nonselective* pulses which can excite the entire object — they are used in three-dimensional imaging. Figure 1-9 shows a simple hard pulse which is of minor importance in magnetic resonance imaging. To improve the quality of pulses we have to shape them, i.e. vary their amplitude with time (Figures 1-9 and 1-10). Widely used pulse shapes include the *Gaussian* and the *sinc* pulses, the latter one giving a better slice profile. The sinc pulse is defined as sinc(x) = sin(x)/x.

Bandwidth is the range of frequencies included in the pulse. A certain bandwidth is needed for any signal. In general, bandwidth is directly proportional to the amount of data transmitted or received per unit time. In analog systems, bandwidth is defined in terms of the difference between the highest-frequency signal component and the lowest-frequency signal component. A typical voice signal has a bandwidth of approximately three kilohertz (3 kHz); an analog television (TV) broadcast video signal has a bandwidth of six megahertz (6 MHz) — some 2,000 times as wide as the voice signal.

Signals, for instance RF pulses, possess a certain waveform. Basically, they vary with time: they are a time function. However, it might be easier and better to work and analyze the properties of the pulse with respect to its frequency components.

This conversion can be achieved with a *Fourier Transform* (FT). The mathematics of the FT is complicated and intricate, but for our purposes it is only the results that count.

While the Fourier transform of a Gaussian pulse is also Gaussian, the FT of a sinc pulse approaches the ideal rectangular slice profile (Figure 1-10). However, even the sinc pulse is not the optimum pulse for a number of MR pulse sequences; thus, many alternatives have been developed.

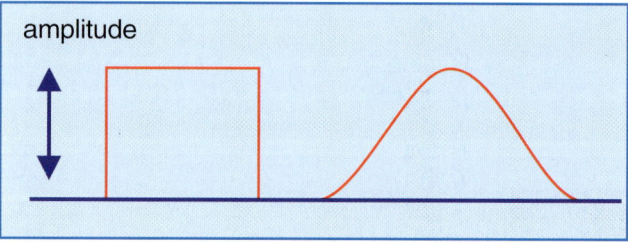

Figure 1-9:
Hard pulse (left) and shaped pulse (right).

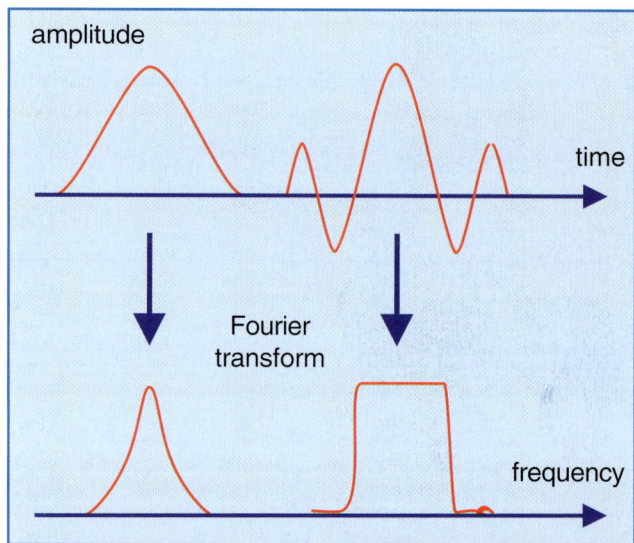

Figure 1-10:
Gaussian (left) and sinc pulses (right). Whereas the Fourier transform of the Gaussian pulse leads to a Gaussian shape, the Fourier transform of the sinc pulse comes close to a rectangular shape. This is more convenient in MR imaging because it allows a better definition of a slice through the human body.

Chapter Two

Nuclear Magnetic Resonance

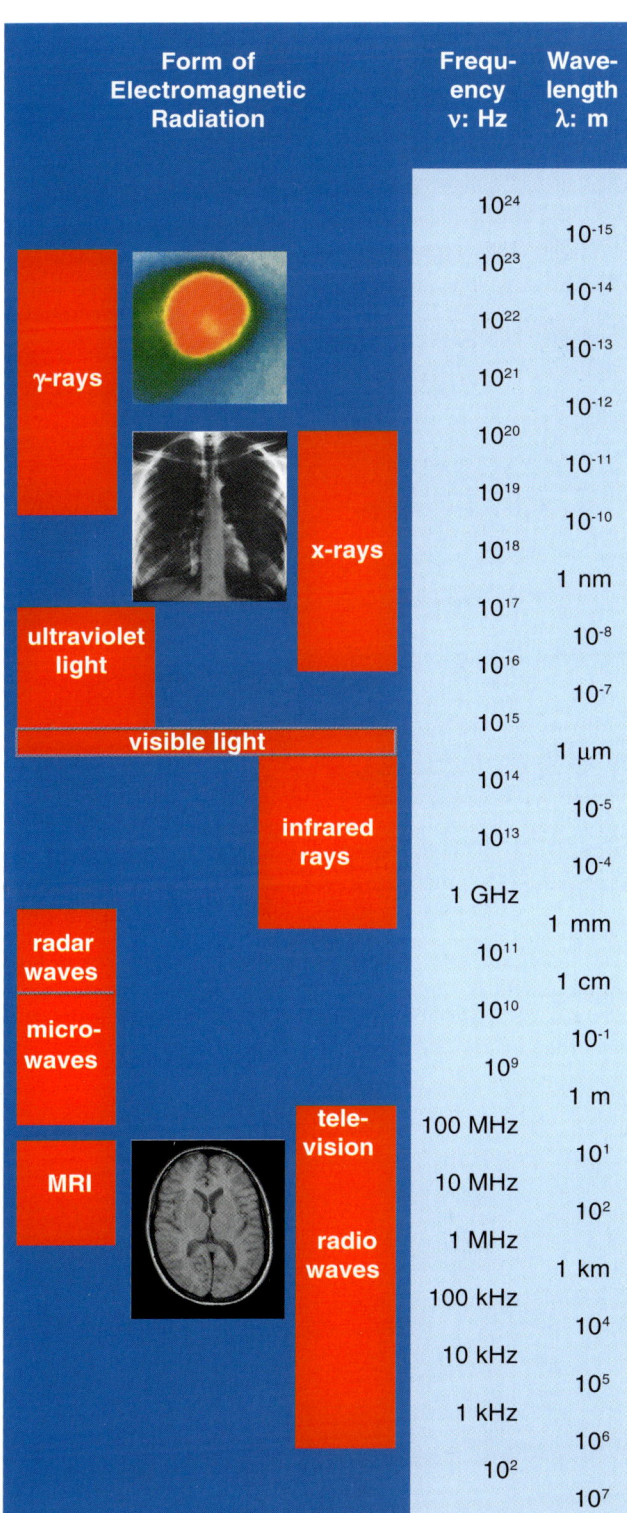

Form of Electromagnetic Radiation	Frequency ν: Hz	Wavelength λ: m
	10^{24}	
		10^{-15}
	10^{23}	
		10^{-14}
	10^{22}	
		10^{-13}
γ-rays	10^{21}	
		10^{-12}
	10^{20}	
		10^{-11}
	10^{19}	
		10^{-10}
x-rays	10^{18}	
		1 nm
	10^{17}	
ultraviolet light		10^{-8}
	10^{16}	
		10^{-7}
	10^{15}	
visible light		1 µm
	10^{14}	
		10^{-5}
infrared rays	10^{13}	
		10^{-4}
	1 GHz	
		1 mm
radar waves	10^{11}	
		1 cm
	10^{10}	
micro-waves		10^{-1}
	10^{9}	
		1 m
tele-vision	100 MHz	
		10^{1}
	10 MHz	
MRI		10^{2}
radio waves	1 MHz	
		1 km
	100 kHz	
		10^{4}
	10 kHz	
		10^{5}
	1 kHz	
		10^{6}
	10^{2}	
		10^{7}

Table 2-1:
Spectrum of electromagnetic radiation.

The Basics

Matter consists of atoms (e.g., ^{1}H, ^{12}C, ^{16}O, ^{31}P, etc.). An atom of one element differs from an atom of another element in its internal structure: the number of protons, neutrons, and electrons. Protons and neutrons comprise the nucleus; protons are positively charged, neutrons possess no charge, and the electrons orbiting around the nucleus are negatively charged. Different nuclear compositions and numbers of surrounding electrons are reflected in different physical properties.

Unlike color or texture some physical properties are not easily perceived. Such is the case with the magnetic properties of the nuclei which are the basis of the nuclear magnetic resonance phenomenon. Although these properties are not easily visualized, they are well defined and obey certain rules. This allows us to make analogies to more well-known phenomena so as to aid our understanding.

With the help of electromagnetic (radio) waves, the magnetic properties permit the production of images of the human body which can furnish information about morphology and function of the human organism. Owing to the stupendous spread of frequencies and wavelengths of electromagnetic waves, their interaction with matter is very dissimilar in the different parts of the spectrum. The radiation used for magnetic resonance imaging is quite different from x-ray and γ-radiation (Table 2-1). It stretches from AM frequencies through mobile, amateur radio and TV to FM radio frequencies, is approximately nine orders of magnitude smaller than the frequencies corresponding to x- or γ-rays (used for radioisotope examinations), and is considered biologically safe (Chapter 18).

Since Röntgen's discovery of the x-rays more than 100 years have passed. He succeeded in generating images of the human body with x-rays which result from interactions between these rays and the electron clouds of atoms.

Nuclear magnetic resonance signals stem from the interaction of radiowaves with the atomic nuclei themselves. This is the reason for the completely different imaging equipment necessary and the different contrast behavior of magnetic resonance imaging as compared with other medical imaging techniques.

Magnetic Properties of Nuclei

Among the most interesting nuclei for magnetic resonance imaging are ^{1}H, ^{13}C, ^{19}F, ^{23}Na, and ^{31}P. All of these nuclei occur naturally in the body, with the proton (^{1}H) being the most commonly used because the two major components of the human body are water and fat, both of which contain hydrogen. They all have magnetic properties which distinguish them from nonmagnetic isotopes.

Nuclei such as ^{12}C and ^{16}O which have even numbers of protons and neutrons do not produce magnetic resonance signals.

The hydrogen atom (^{1}H) consists of a single positively charged proton which spins around its axis. Spinning charged particles create an electromagnetic field, analogous to that from a bar magnet (Figure 2-1).

When atomic nuclei with magnetic properties are placed in a magnetic field, they can absorb electromagnetic waves of characteristic frequencies. The exact frequency depends on the type of nucleus, the field strength, and the physical and chemical environment of the nucleus (Figure 2-2). The absorption and reemission of such radiowaves is the basic phenomenon utilized in MR imaging and MR spectroscopy.

To understand the magnetic resonance phenomenon, two simple macroscopic parallels can be drawn:

First, let us consider a small magnetic needle placed in a magnetic field (Figure 2-3). If the needle is capable of rotating freely, it will orientate itself in the field in such a way that an equilibrium situation is attained. This equilibrium can be maintained indefinitely if no external forces influence the system. If we turn the needle 180° with our finger, an energy-rich, but unstable situation is created. When we withdraw our finger, the needle will jump back to the equilibrium.

A second example illustrates the influence of the external strain on the frequency of the wave absorbed or re-emitted by the system. Imagine three identical guitar strings exposed to different tensions; if we excite the strings, the resultant vibration is dependent on the tension of the strings (Figure 2-4).

In both examples, we have made comparisons between a macroscopic and the microscopic nuclear system. In the first example, we compared the nuclei with small magnet needles and in the second, with strings.

Such parallels provide a mental picture of the phenomenon, but have their shortcomings. One limitation

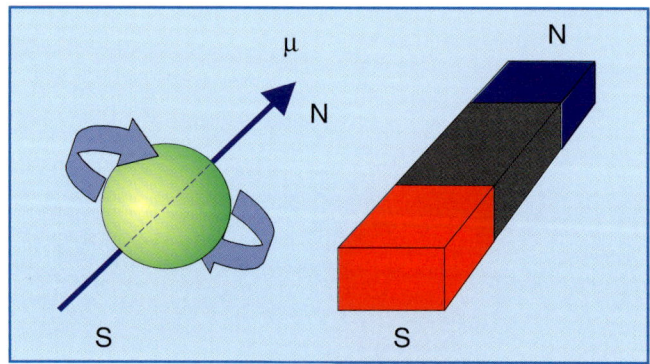

Figure 2-1:
A spinning charged particle possesses a characteristic magnetic moment μ and can be described as a magnetic dipole creating a magnetic field similar to a bar magnet (N = north, S = south).

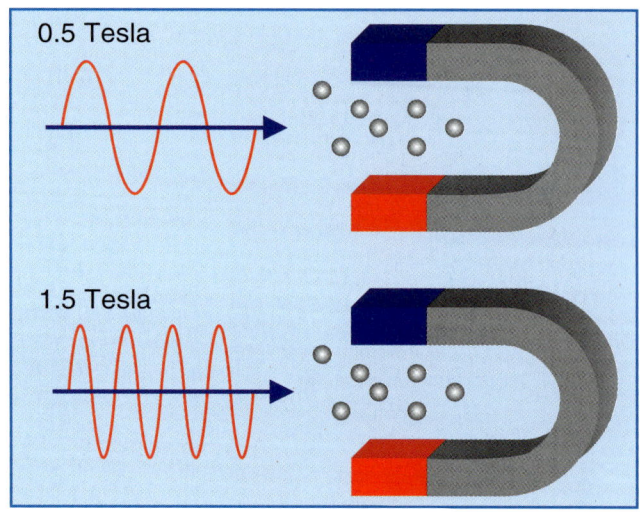

Figure 2-2:
The nuclei are able to absorb electromagnetic waves in both strong and weak magnetic fields. However, the absorption occurs at a field-strength-dependent frequency, which is higher in the strong magnetic field than in the weak magnetic field.

of the models is that all physical phenomena on the molecular scale are quantified. For example, whereas an infinity of different orientations is possible for the magnetic needle, no smooth continuous transitions between the equilibrium state and the unstable, energy-rich state exist for the magnetic nucleus; instead, quantum mechanics predicts that only jumps between these two states are possible for nuclei with a spin of 1/2 such as protons (Figure 2-5).

At equilibrium, we have a slightly larger population in the lower energy level, giving a net magnetization. To observe this population difference we have to provide an amount of energy equal to ΔE (the energy difference between the two levels).

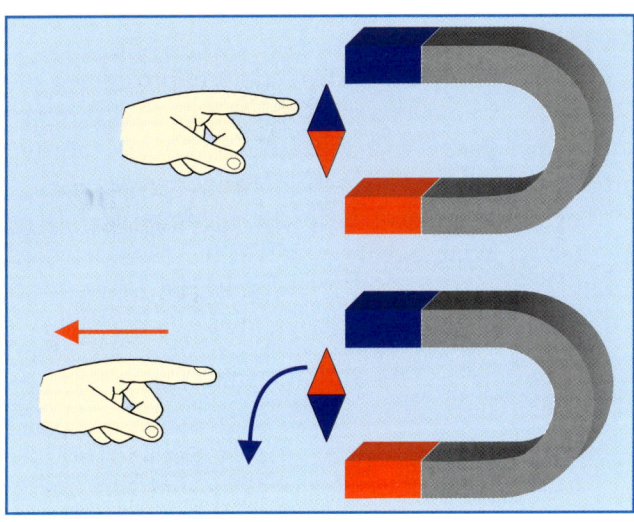

Figure 2-3:
The compass needle will seek the stable equilibrium state. When it is turned around with a finger, energy is brought in and it will be in an unstable energy-rich position. As soon as the finger is taken away, the needle will return to its stable state.

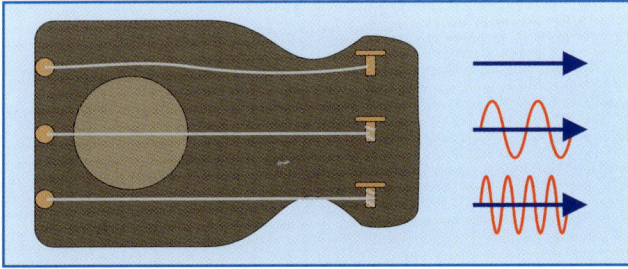

Figure 2-4:
A string (the nucleus) cannot vibrate without being exposed to tension (the external magnetic field). The uppermost string of this instrument has no tension at all, the middle string weak tension, and the lowermost high tension. The higher the tension, the higher will be the frequency of the vibration.

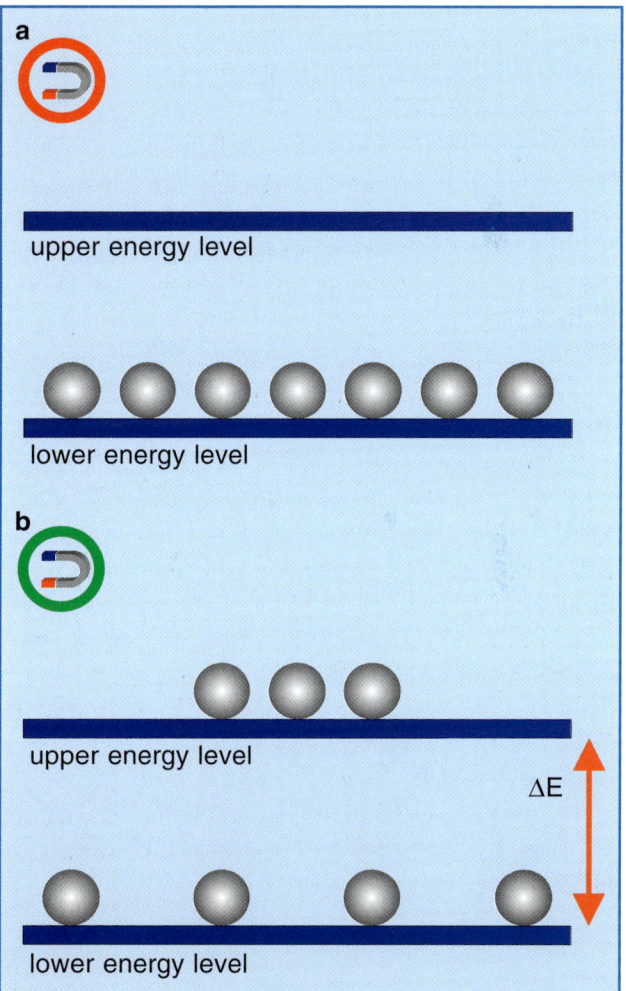

Figure 2-5:
Protons outside a magnetic field (a), and protons in a magnetic field (b). In the presence of a magnetic field, nuclei populate two distinct energy levels. The separation between these levels increases linearly with magnetic field strength, as does the population difference.

The Boltzmann Distribution

When a system has two discrete energy levels, there will be a well-defined probability for finding a particle in the high or the low, more stable energy level.

If we call the number of nuclei in the upper level N_u and that of nuclei in the lower level N_l, then the distribution between the two levels is given by

$$N_l / N_u = \exp(\Delta E / kT)$$

where ΔE is the energy difference between the two levels ($\Delta E = h \times \omega$; $h = 6.62 \times 10^{-34}$ J × s); k is the Boltzmann constant ($k = 1.3181 \times 10^{-23}$ J × K^{-1}); and T is the absolute temperature (Kelvin).

It is clear from this formula that, at equilibrium, the larger the energy difference ΔE becomes, the larger will be the population difference. The energy difference between the two energy levels in the magnetic resonance experiment is proportional to the field strength. Thus, if we increase the field strength we increase the energy difference and hence also the population difference. Since the size of the NMR signal is directly dependent on the population difference, the signal also increases.

This explains the increase of the signal-to-noise ratio with field strength (Chapters 4 and 9).

Radiowaves can be considered as packets of energy. If the energy of such a packet equals ΔE, it will cause a spin to jump to the high energy level. After the radiofrequency (RF) pulse, we no longer have an equilibrium.

To return to this state, the same number of spins which jumped to the higher energy state have to return to the low level. In doing so, they will emit an amount of energy ΔE which corresponds to the signal in an NMR experiment.

The transitions back to the equilibrium do not occur immediately, but over a period of time following the RF pulse.

The Larmor Equation

A mechanical analog to the magnetic nature of the nucleus is that it is a spinning mass with a small net positive charge. Because of the motion of the electric charge, a small magnetic field is created.

In the presence of an external magnetic field, the behavior can be compared with that of a spinning top (Figure 2-6). When the spinning top is submitted to the gravitation of the earth, its motion is complex: it rotates around its own axis and precesses around the direction of the earth. In this later motion, the axis of the spinning top is tilted with respect to the direction of the gravitation. If gravitation were absent, precession would not happen. The magnetic properties of the atomic nuclei make them precess around the external field.

Actually, for protons, two cones of precession exist: one for the nuclei in the state of low energy (represented in Figure 2-6) and another one in the opposite direction for the nuclei in the high energy state.

The frequency ω of this precessing motion is given by the following equation, called the *Larmor equation*:

$$\omega = \gamma \times B_0$$

ω is the angular Larmor frequency (unit: MHz), γ is the gyromagnetic ratio (unit: MHz/T), which describes the ratio of mechanic and magnetic properties of the nucleus and depends on the type of nucleus. B_0 is the strength of the magnetic field in Tesla (T).

In Table 2-2, the values for some nuclei commonly used in MR imaging and MR spectroscopy are given. Note that two nuclei with different gyromagnetic ratios will precess with different Larmor frequencies if placed in the same magnetic field. Consequently, their resonance frequencies will be different.

Table 2-3 shows some typical excitation frequencies in MR imaging and magnetic field strengths for protons and phosphorus nuclei.

Figure 2-7 (right):
The collapse of the Tacoma Narrows Bridge across the Pudget Sound in the US state of Washington in November 1940 was caused by high winds making the structure oscillate. Resonance can contribute to such accidents.

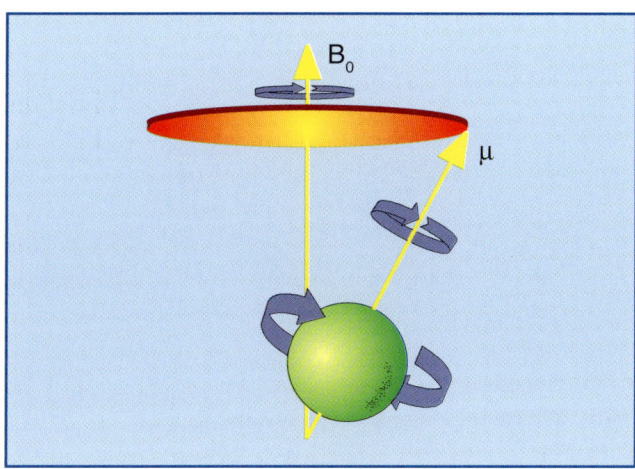

Figure 2-6:
A spinning mass with a net positive charge will create a small magnetic moment μ (or 'spin'). In the presence of an external field, B_0, the moment will precess around the direction of the external field.

Resonance

Let us look at another example. A bridge can oscillate and collapse when the marching rhythm of a column of soldiers or the undulation caused by strong winds correspond to its own structural resonance frequency; similarly a glass can be broken by the voice of a singer (Figure 2-7). This correspondence of frequencies allows energy to be transferred from the external world (the soldiers' legs) to a given physical system (the bridge) and it is called *resonance*.

Similarly, a resonance phenomenon will occur when an electromagnetic wave of appropriate frequency (equal to the Larmor frequency) reaches the nuclei; then, nuclei located in the state of lower energy will be transferred to the state of higher energy.

Table 2-2:
Magnetic values for some common elements.
Magnetogyric ratio: MHz/T; to obtain the resonance frequency, one multiplies the cited number with the field strength.
Relative abundance: percentage.
Sensitivity: at constant field, taking into account their relative abundance.

Nucleus	Magneto-gyric ratio	Relative abundance	Sensitivity
^1H	42.58	99.98	1.00
^{13}C	10.71	1.11	0.01
^{19}F	40.05	100	0.83
^{23}Na	11.26	100	0.093
^{31}P	17.23	100	0.066

Table 2-3:
Dependence of field strength and frequency for some commonly used fields (rounded values).

Field Strength (T)	Frequency (MHz)	
	^1H	^{31}P
0.1	4.3	1.7
0.3	12.8	5.1
0.5	21.3	8.6
1.0	42.6	17.2
1.5	63.9	25.9
2.0	85.2	34.5
3.0	127.8	51.8
4.7	200	81

Magnetization

Until now, only the microscopic behavior of the nuclei has been considered. However, any real sample contains a tremendous number of nuclei - a drop of water, for instance, contains about 10^{21} nuclei.

Macroscopically we observe the net effect of all microscopic events. In the presence of an external field, nuclei with magnetic properties similar to those of protons can occupy either the stable (and lower) energy state or be in the excited (upper) state, which has a slightly higher energy.

The energy difference between these two states is so small that the number of nuclei occupying each of the states is almost identical. Out of a million nuclei, there are only a few more low-energy nuclei than high-energy nuclei. Since the signal obtained in a magnetic resonance experiment depends on the difference in population between the two states, the resulting signal is very weak.

Figure 2-8 shows that the sum of the nuclei precessing around the direction of the external field is equivalent to a single magnetic moment, called *net magnetization*. This represents the total magnetization of the population difference (those excess nuclei in the lower energy level at equilibrium). To detect this net magnetization, it is necessary to tip it away from the axis of the main field. This is achieved with an electromagnetic pulse at the resonance frequency.

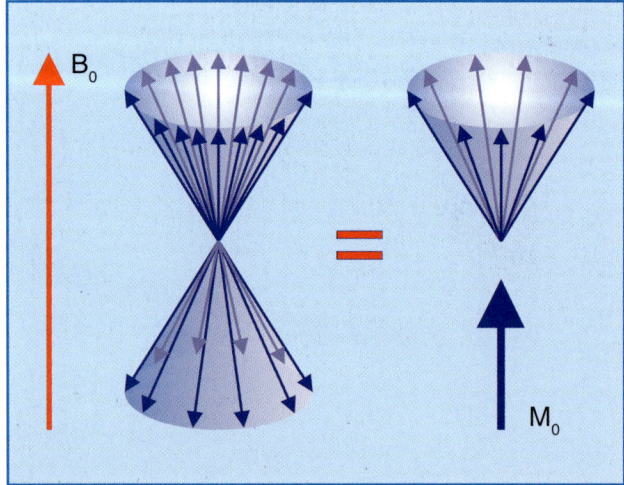

Figure 2-8:
Alignment of the individual magnetic moments with the external field. The population of the upper energy state is slightly higher. The sum of all individual components is the macroscopic magnetization or net magnetization M_0.
According to the Boltzmann distribution, the number of spins in the upper energy level rises with increasing field strength.

The Rotating Coordinate System

Before we try to explain the effect of an RF pulse on a spin system, we have to change our way of looking at this system. It is very difficult to understand the movements within this system because the spin rotates around its own axis and precesses around the axis of the magnetic field B_0 at the Larmor frequency. The RF pulse adds another movement. To follow these movements requires a lot of imagination (Figure 2-9).

However, if we try to look at the system from the perspective of a coordinate system rotating at a frequency equal or close to the Larmor frequency, the situation become less complicated.

This may not seem obvious, but in fact we are doing this all the time. If a person walks by us, it is very easy for us to assess how much faster than us he is walking. However, if we view the same action from outer space and have to take into account the rotation of the earth the problem would be much more complicated (Figure 2-10).

We are in fact using a rotating frame of reference since we are rotating at the same rate. In NMR we can achieve this by using a frame of reference which is rotating at the resonance (Larmor) frequency.

On resonance, spins will be stationary in this frame whereas off-resonance spins will rotate at a frequency which is the difference between their frequency and the resonance frequency.

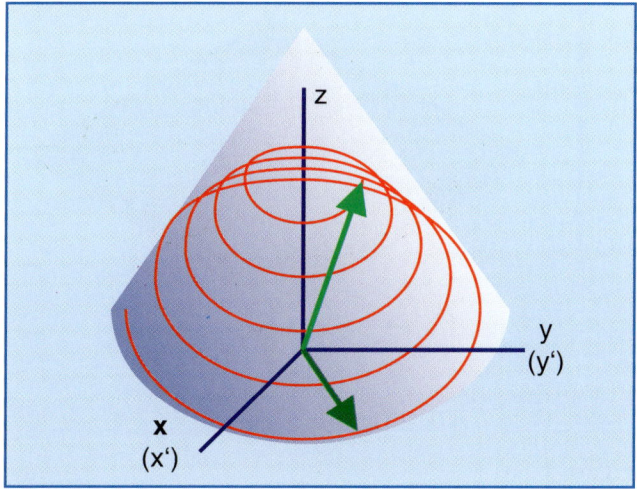

Figure 2-9:
The spiral motion of the tip of the magnetization during RF excitation shown in a stationary frame. The rotating frame depicts only the blue axes and the green arrows.
To distinguish when we are using the rotating frame of reference, we denote the x- and y-axes as x' and y'.
The same procedure can be used for z, but z and z' are identical.

Figure 2-10:
Another example of a rotating (coordinate) system: Watching the horses on a merry-go-round from (a) far away and (b) from the merry-go-round itself. From outside, the horses look blurred and, if they are moving fast, it is not easy to distinguish them. If one's point of reference is on the merry-go-round, it is much easier to distinguish the horses and their features.

The Magnetic Resonance Signal

To excite a spin system, one can expose the spins to a continuous electromagnetic wave of the right frequency. However, the method most commonly chosen for excitation of atomic nuclei in a magnetic field is to apply radiowaves of high intensity during a short period of time (*pulsed magnetic resonance*).

The frequency of these RF waves should be equal to, or close to, the Larmor frequency of the nuclei. Viewed from the rotating coordinate system (Figure 2-11 top), this results in a rotation of the magnetization away from the direction of the external field (Figure 2-11 center and bottom).

To understand this, we have to remember that spins at the resonance frequency are stationary in the rotating frame, implying a zero effective magnetic field. Therefore, the only field the spins experience is the B_1 field, which is the field created by the RF pulse. The spins rotate about B_1 in the same way as they rotate about B_0 in the stationary frame of reference.

In other words, prior to the RF pulse, the spins rotate about B_0 which is aligned along the z-axis. At this point, there is no net magnetization in any direction within the x'-y' plane. The RF pulse then tips the net magnetization away from the z-axis, towards the x'- and y'-axes of the rotating frame.

Following the pulse, the spins are still precessing about B_0, but their precession is no longer random; they precess in phase, and a net magnetization is produced in the x'-y' plane. This magnetization is aligned along the y'-axis following a 90° pulse along x'. For a given RF intensity, the pulse angle is determined by the duration of the RF pulse. The duration of a 180° pulse is twice as long as that of a 90° pulse.

In the standard stationary frame of reference, we now have a component of magnetization rotating at the Larmor frequency perpendicularly to B_0 (= longitudinal magnetization) in the x'-y' plane (= transverse magnetization). According to Faraday's law of induction, this transverse magnetization can induce a voltage in the receiver coil surrounding our sample.

When the excitation pulse is switched off, the spins start returning to their equilibrium and emit a signal.

The signal that is received from a homogeneous sample in a homogeneous magnetic field typically appears as shown in Figure 2-12a. It is called *free induction decay*, or *FID*, of the system. It looks like a damped oscillation.

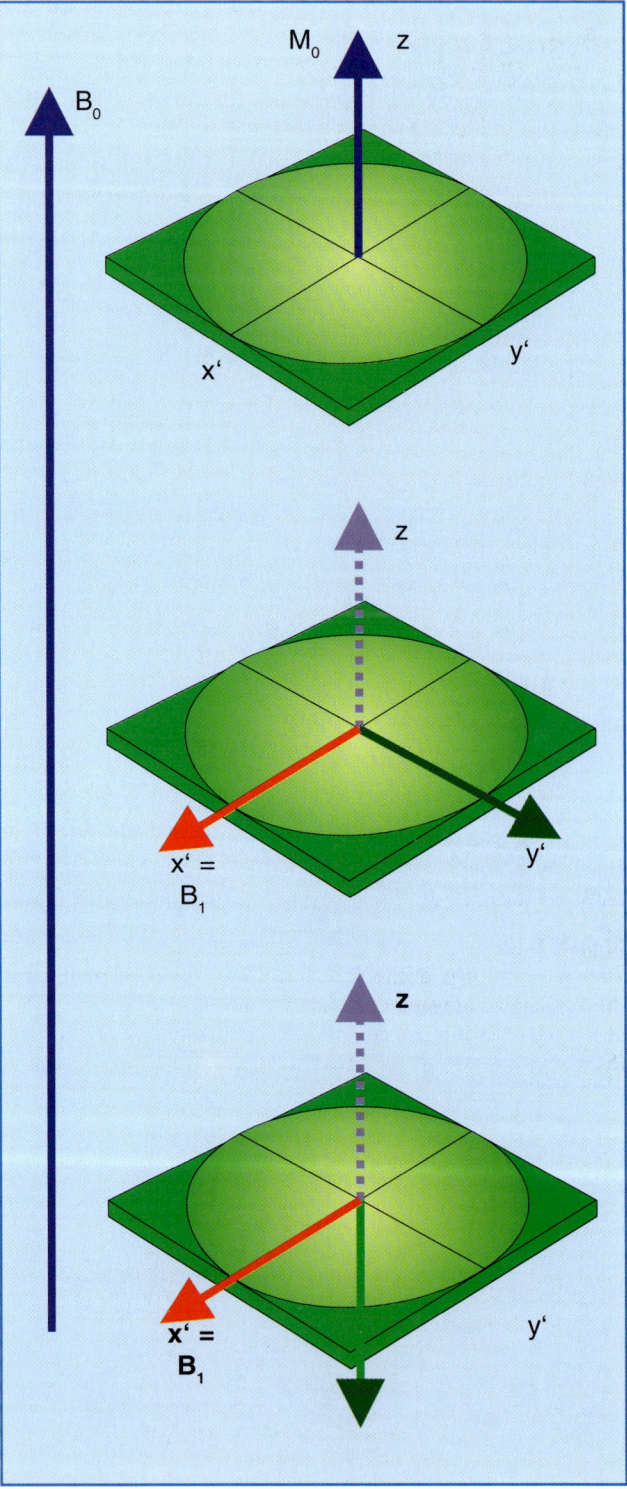

Figure 2-11:
Top: At equilibrium there is one stationary magnetic moment, M_0, directed along B_0.
Center: After being exposed to an RF pulse of an appropriate frequency, the magnetization (M_0) is tipped away from its equilibrium situation; in this case the pulse has tipped M_0 by 90° and is consequently called a *90° pulse*.
Bottom: If a pulse lasting twice as long is applied, a *180° pulse* results, which inverts the magnetization.

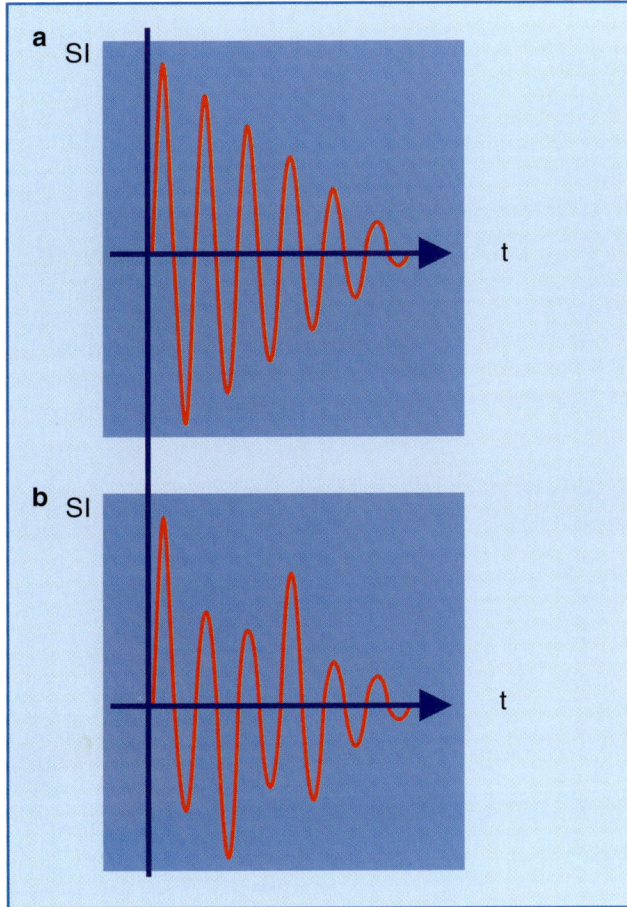

Figure 2-12:
The free induction decay (FID) (a) of a sample of pure water, (b) of a sample of water containing additional components. Usually, FIDs are far more complex than shown in these examples (SI = signal intensity; t = time).

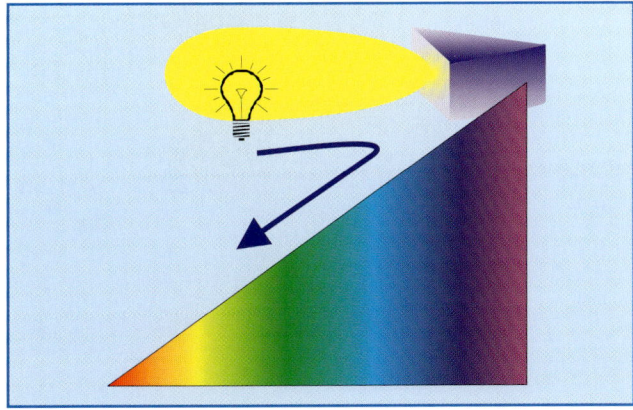

Figure 2-13:
Frequency analysis: visible light is broken down into its components by a prism. The constituent colors are those of the rainbow which have different frequencies.
The components of a magnetic resonance signal can be analyzed by subjecting the signal to a Fourier transform. For this purpose the prism is replaced by a computer.

If the magnetic field is not homogeneous, different parts of the sample experience different field strengths, and thus different parts of the sample will show different Larmor frequencies, leading to a more complicated FID (Figure 2-12b).

The deconvolution of a FID is essential for the analysis of the information hidden in it. The idea of separating electromagnetic radiation into frequency components is familiar to us through the use of a prism to split white light into its constituent colors, each of which has a different frequency (Figure 2-13).

Frequency Analysis: Fourier Transform

It is more difficult to analyze the components of sound or, in our case, the magnetic resonance signal.

The physical phenomenon of a FID can be compared to the sound received from a bell. This sound is initialized by a short-lived impulse produced by the action of the clapper. The resonating sound resulting from it is of high intensity at the beginning, but decays in intensity as a function of time (Figure 2-13).

The signals emanating from an inhomogeneous magnetic field can be compared to several bells tolling at the same time, creating a sound pattern which does not easily allow the differentiation of a single bell.

Before it is possible to separate the received oscillating signal into its components, it must be digitized, converted into a binary sequence, and stored in a computer.

Applying a *Fourier Transform* (*FT*), one then can analyze the signal for its frequency components and determine the intensity of each frequency. With a *Fast Fourier Transform* (*FFT*) algorithm, this may take only a few milliseconds.

It is possible to analyze the frequencies of a number of bells tolling at the same time by measuring their combined response at a particular frequency and then proceeding to the next frequency until the whole frequency range has been covered. This is analogous to the original magnetic resonance experiment. However, if we excite and then measure all of the frequencies at the same time in the same experiment, we obviously greatly increase the efficiency of the experiment. In pulsed magnetic resonance, all of the frequencies of interest are excited by the pulse, and we rely on the Fourier transformation to sort out the intensity of response at each frequency (Figure 2-14).

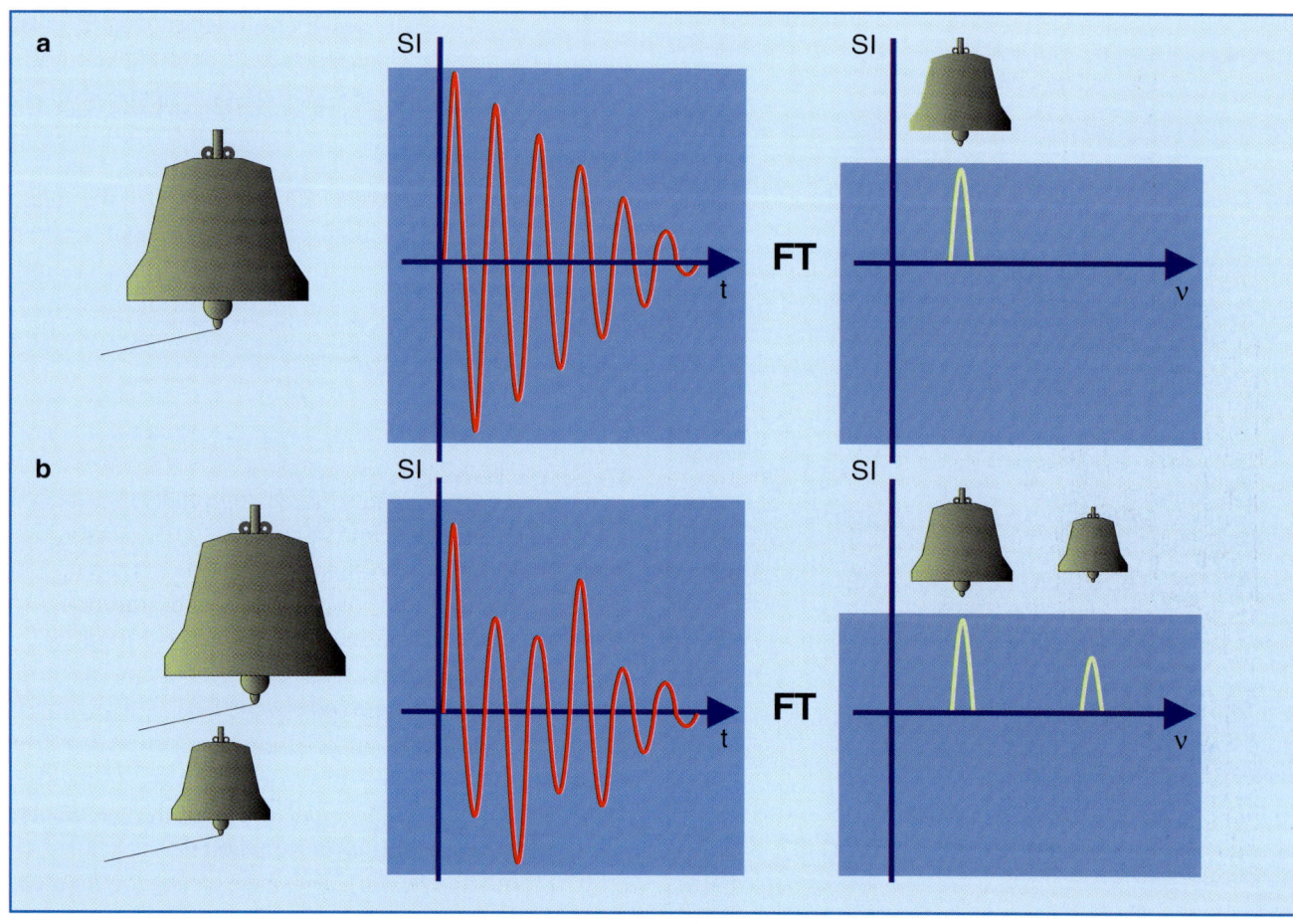

Figure 2-14:
The equivalent example: the sound wave originated by a single bell and by two bells of different sizes. Tolling a single bell once gives a clean sound (not in reality, but for the sake of our example), while tolling several bells at the same time gives a mixed sound. Fourier transform (FT) of the signal emitted by a single bell (or a pure water sample in NMR) and two different bells (water containing different components in NMR) provide us with spectra showing the frequency content of the sound. The same mathematical analysis can be used for the FID (SI = signal intensity; t = time; ν = frequency).

Chapter Three

Instrumentation

Introduction

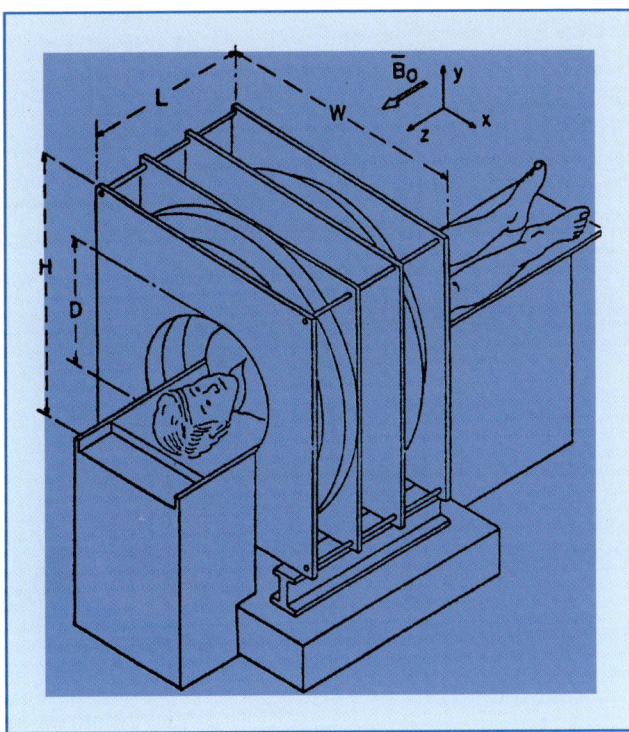

Figure 3-1:
'A sketch of a possible magnet configuration for medical zeugmatography.' Graphic depiction from 1978 of what would become the first whole-body imager at Paul C. Lauterbur's laboratory.[1]

There are a wide variety of MR imaging systems and technologies. The extensive range of MR systems can be confusing for the potential buyer. Therefore, users should identify their specific needs[2].

Analytical NMR and MR imaging systems are very similar in their basic components.

However, imaging machines additionally require gradient coils, the implications of which will be discussed later, and Faraday shielding which protects the instrument against undesirable interference by radio waves from broadcasting stations transmitting on, or close to, the resonance frequency.

Any MR imaging equipment includes the following elements (Figures 3-1 and 3-2):

- magnet large enough to accommodate the sample to be examined (mouse or patient);
- gradient coils and electronics;
- RF-pulse transmitter and RF receiver;
- power supplies and cooling systems;
- data acquisition and processing system, including a powerful computer;
- operation and evaluation console(s).

A typical layout of an imaging system is depicted in Figure 3-3; in this case a mobile MR imaging unit is shown.

The central part of the MR machine is the magnet, which should create a static, stable, and homogeneous magnetic field. The field strength of these magnets can differ by several hundred percent according to the purpose of the equipment.

MR imaging systems are generally classified according to their magnetic field strength. Table 3-1 gives an overview of this classification.

Ultralow equipment (< 0.1 T) is hardly used any more, but there is a trend towards ultrahigh research machines, operating between 3 T and 8 T.

For further details on possible hazards and side effects of magnetic fields, especially at field strengths beyond 2 T, refer to Chapter 18. Field strength is also discussed in Interlude VI.

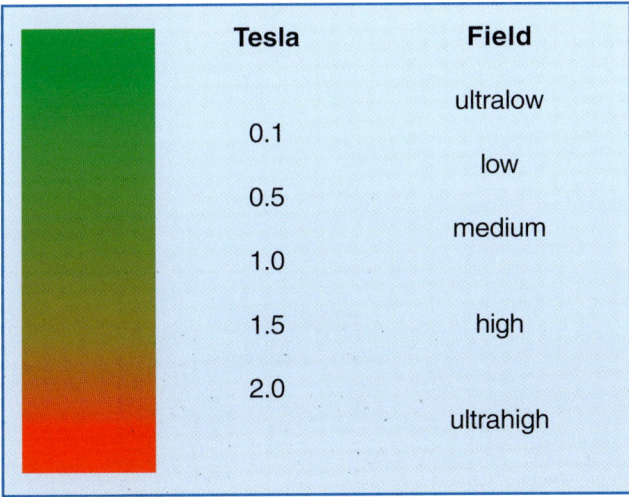

	Tesla	Field
		ultralow
	0.1	
		low
	0.5	
		medium
	1.0	
	1.5	high
	2.0	
		ultrahigh

Table 3-1:
Definition of field strength: magnetic resonance systems are called *ultralow-field systems* below 0.1 T, *low-field systems* between 0.1 and 0.5 T, *medium-field systems* between 0.5 and 1.0 T, *high-field systems* up to 2.0 T, and *ultrahigh-field systems* above 2.0 T.

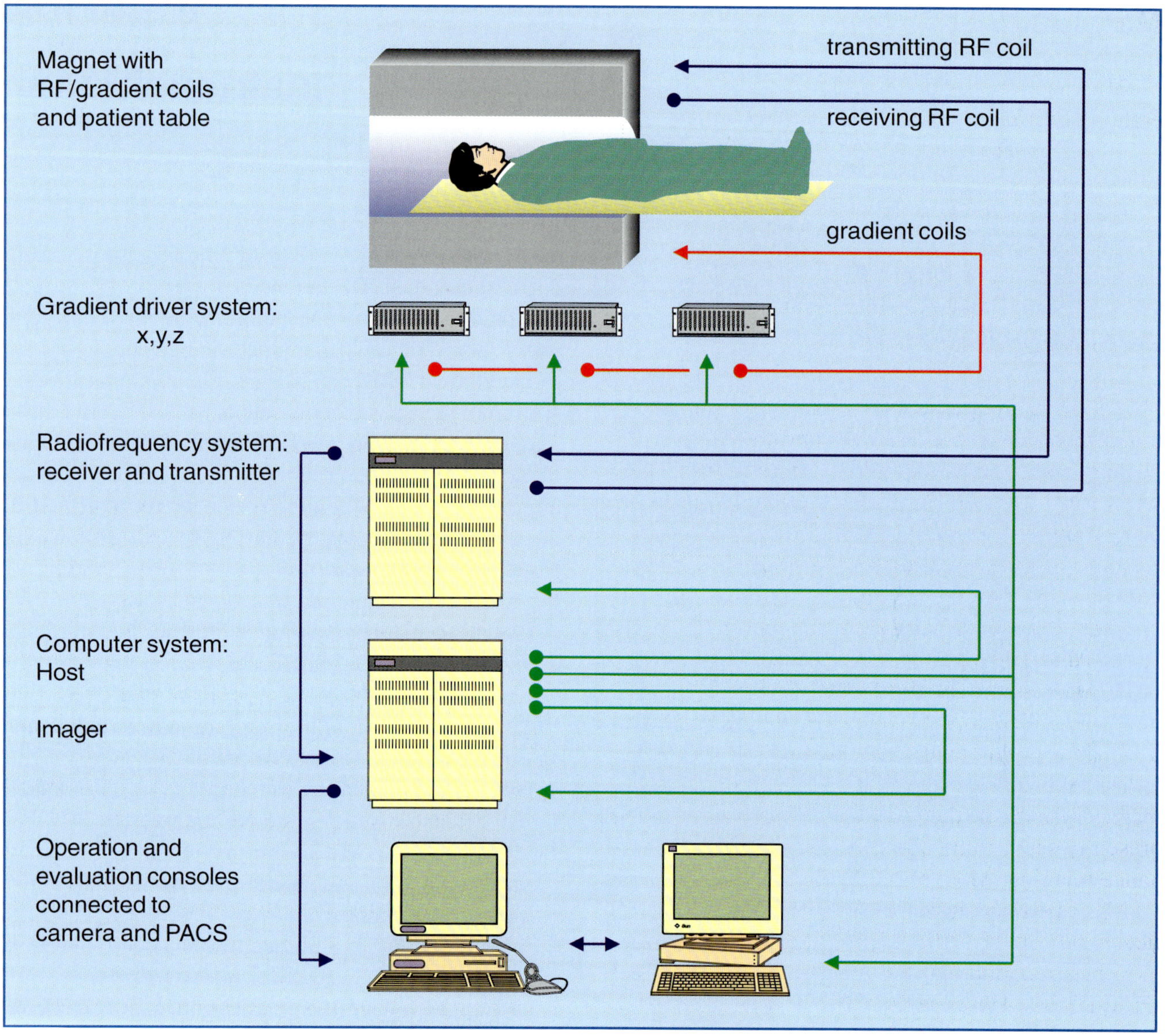

Magnet with
RF/gradient coils
and patient table

transmitting RF coil

receiving RF coil

gradient coils

Gradient driver system:
x,y,z

Radiofrequency system:
receiver and transmitter

Computer system:
Host

Imager

Operation and
evaluation consoles
connected to
camera and PACS

Figure 3-2:
The main components of a MR imaging system.

Figure 3-3:
Complete superconducting magnetic resonance imaging system (in a trailer). All necessary systems and subunits have been accommodated in limited space.

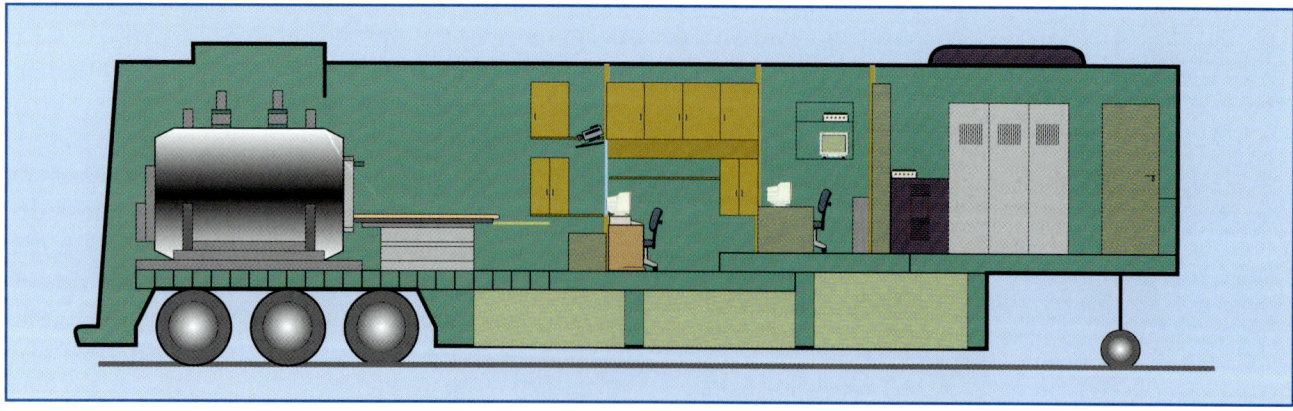

Magnet Types

The magnetic field of an MR system can be generated by different magnet systems:
- permanent magnets;
- resistive magnets;
- superconducting magnets.

Table 3-2 summarizes advantages and disadvantages of different magnet types.

Permanent Magnets

Certain alloys possess ferromagnetic properties. A magnet built of such materials has the advantage of needing no power to maintain the field strength. Likewise, it needs no cooling because there is no power dissipation.

Such systems have small fringe (stray) fields when compared to the other magnet systems. Capital and operational costs of permanent magnets are low.

The disadvantages are the weight of the currently produced systems for whole-body imaging, although new alloys developed during recent years have cut down the weight of permanent systems from 100 tons to less than 20 tons. Another drawback of permanent magnet systems are the field-strength limitations, which presently seem to be about 0.3 T for magnetic resonance imaging. Most of them operate at about 0.2 T.

Many permanent magnets have a vertical magnetic field which distinguishes them from some resistive and most superconducting systems with horizontal fields (Figure 3-4a). The field direction has an impact on the use of certain transmitter and receiver coils.

Electromagnets or Resistive Systems

Resistive systems consist basically of a suitable coil or collection of coils through which a strong electric current is passed. If these coils are set up in a proper geometry, a homogeneous magnetic field can be created, as shown in Figures 3-1 and 3-5. Such systems have a high power consumption (e.g., a 0.1 T unit requires about 20 kW), create a lot of heat, and therefore need large-capacity cooling systems.

The practical upper limit for large-bore magnets is about 0.7 T, but usually 0.3 T is considered the upper limit for commercially available machines. Fringe fields are present around such systems. The weight of these systems is typically below 5 tons. They are the lightest of all MR imaging systems. Resistive magnets have the advantage that they can be switched off when the system is not being used or during emergencies.

Hybrid Magnets

Some companies have developed magnets which are hybrids between permanent and resistive systems. They are iron-cored electromagnets in which the magnetic energy of the resistive magnet is concentrated in the gap between the soft-iron pole pieces (Figure 3-4b). These systems reach field strengths up to 0.4 T and are the most commonly used. Their weight is between 10 and 15 tons.

Superconducting Magnets

When certain alloys are cooled down to temperatures close to absolute zero, they show drastically reduced resistance to electric current: they become superconducting. Thus, when superconducting alloys are placed in liquid helium (at temperatures below a critical value of between -263° C and -269° C or 4 to 10 K), high currents can be driven in a coil built of that alloy, and an extremely stable magnetic field of very high field strength can be produced. The basic design for superconducting magnets involves a double cooling system using liquid nitrogen as cryogenic liquid in the first thermos container (*cryostat* or *dewar*) and liquid helium in the second inner dewar (Figure 3-6). Recently, single-dewar systems have been developed.

When charged with current, the superconducting magnet uses virtually no electrical power, but consumes cryogenic liquids. Helium must be replenished by refilling or through a compreessor connected to the MR system which reliquifies cryogens. Superconducting magnets have large fringe fields and are usually shielded so that the environment is protected.

The field-strength limitations for superconducting magnets are not yet established. For imaging purposes, small systems up to 9.4 T and whole-body systems up to 8 T have been used, and for spectroscopy fields of up to 14.1 T are in use. Only superconducting magnets can be used for *in vivo* spectroscopy and functional imaging because of their high field strengths.

The magnetic field of a superconducting magnet can be discharged when the coil accidentally loses its superconductivity. This creates a sudden increase of temperature which, in turn, heats the liquefied coolant gases. They start boiling, increasing in volume, and helium is set free. Such an incident is described as a *quench* (see Chapter 18).

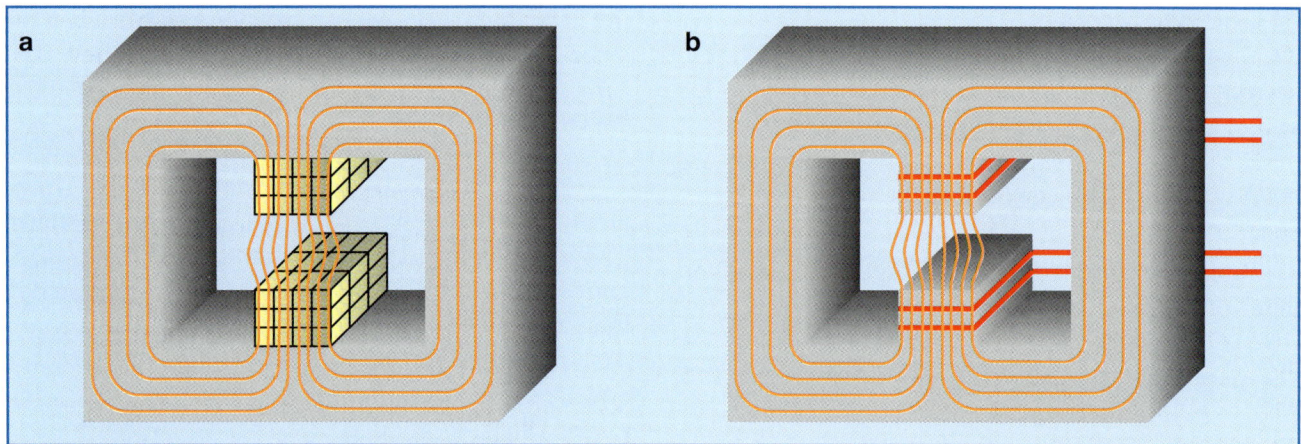

Figure 3-4:
Schematic drawings of (a) a permanent magnet and (b) a hybrid magnet.

Permanent magnets can be designed in different ways, from a Greek temple shape to a C-shaped open system. In this case the field is produced by magnetized ceramic bricks; the outside consists of iron, which provides structural support to the system, contains the stray field, and thus intensifies magnetic field strength. The field strength of permanent magnets is easily influenced by the surrounding temperature, therefore temperature-stabilizing air conditioning is necessary for the magnet room.

Hybrid magnets combine permanent magnets with electromagnets. The power consumption is high, but field strength can be increased compared to a permanent or purely resistive magnet system. These magnets are also described as 'iron core' electromagnets.

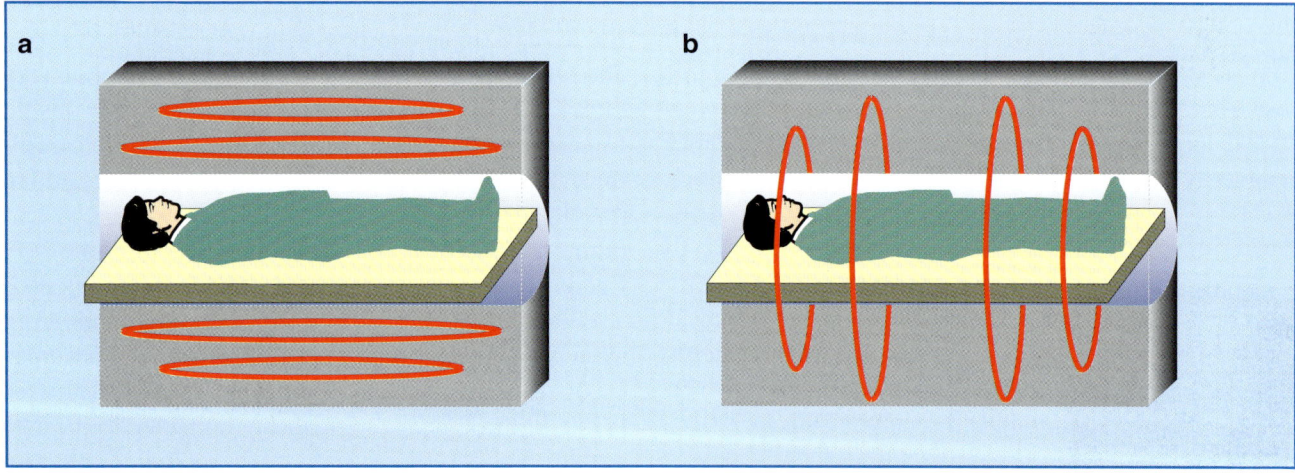

Figure 3-5:
Cuts through two different kinds of air core resistive electromagnets. As shown in Figure 3-1, resistive magnets commonly consist of four loops of wire creating the static magnetic field. They can be arranged (a) parallel or (b) perpendicular to the patient table; the perpendicular (head-to-foot) orientation is more common.

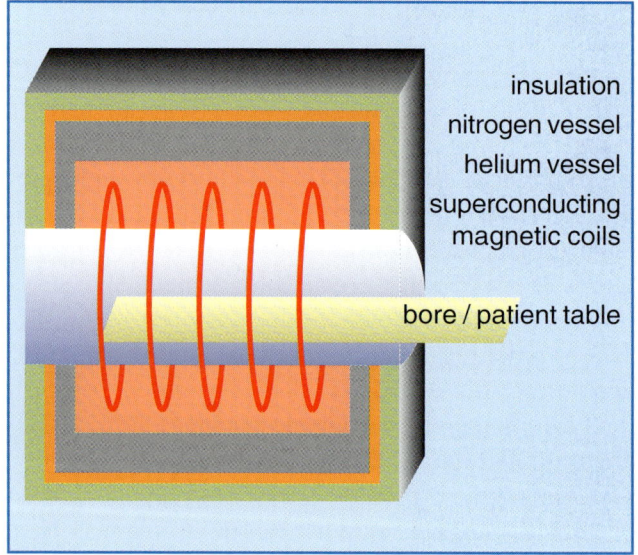

insulation
nitrogen vessel
helium vessel
superconducting
magnetic coils

bore / patient table

Figure 3-6:
Schematic drawing of a superconductive MR imaging system. The magnetic field is produced by electric current flowing in wire loops cooled by the surrounding liquid helium. The power supply is disconnected once the system is charged and running at the desired field strength.

Usually no permanent damage to the magnet is induced but the magnet has to be refilled with helium and cooled down to superconductivity, which may last several days.

During the last few years new superconducting materials have been developed, which allow superconductivity to occur at higher temperatures (up to 100 K). However, the majority of the materials developed to date are rather brittle and unsuited to wire (and hence magnet) production. In addition, many of the materials lose their superconductivity in the presence of strong magnetic fields. For these reasons the midterm impact of these developments upon magnetic resonance systems is expected to be limited.

Shimming

None of the above mentioned magnet systems will produce a perfect homogeneous field, but careful design may allow for fields where the inhomogeneities are far better than 100 parts-per-million (ppm) within the region of interest. Field inhomogeneities reduce the efficiency of imaging experiments and prohibit spectroscopic investigations. To improve the field characteristics, most magnet systems are delivered with *shim* coils. When currents are passed through these coils, correctional fields of known geometry are produced and can compensate for the inherent inhomogeneity of the magnet.

Homogeneities better than 0.01 ppm can routinely be achieved with high-field analytical magnetic resonance magnets over small sample volumes (less than 1 cm^3). Using *in vivo* MR spectroscopy with localized shimming, homogeneities of less than 1 ppm can be achieved for small volumes. For MR imaging where larger volumes are used, poorer homogeneity is acceptable.

The shim coils can be placed in liquid helium inside the superconducting main magnetic field and adjusted one-by-one to shape the field (*active shimming*).

A similar effect can be achieved by mounting small ferromagnetic metal pieces at the appropriate locations inside or outside the magnet bore. Each of these pieces will contribute to the magnetic field and, if the symmetry of the field is kept, a very homogeneous field can be obtained (*passive shimming*).

Table 3-2:
Properties of different magnet types.

Magnet Type	Advantages	Disadvantages
Permanent	• no electric energy necessary; • good patient acceptance due to open architecture; • limited fringe field; • no refrigerants necessary; • sufficient image quality for many routine examinations; advanced methods might not be applicable.	• cheaper but usually not competitively priced compared to medium and high-field systems; • sensitive to temperature changes; • systems cannot be switched off; • limited field strength, thus: • low signal-to-noise ratio.
Resistive	• good patient acceptance due to open architecture; • no refrigerants necessary; • easy installation in difficult sites; • many advanced imaging methods possible; • machines can be switched off.	• cheaper but usually not competitively priced compared to permanent and superconductive systems; • limited field strength, thus: • limited signal-to-noise ratio; • high electric power consumption.
Superconductive	• high signal-to-noise and homogeneity, making • some advanced imaging methods more easily applicable; • only systems available for spectroscopy.	• high capital and running costs; • special sites necessary; • prone to artifacts; • higher claustrophobia rates and more noisy than other systems; • systems cannot be switched off.

Shielding

Shielding is applied to limit the fringe field of the magnet, to compensate for inhomogeneities of the magnetic field, partly to increase the field strength, and to protect the environment (Table 3-3). *Passive shielding* involves large quantities of iron, easily 30 tons, symmetrically placed around the magnet. *Active shielding* is accomplished by additional superconductive coils. Whereas the inner set of coils produces the main magnetic field, the outer set contains and reduces the fringe field which surrounds the magnet. Commonly, both sets are electrically coupled for fail-safe operation.

Gradients

Magnetic field gradients are necessary to create MR images. The reasons for this will be explained in detail in Chapter 6. As shown in Figure 1-4, two wires with electric current moving in the opposite direction can create a linear change of the magnetic field between each other when placed at the correct distance. This is also known as a 'magnetic field gradient'.

Three sets of gradient coils are necessary to be able to create a weak magnetic field in any direction in space (Figure 3-7). With mid- and high-field systems, the strength of the magnetic field created by gradient coils is approximately 100 times lower than the main field.

The performance of gradients coils is measured in mT/s, their *peak amplitudes*. Common peak amplitudes are 10 mT/m; high performance systems require up to 30 mT/m. A second important property of gradients is their *rise time* (ms) or *slew rate* (mT/m/ms). The faster the rise time or the higher the slew rate, the better the system performance will be;. i.e. the faster image data can be acquired (cf. Figure 6-6).

Eddy Currents

These currents are introduced by the gradients' time varying magnetic field. They can degrade both the homogeneity of the static magnetic field and distort the gradient pulse profiles. If they are not compensated for, image quality will significantly decrease. One way to counteract their influence is by shielding the gradient so that their fields are restricted to the interior of the patient bore.

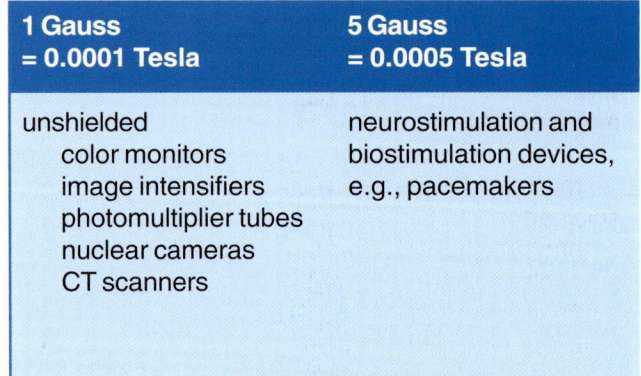

1 Gauss = 0.0001 Tesla	5 Gauss = 0.0005 Tesla
unshielded 　color monitors 　image intensifiers 　photomultiplier tubes 　nuclear cameras 　CT scanners	neurostimulation and biostimulation devices, e.g., pacemakers

Table 3-3 (continued on next page):
Minimal field strengths at which certain devices may start malfunctioning. Shielding can be necessary to protect the hospital environment from the magnetic field emanating from the MR system. Certain equipment must not be exposed to magnetic fields.

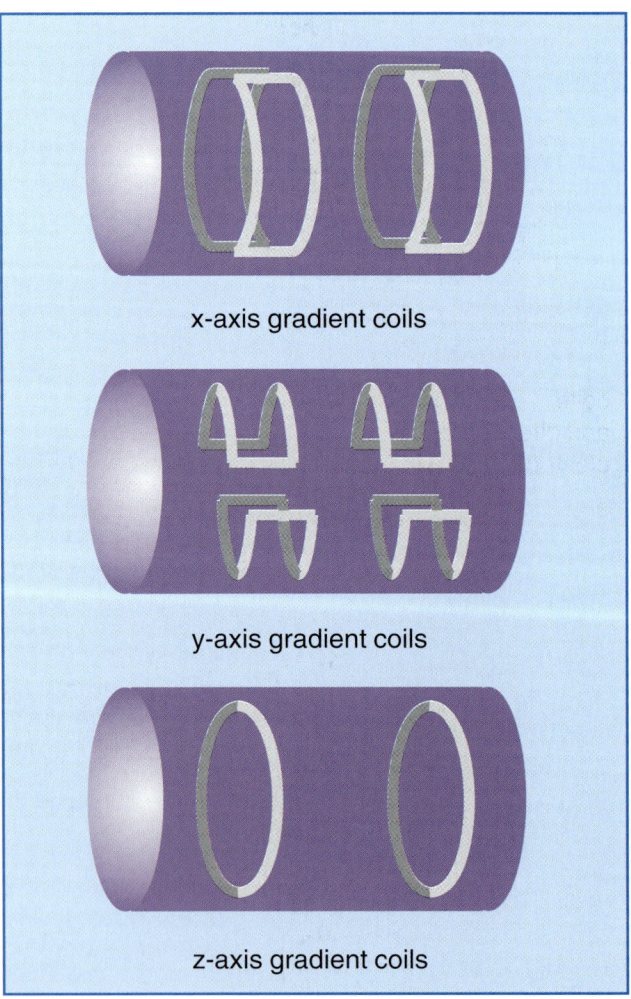

x-axis gradient coils

y-axis gradient coils

z-axis gradient coils

Figure 3-7:
Schematic depiction of gradient coils in a superconducting or resistive magnet system. To cover all three spatial dimensions (x, y, and z), three sets of gradient coils are installed in the bore of the MR scanner. By changing the current flow relative to each other, planes in any direction can be laid through the patient body.

10 Gauss = 0.001 Tesla	20 Gauss =0.002 Tesla

magnetic data carriers:
    disks, diskettes,
    tapes, credit cards
shielded
    photomultiplier tubes,
    nuclear cameras,
    CT scanners
    color monitors
	computers
	disk drives
	mechanical watches

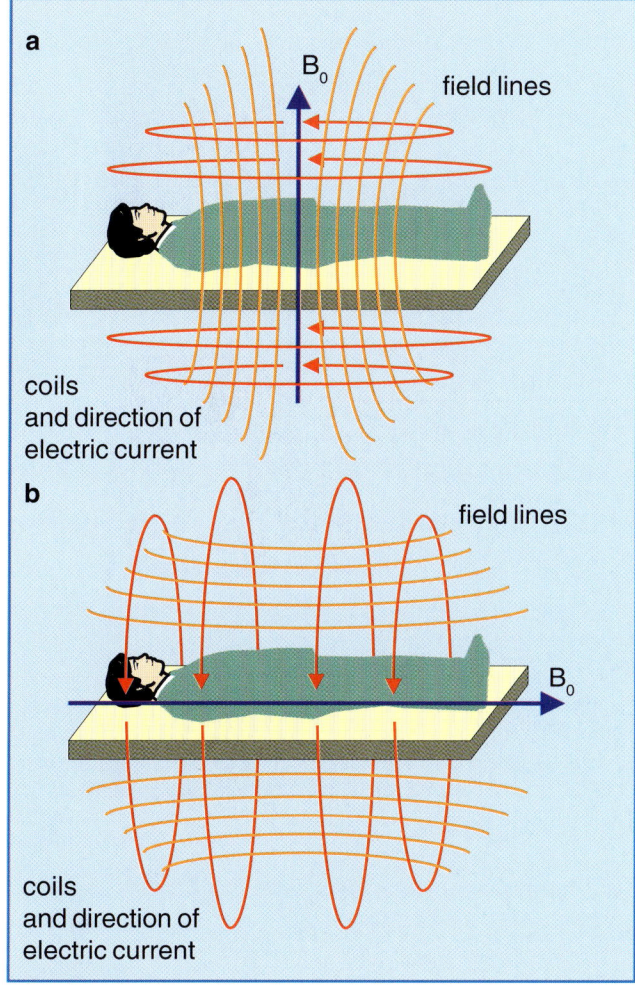

Figure 3-8:
The direction of the main magnetic field depends on the orientation of the coils of the magnet. The field can either be vertical as in (a) or horizontal as in (b). In superconducting and some resistive systems, the field is usually horizontal and thus parallel with the patient in the magnet.

Transmitter and Receiver

As mentioned earlier, the initial high power excitation of the nuclei is accomplished with a shortlasting RF pulse at a frequency close to or at the Larmor frequency. Both the radio waves and pulses are generated in the transmitter section of the magnetic resonance equipment.

The desired frequency is produced by a frequency synthesizer. The synthesizer output is then modulated with an 'envelope' to provide the required pulse shape for the RF excitation.

The receiver is basically a highly sensitive low-noise detector of signals in the high- and very-high-frequency (HF and VHF) range. The magnetic resonance signals are typically a few microvolts in amplitude. In the receiver, the signal is improved by a factor of 500 to 1,000.

After this stage, the signal is converted from an HF resonance signal (MHz) to an audio frequency signal (kHz).

Transmitter and Receiver Coils

The object to be studied, whether it is a 1 mm^3 sample for chemical analysis or a whole human body, is placed *inside* an antenna or coil. It should fill the coil as much as possible — the coil should have a good filling factor (> 70%).

The oscillating magnetic field B_1 of the RF coil has to be perpendicular to the main magnetic field B_0 generated by the magnet if the spins are to be excited. The most common configuration is to have the main field orientated along the bore of the magnet, so the coil must produce a field perpendicular to the bore of the magnet (Figure 3-8b).

Separate coils can be used for transmitting and receiving, but in most cases a single coil is used for both excitation and detection (*transceiver* coil).

Since the excitation pulse is many orders of magnitude stronger than the responding magnetic resonance signal emitted by the human body the receiver will be damaged if it is subjected to all or part of the RF pulse. To overcome this problem, a device known as a transmit/receive switch is used which can very rapidly switch the routing of signals.

Coils consist of one or more windings of low-resistance wire, usually copper. The geometry of the single or multiple windings is crucial for the proper excitation and subsequent signal detection.

Coils come in different categories and for different magnet types. Some of the most popular coil geometries are the solenoid, saddle-shape (Helmholtz), birdcage, and slotted resonator. This group of coils is identified as *volume coils*. Figure 3-9 illustrates the form of some commonly used coils in MR imaging. Typical examples are head and body coils, or coils for studying the knees or the neck.

Since the solenoid coil generates an oscillating magnetic field parallel to the coil's axis it has to be aligned across the bore of the magnet. This limits the use of this kind of coil to magnetic fields that are orientated perpendicular to the patient couch.

All coils except the surface coil are designed to produce a very homogeneous RF field such that all of the sample experiences the same degree of excitation.

Surface coils are used to detect magnetic resonance signals from a small region close to the coil when it is placed, for instance, on a patient's spine, orbit, or temporo-mandibular joint. The advantage of using surface coils is that in the small volume of tissue close to the coil, we can obtain a better signal-to-noise ratio than that obtained by a standard volume RF coil.

Surface coils receive a high signal only from an approximately hemispherical area below the coil with a depth of half its diameter (Figure 3-10). Thus, both the RF fields and the detection sensitivity of a surface coil are highly inhomogeneous, leading to a position-dependent excitation. The variation of RF intensity with depth causes the flip angle to vary with depth when the surface coil is used as a transmitter. The problems of varying pulse angle with depth in surface coils can be overcome either by using special RF pulses or by transmitting the RF pulse on the standard body coil and only detecting the signal with the surface coil.

An increasingly common modification to the standard coil design is the *quadrature* (or *circularly polarized*) coil which uses at least two RF fields orthogonally to each other and improves both the efficiency of the coil and the signal-to-noise ratio of the resulting signal by $\sqrt{2}$.

Phased-array (or *synergy*) *coils* consist of a number of small surface coils. The signals received by these coils can be collected simultaneously and the data can

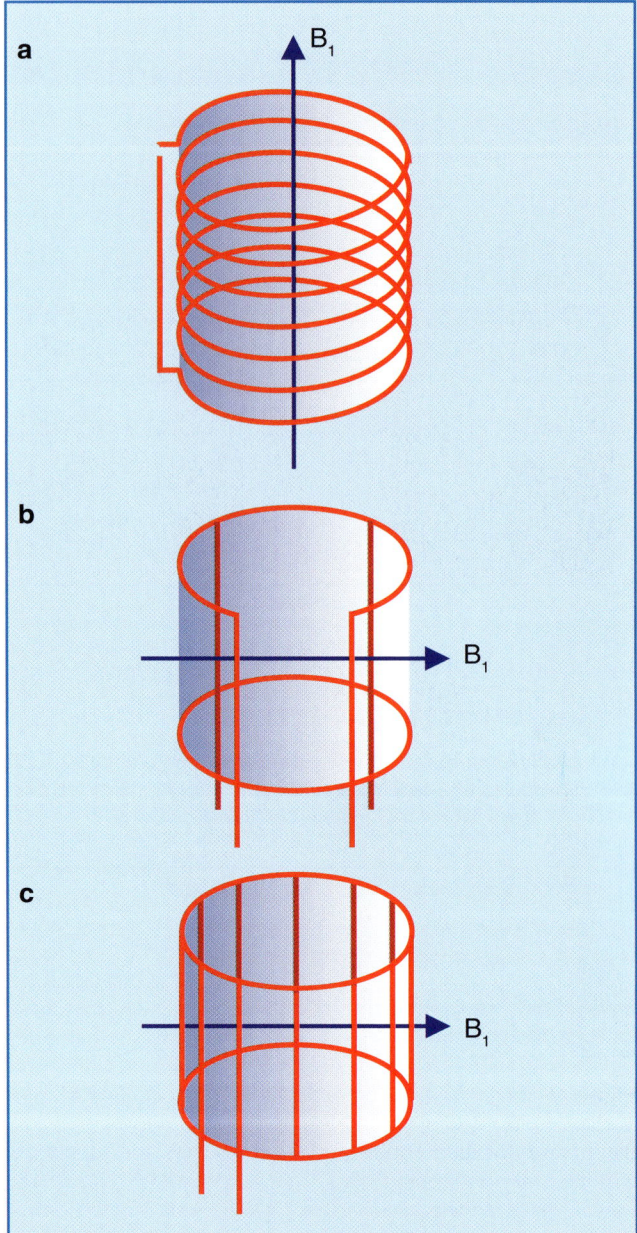

Figure 3-9:
Three different types of coils: (a) solenoidal coil; (b) Helmholtz or saddle-shape coil; (c) birdcage coil. Their oscillating magnetic field B_1 must be perpendicular to the main magnetic field B_0. Their RF field is uniform within the volume.

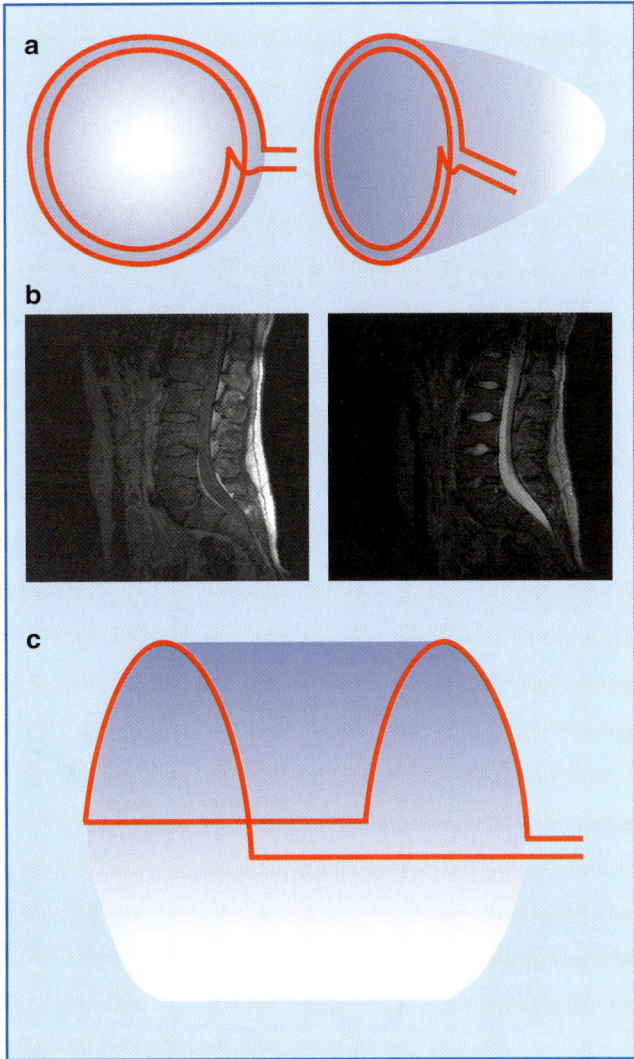

Figure 3-10:
Surface coils. (a) Graphic of a simple surface coil design. Since the intensity of the RF field varies with depth, the pulse angle will also vary with depth unless special ('adiabatic') pulses are used. Similarly, the detection sensitivity will also decrease with increasing depth. (b) T1- and T2-weighted images of surface coil acquisitions of the lumbar spine. Spinal cord and spine are well depicted, but there is hardly any signal from the anterior parts of the pelvis. (c) Wrap-around or half saddle-shape surface coil.

be combined to construct a single image of the body region covered.

Compared to a large surface coil with poor signal-to-noise ratio, phased-array coils have the advantage of better signal-to-noise; however, each single subunit requires its own receiver channel, which makes reconstruction difficult and coils expensive.

Faraday Cage

Table 2-1 shows that the resonance frequencies of all MR scanners overlap with commercial, military, and amateur radio and television frequencies. Electric machines can also create electromagnetic waves. It can easily happen that the receiver of the MR imager picks up such radio signals from the outside world, which then interfer with the signals from the examined sample or patient. This leads to noisy images or, in the worst case, the complete loss of images.

Faraday shielding is used as a protection against electromagnetic interference. High-field systems require a complete Farday cage (usually a copper cage with windows, including an electrically conducting screen), which has to be grounded. Connections from the inside of the cage to the outside have to be very carefully made and shielded.

Data Acquisition System and Computer

The analog signal emanating from the spins must be converted into a digital (numerical) form suitable for storing and processing on a computer. This digitization of the signal is achieved by using an *analog-to-digital converter* (ADC). The output of the ADC is a digital version of the FID for each data point. After the recording is complete, the digitized FID is stored on a magnetic disk or another storage medium.

A specialized computer (*the image processor*) is used to pre-process and process the raw data into images. The Fourier transform of the raw data can be greatly accelerated by array processors of dedicated hardware.

The *host computer* is responsible for controlling and monitoring the entire MR system. Additional consoles can be connected to the host computer for such functions as patient management, display of image data and calculation results, image post-processing, and documentation and archiving.

The Right Choice

There are some 20,000 MR imaging systems operating all over the world (Figure 3-11).

MR systems are available in the well-known tunnel shape (Figures 3-1 — 3-6) or as open systems (Figure 3-12). The latter reduce patient claustrophobia and are better suited for interventional purposes because there is access to the patient from all sides. Open systems are available between 0.3 T with permanent magnets to 1.5 T with superconducting magnets. Recent development focus on machines between 0.7 and 1.0 T.

Dedicated systems are aimed at, e.g., orthopedic imaging. Examinations of knees, elbows, hands, and shoulders are the main indications. These systems are smaller and easier to install.

The choice of an MR system might be quite agonizing. Field strength is one, but not the only and most important, parameter; higher field strength does not necessarily guarantee a better quality system or better diagnostic outcome for the patient.

Needs must be carefully assessed in the choice of an MR imager. The quality of the component parts, both in terms of hardware and software, makes a considerable. Service, maintenance, and knowledge of how to run the system are of pivotal importance for image quality and assessment (see also Interludes One and Six).

References

1.   Lai C-M, House WV, Lauterbur PC. Nuclear magnetic resonance for medical imaging. In: IEEE/ERA: Technology for noninvasive monitoring of physiological phenomena. Electro/78. Proceedings. Boston, 25 May 1978. 1-15.
2.   Saint-Jalmes H, Bittoun J. Hardware development. In: Rinck PA, de Francisco P (eds.). The rational use of magnetic resonance imaging. Oxford: Blackwell Science 1995; 307-322.

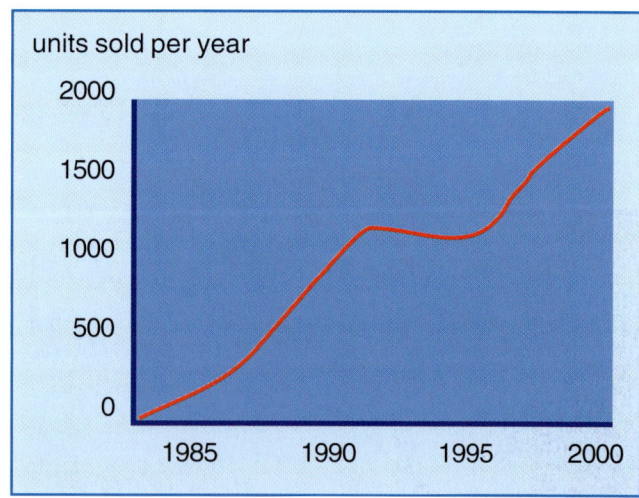

Figure 3-11:
Whole-body MR units sold worldwide since 1983. The biggest markets are the United States of America and Japan; Europe buys approximately one quarter of all units. In 2001 there will be an estimated 20,000 units worldwide.

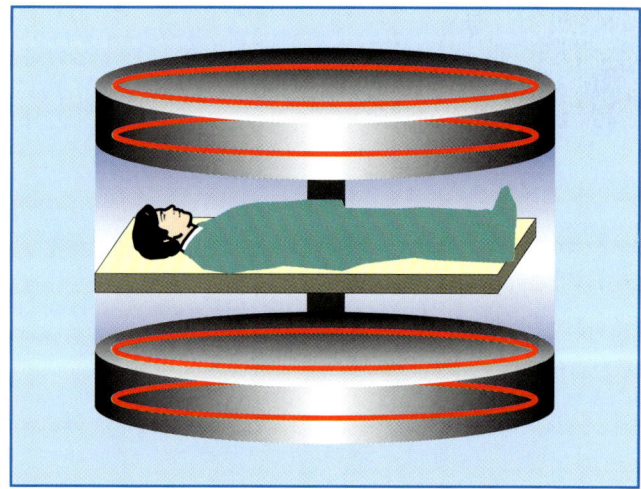

Figure 3-12:
Open MR system. The magnet is separated into two sections above and below the patient couch.

How to Purchase an MR Machine
A Guide in Ten Easy Lessons

Murphy's Law is the most reliable guideline when buying an MR machine: anything that can go wrong usually does.

This is what you have to learn before you start diving into this adventure.

Lesson One

Know what you want to buy. If you do not know anything about magnetic resonance, you can go straight to any company. The salesperson will know as much as you — possibly slightly less — so this is the perfect arrangement.

The chances are that it will not be you who chooses the machine: you will choose what this company tells you is the right choice. Or, more positively, the better the education of the salesperson, the easier and more efficient will be the collaboration between user/buyer and company.

Sales and marketing people hardly ever lie, but they would not dream of telling you the truth. Their claims for performance should be multiplied by a factor of 0.25.

Every company will say that it has the best equipment available in the world, and that all other companies have outdated equipment that will not perform properly. Marketing people use a special lingo; for instance, they use the adjectives 'ultimate' and 'optimum' instead of 'just another'. The 'ultimate MR machine' translates into 'just another MR machine' (it is unlikely that they mean 'ultimate', which equals 'the last one you need before your death').

Lesson Two

There is no such thing as a free lunch. A company may invite you, the hospital administrator, five local politicians, and several others who do not understand anything about magnetic resonance to travel around the world in a chartered jet. You (the taxpayer) or the next customer (also the taxpayer) will pay for it. However, even if you do not accept such invitations, the price does not drop.

The price will only drop the day you sign a contract with company A. One hour before signing, you will get a telephone call from company B, stating that it will reduce its price by 50%.

Lesson Three

After buying an MR machine, you will find out that the company has not delivered what you thought you had purchased.

Even if you have a detailed written contract, certain parts of the hardware or certain software programs only existed on the drawing board of the company's development department. They are not part of the delivered equipment because they do not exist. But you have already paid for them. This is usually called 'works-in-progress'.

The identical unit that you have seen at the company's headquarters or at a showcase performs differently from the unit that has been delivered to you. At exhibition booths, you always see 'typical' images that look great. No one tells you that the patient was dead when imaged; thus, there are no motion artifacts.

The salesperson who have negociated with will have left the company at this time. The company itself will have merged with another company, which considers the contract signed with you null and void.

Lesson Four

Never expect functioning equipment. Wherever computers are involved, things will go wrong. Think twice when you start planning. There are a lot of fantastic ideas to solve the problems of the world, medical imaging included. But if no one is using these ideas (or equipment), there is probably a good reason.

By the way, you should have thought three times.

Lesson Five

If something is wrong with the magnet, the responsible company people will tell you that all the troubles are caused by Eddy Current. This impertinent guy interferes and messes up everything. No one understands either where he comes from or where he can be found.

Trying to operate the equipment will be nearly impossible. Manuals are written in such a way that even their writers will not be able to operate the machine. Much space is given to unnecessary details, but there is no description of how to switch on the computer.

Lesson Six

When the MR machine is delivered, you will find out that within the next two months, a new version will replace the one you purchased. You have bought one of the last models of a version which will be discontinued and cannot be upgraded in the future.

Guarantees and warranties do not exist and are voided immediately after the first installment of your payment. Anything in writing is not worth the paper it is written on.

Similarly, deadlines only mean that the company acknowledges that the Gregorian calendar has replaced the Julian calendar some time ago. The dates given are meaningless.

There is one basic rule, however: everything takes longer than you think.

Lesson Seven

Manifold options exist. Some of them are necessary to run the equipment properly. You have to buy them at horrendous prices (value-added tax not included). This often happens with car manufacturers, who sell cars without tires (see 'Options').

Options bought at a later stage will be even more expensive (see 'Options').

Among these options is the Faraday cage. Without a Faraday cage, all images produced after 10 a.m. will have a central artifact caused by Radio Vatican, which starts broadcasting at exactly this time on exactly the frequency you use as the resonance frequency.

Service and maintenance are not included in the purchasing contract (see 'Options'). To avoid unpleasant surprises, you should discuss and include them before signing the delivery contract. The service people will not be trained to cope with the problems they have to face, anyway.

Downtime is not what you think it is, but what the company defines. If the machine does not produce images, it is not necessarily out of order. Some companies even try to change the groundrules. Instead of paying penalties when the equipment malfunctions, they try to make the customer pay by installing a control clock. If the MR imager is used more than eight hours per day and five days per week, additional service charges apply (see 'Options').

Lesson Eight

The multiformat camera / workstation / whatever you have bought to be connected to your MR system cannot be connected.

As soon as it is connected, no information will ever leave this piece of equipment, because either its ports do not conform to any standard or there are no ports at all. Laser cameras do not work. Film developing units eat films. PACS links send pictures everywhere except to where you want them to be delivered to. DICOM is not a unified standard of image data transmission but an in-house company format that is changed on the first day of every month.

If two companies are involved, such as an MR manufacturer and a camera producer, you, the customer, are lost. If something goes wrong, one company will blame it on the other one and nothing will happen. If something finally does happen, you will pay dearly (most likely to both companies).

Lesson Nine

Something will be wrong with your building plans but you will detect it too late. The bigger the hospital, the more people will be involved in the planning and the bigger the mistakes will be.

For instance, the sewer system of your patient toilets will be connected straight to the emergency water evacuating system of your computer room. One day the pipes will be clogged, but there will be a patient who flushes the toilet anyway.

Lesson Ten

Do not believe what your colleagues in the next town tell you about their machine. They either hate it because they just went through lessons one to nine, or they love it because they do not know better. If they have the highest patient throughput in the country, it is because their machine is directly connected to a cash register.

Soon you will be part of this club: either looking forward with dismay or backward with anger.

Chapter Four

Relaxation Times and Basic Pulse Sequences

Figure 4-1:
Spinning away: You may need relaxing times to understand relaxation times — but how long do you stay in this position of low energy when the waves start hitting you ?

Relaxation Processes

T1 ('T-one') is the spin-lattice or longitudinal *relaxation time*; the characteristic time constant for *spins* to tend to align themselves with the external *magnetic field*. Starting from zero *magnetization* in the *z* direction, the *z* magnetization will grow to 63% of its final maximum value in a time T1.

T1-ρ ('T1-rho') is the spin-lattice relaxation time in the rotating frame; the characteristic time constant for loss of magnetization of spins under the influence of a *spin-locking* B_1 field. Despite its name T1-ρ relaxation is more closely related to T2 relaxation than T1. It will not be discussed in detail in this book. If one applies a long lasting B_1 magnetic field immediately after a 90° pulse, the dephasing of the spins in the *x-y* plane is stopped while the B_1 field is on. This is called spin-locking and the B_1 field is called a spin-locking pulse even though it may last hundreds of milliseconds.

T2 ('T-two') is the spin-spin or transverse *relaxation time;* the characteristic time constant for loss of *phase coherence* among *spins* oriented at an angle to the static *magnetic field*. It arises from interactions between the spins, with a resulting loss of *transverse magnetization*. The *x-y* magnetization will decay so that it loses 69% of its initial value in a time T2.

Table 4-1:
The different relaxation processes.

The Spin-Lattice Relaxation Time

Excitation of an equilibrium system always transfers the system to an unstable state of high energy. The length of time the system will remain there depends on the local conditions, as illustrated in Figures 4-1 and 4-2. Table 4-2 gives an overview of the different phenomena the system is or can be exposed to while returning to its equilibrium.

For a system of spin nuclei in a magnetic field, an unstable situation is created by the excitation pulse — the system is 'pumped up' with energy supplied by the RF pulse. At the molecular level, the return to equilibrium depends on the local magnetic and electric conditions at the excited nuclei.

In the same way, we need a resonance condition to exchange energy from the external world to the spin system. The excited spin system needs to be exposed to electromagnetic fields oscillating with a frequency at or close to the Larmor frequency of the nuclei before it can relax. The relaxation corresponds to the excess nuclei, which were transferred to the upper energy level returning to the lower energy level (Figure 2-5). If an isolated proton is left excited in absolute vacuum in the absence of any sort of electromagnetic fields, several years might be needed before the nucleus could, by itself, spontaneously return to the equilibrium state of low energy. However, if the proton is surrounded by water, this process can be 'stimulated' by the surrounding nuclei and will then require only a few seconds.

The process of returning to the equilibrium state from an excited state is called the *spin-lattice relaxation process* or *longitudinal relaxation process*. It is characterized by the T1 relaxation time. The T1 relaxation time is the time required for the system to recover to 63% of its equilibrium value after it has been exposed to a 90° pulse. For a given kind of nucleus, T1 depends on several parameters:
- type of nucleus;
- resonance frequency (field strength);
- temperature;
- mobility of observed spin (microviscosity);
- presence of large molecules;
- presence of paramagnetic ions or molecules.

The two latter factors are of special interest. In pure water, the process of reorientation (translational movement, rotation, etc.) of a single small water molecule occurs very rapidly. Each molecule has its own magnetic field, and in pure water the orientation of the molecule is rapidly changing, resulting in a fluctuating magnetic field at neighboring nuclei.

To promote relaxation, the frequency of the reorientation must be at, or close to, the resonance frequency in pure water. If the frequency of this reorientation is much higher than the Larmor frequency of the protons the relaxation is inefficient.

However, if we add more slowly moving *large molecules* such as proteins to the water, the water molecules will interact with them. This interaction involves temporary attachment of the water to the proteins and subsequent release. This temporary bonding radically reduces the frequency with which the water molecules reorientate themselves (Figure 4-3). To characterize the motion of a molecule, the correlation time (t_c) is used which measures the minimum time required for a molecule to re-orientate itself.

Because of the presence of protein surfaces, the T1 relaxation times of water in living tissue are always shorter than those obtained for pure water. In Table 4-2, some representative T1 values of normal tissues are listed.

T1 values vary with field strength. This influences image contrast in MR imaging so that it is not possible to make direct quantitative comparisons between T1 values at different fields. Thus, it is necessary to always mention the field strength when quoting T1 values. T1 data of brain tissues at different fields are shown in Figure 4-4.

The explanation as to how the presence of *paramagnetic ions* or *molecules* can enhance the relaxation rate of water is highly complex. Electrons produce a much stronger magnetic field than the nuclei, but when pairing of electrons occurs, there is only a weak net field. In paramagnetic ions we have unpaired electrons, the reorientation of which produces a very strong fluctuating magnetic field, resulting in a significant reduction in the relaxation time (Figure 4-5). Typical paramagnetic substances include Mn^{2+}, Cu^{2+}, Fe^{2+}, Fe^{3+}, Gd^{3+}, as well as molecular oxygen and free radicals. In certain circumstances, the ability of paramagnetic compounds to alter relaxation rates can be utilized to change the contrast in magnetic resonance images; in this capacity they are used as magnetic resonance contrast agents (see Chapter 13).

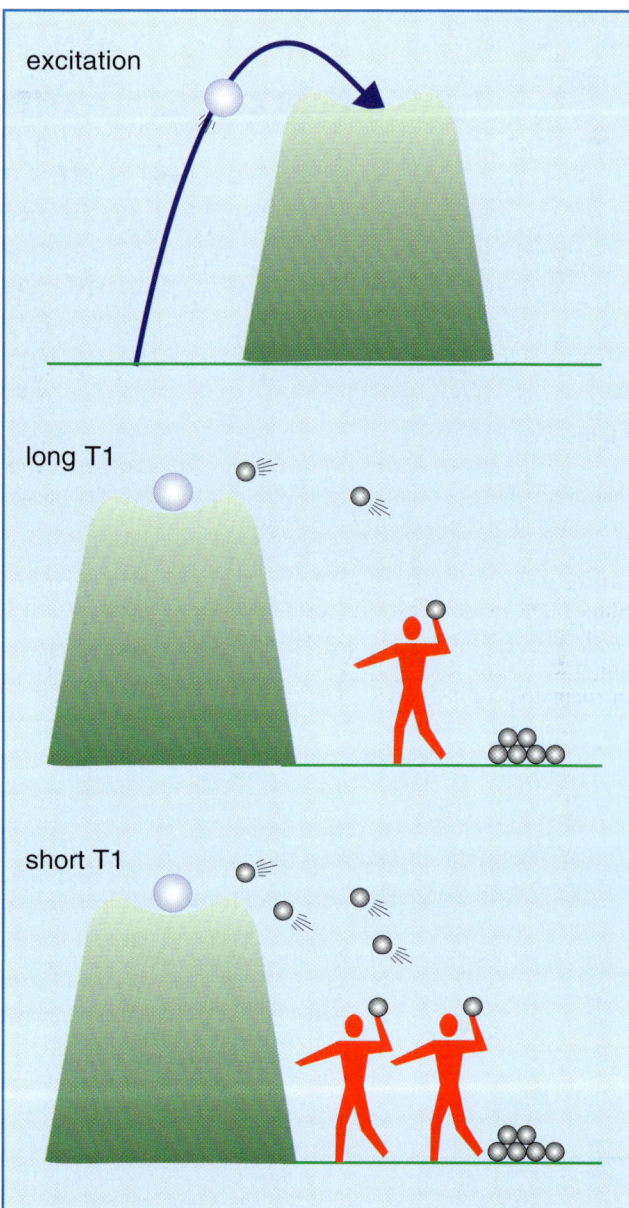

Figure 4-2:
A ball is stuck on top of a small hill (unstable high state of energy) and two boys are trying to get it down by throwing rocks at it. On statistical grounds, it will take less time for two boys to achieve their goal compared with one boy.

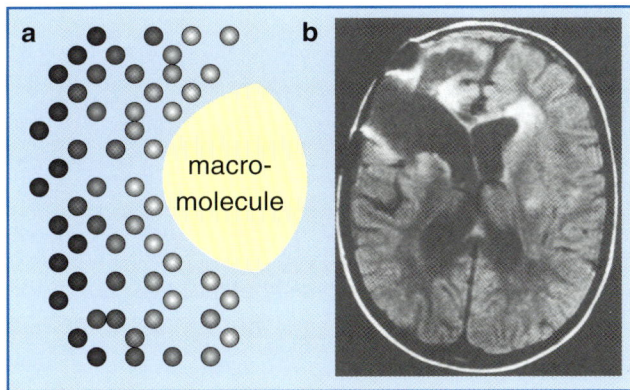

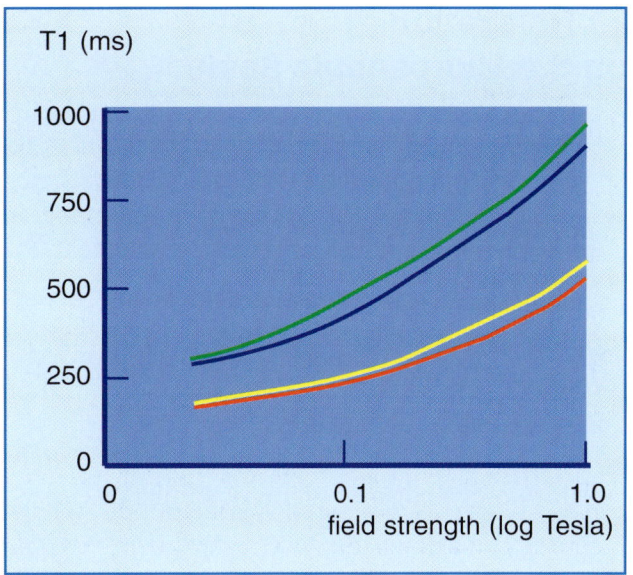

Figure 4-3:

(a): Pure water, i.e., water in the bulk phase, moves much faster than water close to macromolecules or membranes. The slower the molecular motion, the shorter the relaxation time T1 (and T2), here shown as an increase in brightness.

(b): T1-influenced MR image of a brain after a tumor operation. Fluid-filled areas are dark (long T1), while edematous areas are bright. The reason for this signal behavior is explained in the graph to the left: bulk water moves faster, protein-bound water in edematous tissue moves slower (short T1).

Figure 4-4:

Change of T1 relaxation times of gray and white matter versus field strength. Temporal gray matter (GM) is depicted green and parietal GM is blue; frontal white matter (WM) is depicted yellow, temporal WM is red.

Relaxometry deals with the relaxation behavior of different substances. An area of relaxometry which developed specific medical applications in the second half of the 1980s is field-cycling relaxometry. This technique requires a special NMR machine, the field-cycling relaxometer. With this machine, *ex vivo* or *in vitro* measurements of the relaxation behavior of tissue samples or contrast-enhancing compounds can be performed at any field strength; thus identical samples can be examined under identical conditions.

Field-cycling relaxometry showed that T_1 increases nonuniformly with field, leading to specific fingerprints of T_1 increase for different tissues [11].

Figure 4-5:

The presence of another boy (called 'Gadolinium') on top of the hill kicking the ball down would significantly shorten the time the ball would stay in the unstable state. Paramagnetic compounds influence excited spins in a similar way and shorten T1.

Table 4-2:

Tissue/Organ T1 relaxation time values (ms)			
Brain:			
Gray matter	450	Kidney	
White matter	350	Renal medulla	680
CSF	1500	Renal cortex	570
Heart:		Bowel (wall)	300
Myocardium	380	Testes	880
Abdominal organs:		Muscle	500
Liver	380	Fatty tissue	230
Pancreas	460	Bone marrow	490
Spleen	650	Skin	320

Some T1 values of some human tissues measured on an MR imaging system at 0.15 Tesla. The standard deviation of these values can be between 10 and 30%; therefore, these values are not very reliable.

T1 Relaxation
on the Microscopic Scale

There are good theories for the explanation of re-laxation times of pure substances, for instance water. A living system, however, contains a large number of chemical components, all of which contribute to the observed proton magnetic resonance signal. These components possess different relaxation times. Thus, the analysis of the observed signal in terms of the different subsystem parameters (concentration and relaxation times) is complex but important.

For the sake of simplicity, we will deal with T1 only in two-component systems. A similar discussion is possible for T2.

For example, T1 of muscle tissue protons obtained at 0.1 Tesla is about 300-400 ms, but more than three-quarters of the received proton signal stems from water protons, which in the pure liquid show a T1 of several seconds. Using an example from clinical routine, cerebrospinal fluid (CSF) has similar relaxation times as water. Brain edema, which reflects pathologically high water content in brain tissue, possesses relaxation times that are closer to brain tumors than to CSF (Figure 4-3).

What is the reason for this discrepancy ?

This is best explained using the relaxation rate R1. R1 equals 1/T1. Different R1 components can be added to each other to create a new R1.

It has to be emphasized that *any* biologic sample is a complex mixture of different chemical compounds. The T1 of such a sample is a parameter reflecting the physical and chemical properties in the environment of the observed nuclei. If the environment is not the same throughout the sample, then the obtained T1 will only reflect the mean properties of the sample. In most tissues, one component, usually water, dominates the relaxation behavior. In special cases, where two components having significantly different T1 values are present in comparable amounts, a complex situation arises, which makes a quantitative interpretation more difficult.

Let us consider two systems containing two different groups of protons, one moving fast, one moving slower. Both possess different T1 relaxation times and thus different R1 relaxation rates. We can compare them with the example given in Figure 4-6, in which there

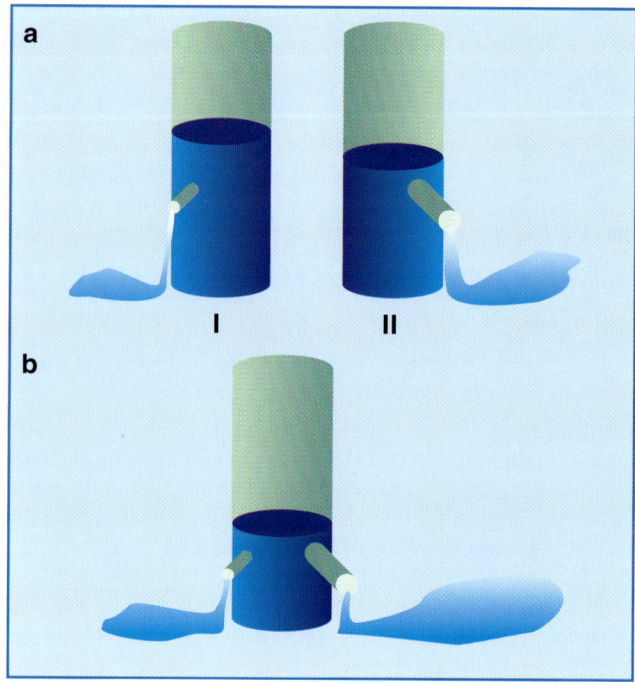

Figure 4-6:
The container example explains the use of relaxation rates instead of relaxation times in complex systems. (a) Two containers I and II with different sized outlets; (b) one container with two different sized outlets.

are two containers, I and II, filled with water. Both of them have an outlet, but the outlet of II is larger than that of I (Figure 4-6a). The rate, R, at which water is leaving I and II can be expressed in milliliters per second, and the time needed to empty the containers is given by V/R, where V is the volume of the water (assuming that the water pressure is constant).

If we construct a container with volume V and equip it with two outlets (Figure 4-6b), one similar to the outlet of container I and one similar to the outlet of II, then the water in this container will leave at a rate which is the sum of the two outlet rates.

This reflects the relaxation time of a tissue composed like our example in Figure 4-3. Although we have two different components, we only measure one common relaxation time for this tissue.

If the exchange rate between the two groups of protons is very slow or absent, it is possible to distinguish the two different contributions to the relaxation behavior. A physical reason for such a behavior can be found, for example, in samples containing both fat and muscle tissues. The fat cannot exchange protons with the water in the muscle tissue. In the case of slow proton exchange, the system will show double exponential relaxation. Other biological systems can show a single exponential relaxation behavior, as if they were relaxing governed by a single relaxation time.

It is possible to distinguish the data, provided that enough data points are available. However, the accuracy actually needed for such measurements is often underestimated.

Cross Relaxation

Solids, such as proteins and membranes, have a wide range of resonance frequencies, which allows for energy exchange between different parts of the solid. The process of energy exchange in a solid is referred to as *spin diffusion*.

Thus, if part of the solid relaxes more rapidly than the rest it can enhance the relaxation of the whole solid.

A similar process can occur between solids and bound water molecules, with the presence of solids (such as proteins and membranes) in tissue acting to reduce the observed relaxation time for the water.

This process is described as off-resonance irradiation and can be exploited to enhance contrast (magnetization transfer contrast; Chapter 11).

The T1 Process
on the Macroscopic Scale

To recapitulate, the T1 relaxation time is the time required for the system to recover to 63% of its equilibrium value after it has been exposed to a 90° pulse. To measure this time, several radio frequency pulse sequences can be employed.

Partial Saturation Pulse Sequence

This pulse sequence is the most simple sequence in magnetic resonance. Sometimes it is also called 'saturation recovery' pulse sequence, although the latter sequence differs from partial saturation by longer repetition times.

If at time zero, the equilibrium magnetization M_0 is exposed to a 90° pulse, it will be tipped down into the x'-y' plane.

After a delay time, called repetition time TR, the spin system is exposed to a second 90° pulse (Figure 4-7), which brings the magnetization down in the x'-y' plane where the FID can be monitored.

If TR is equal to, or greater than, 5 × T1, the mag-

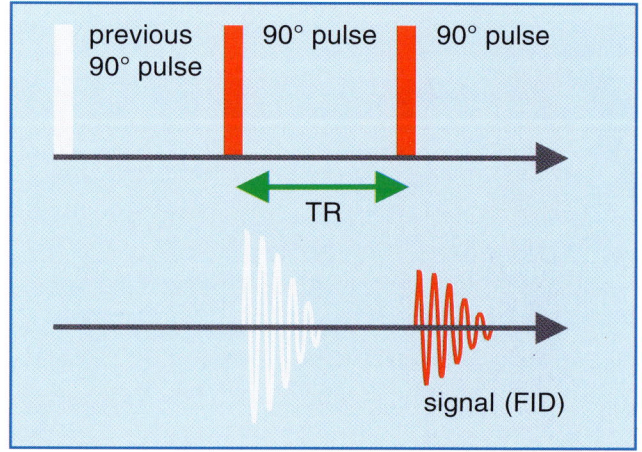

Figure 4-7:
Pulse sequence diagram of a partial saturation sequence, consisting of 90° pulses. The time between the pulses is called repetition time, TR. When TR is not long enough for the spins to return completely to the equilibrium (i.e., TR<5×T1), the signal intensity of the FID is lower than the maximal signal intensity possible.

Figure 4-8:
Partial saturation sequence. The magnetization M_0 is tipped by a 90° pulse. During the repetition time, TR, the system will relax and magnetization will start its return to the equilibrium state. To monitor the size of the magnetization, the system is exposed to a second 90° pulse.

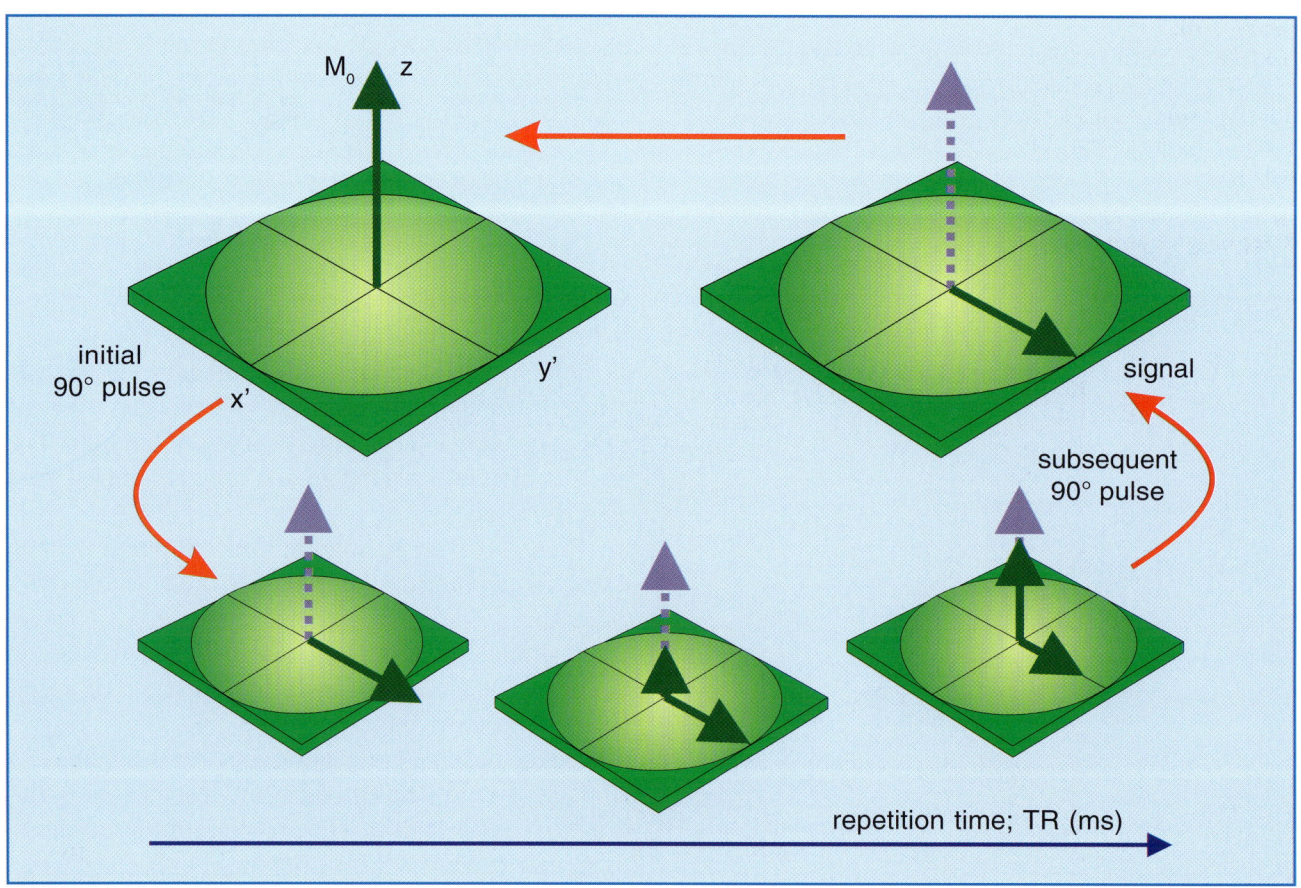

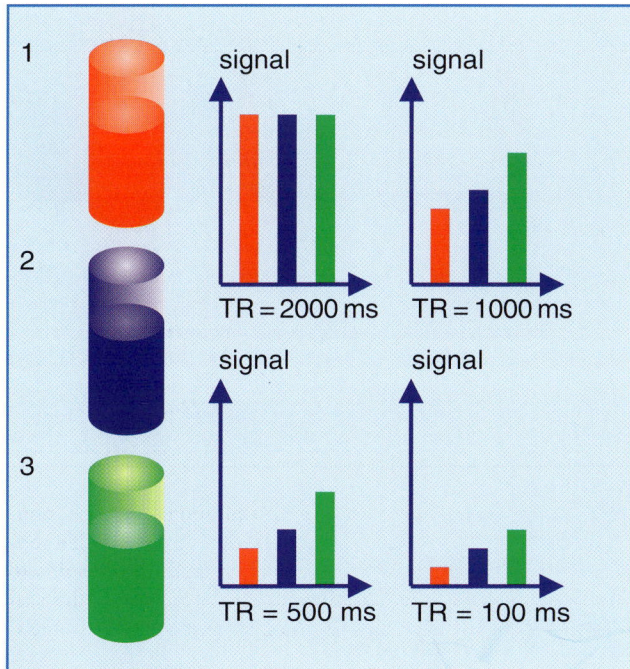

Figure 4-9:
Three different samples — (1) blood, (2) muscle, and (3) fat — assumed to have identical amounts of hydrogen but decreasing relaxation times, are exposed to a train of pulses with different repetition times TR. Note that the blood sample shows the most pronounced saturation behavior, since it has the longest T1.

Figure 4-10:
The relative signal intensity (SI) in a partial saturation experiment. TR is the repetition time between two 90° pulses. Two different tissues with T1 relaxation times of 500 and 1,500 ms, respectively, are shown. The signal recovery is 63% after a period of T1.

netization in the x'-y' plane is equal to M_0. However, if TR is comparable to T1, incomplete relaxation will take place, leading to an observable magnetization smaller than M_0 (Figure 4-8).

The time dependence of Mz (the z-magnetization which equals the signal intensity) on TR, Mz(TR) can be studied by introducing a range of fitting repetition times TR. In the simplest case, the return to equilibrium is a monoexponential function:

$$Mz(TR) = Mz(0) \left(1 - \exp[-TR/T1]\right)$$

Thus, it is understandable that if the system is being re-excited at a repetititon time TR smaller than 5 × T1, the recorded magnetization is less than the maximum value M_0. How much less depends on the ratio of TR over T1.

This effect can be utilized to great advantage if different substances in a given sample have different T1 values. It is possible to reduce part of the signal emerging from the sample, for instance to suppress the signal emerging from fatty tissue.

Also, different samples respond very differently to a train of equidistant 90° pulses (Figures 4-9 and 4-10). This is the basis for TR-influenced contrast behavior in MR imaging.

Partial saturation as described here is only used in a modified form in clinical imaging, namely with the incorporation of a gradient echo. This corresponds to a FLASH imaging sequence which will be discussed at a later stage.

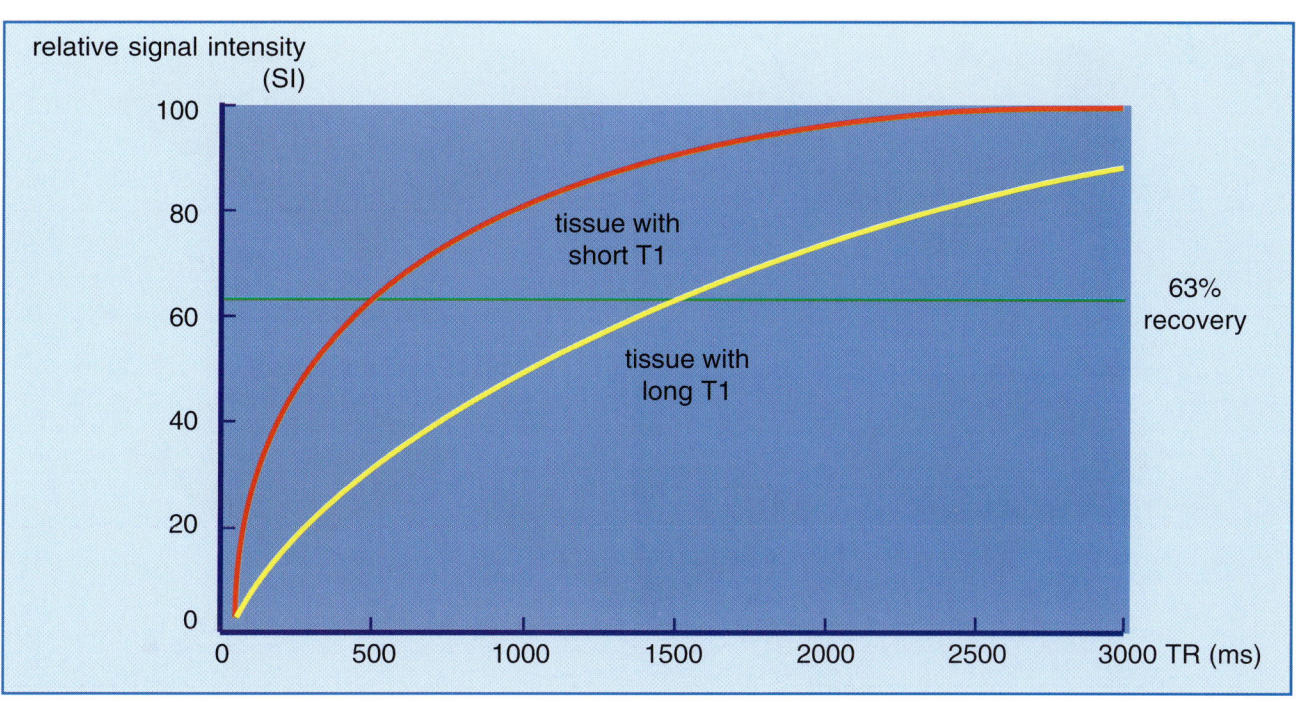

Inversion-Recovery Pulse Sequence

If a spin in equilibrium is subjected to a 180° pulse, the sum magnetization M_0 is inverted with respect to the direction of the external field and becomes antiparallel to the main magnetic field.

Following the inversion, the magnetization starts to recover towards its equilibrium state, the rate of recovery being determined by T1. If after a certain delay time, the inversion time TI, we expose the system to a 90° pulse, the actual magnetization $Mz(TI)$ will become observable in the x'-y' plane as an FID.

By applying a range of different delay times, the time dependence of magnetization, and thus the signal of the inversion time, can be studied in detail. After a delay time of approximately $5 \times T1$, the magnetization is back to equilibrium.

This 180°-90° pulse sequence is called an *inversion-recovery sequence* (Figures 4-11 and 4-12).

In the simplest case, the return to equilibrium can be expressed mathematically as a monoexponential function:

$$Mz(TI) = Mz(0) \; [1 - 2 \times \exp(-TI/T1)].$$

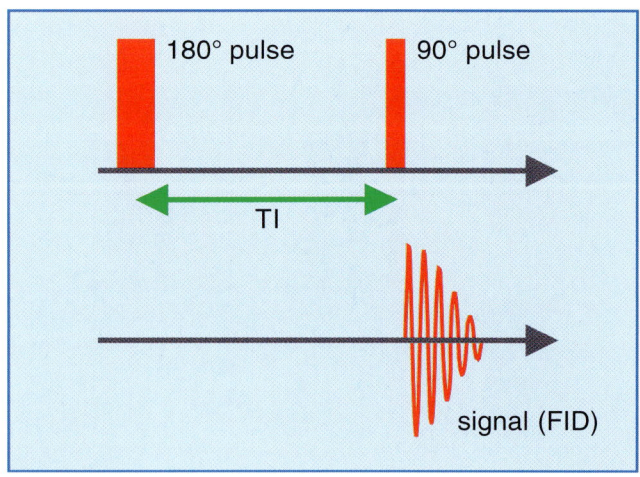

Figure 4-11:
Pulse sequence diagram of an inversion-recovery sequence. The 180° inversion pulse inverts the magnetization. During the inversion delay (TI), the magnetization recovers at a rate determined by the T1 of the sample. At a certain point during recovery, a 90° pulse is applied and the resulting signal is measured.

Figure 4-12:
Inversion-recovery sequence. The magnetization M_0 is inverted by a 180° pulse. During the delay time TI, the system will relax and magnetization will start its return to the equilibrium state. To monitor the size of the magnetization, the system is exposed to a 90° pulse, which tips the magnetization into the x'-y' plane and converts the magnetization into signal.

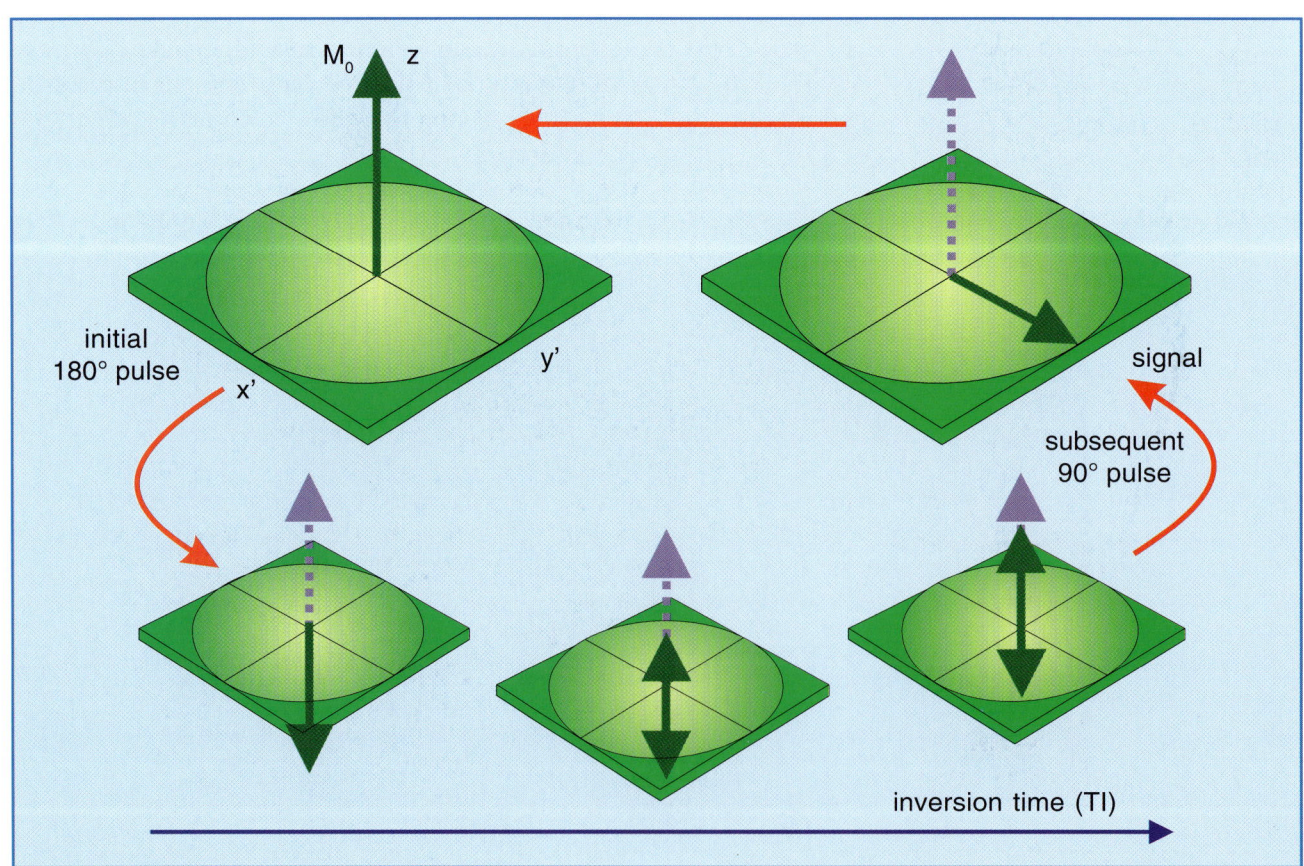

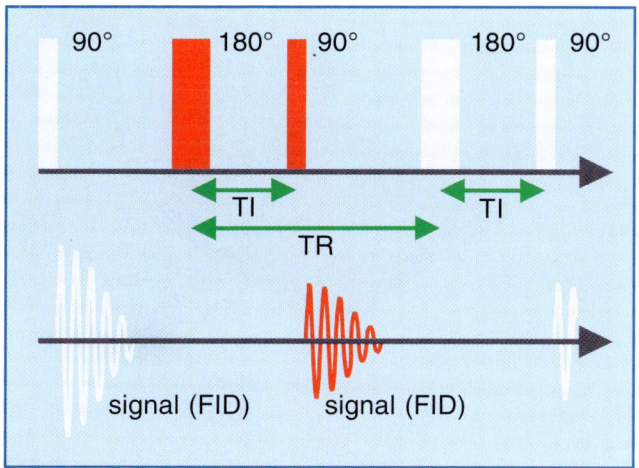

Figure 4-13:
Pulse sequence diagram of an inversion-recovery sequence. The first 90° pulse is from the previous cycle. Data collection follows after the second 90° pulse. The time between the 180° pulses is called repetition time, TR. When TR is not long enough for the spins to return completely to the equilibrium (i.e. TR<5× T1), the signal intensity (SI) is lower than the maximal intensity possible.

Figure 4-14:
The relative SI measured in an IR experiment as a function of TI, the time between the 180° pulse and the 90° pulse. Note that $Mz = 0$ for TI = 0.69×T1 (in this example two tissues with T1 = 500, ρ = 72% and 1500 ms, ρ = 100%; TR = 2000 ms).

The development of the signal intensity is depicted in Figures 4-13 and 4-14.

As with the partial saturation pulse sequence, signal intensity depends on the repetition time TR. In the case of the IR sequence, the repetition time is the time between the 180° pulses.

It is advisable to choose a TR at least 3 × T1 of the tissue of interest to allow for recovery of the longitudinal relaxation of that tissue, thereby avoiding a reduction in signal intensity.

In analytical chemistry, inversion-recovery is applied as a 180°-90° pulse sequence; the initial amplitude of the FID is proportional to the value of the net magnetization at the time of the measurement.

In MR imaging, the sequence is commonly adjusted to the needs of creating an image and, for instance, combined with a spin-echo pulse sequence.

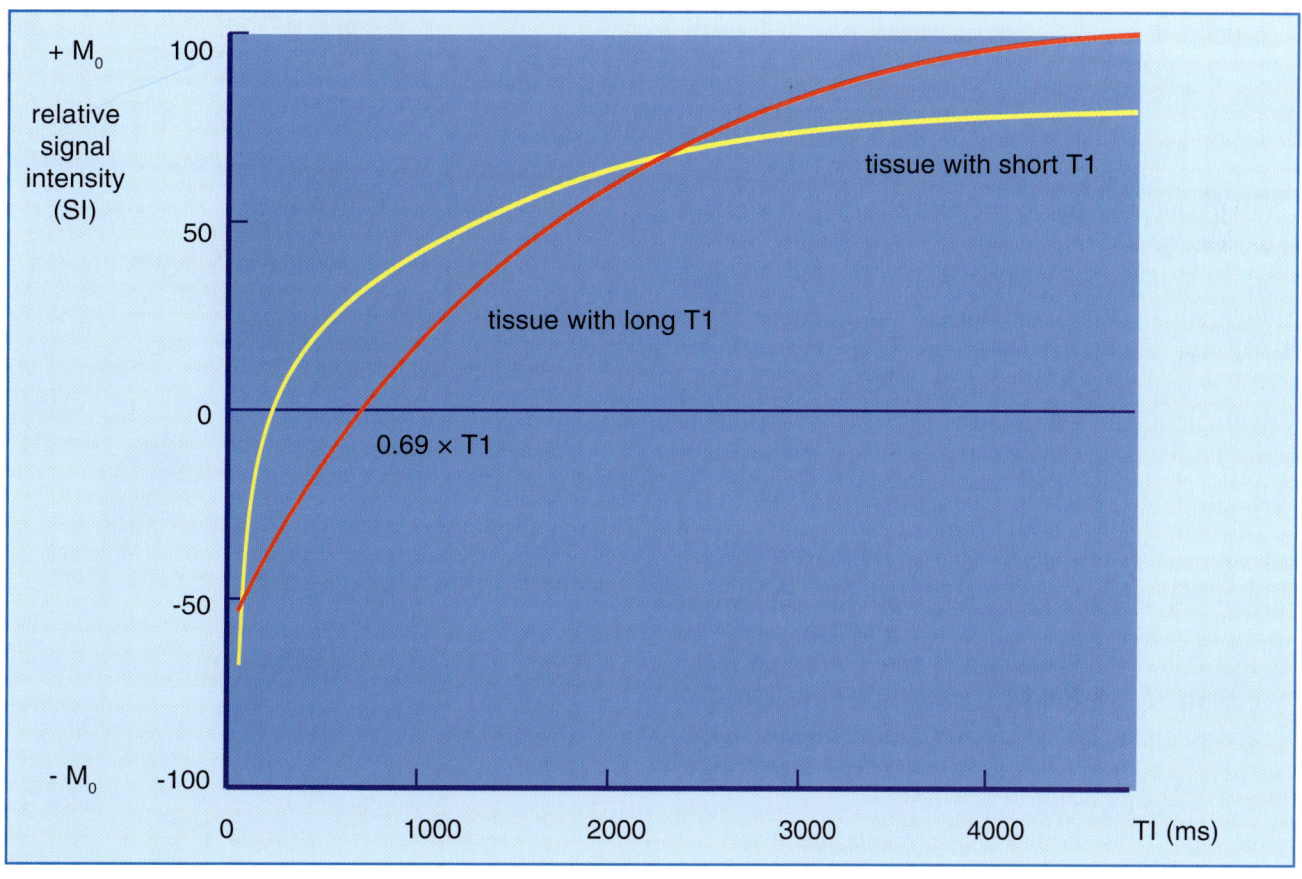

T2, the Spin-Spin Relaxation Time

After a spin system has been excited by an RF pulse, it initially behaves like a coherent system; i.e., all microscopic components of the macroscopic magnetization precess in phase (all together) around the direction of the external field.

However, as time passes, the observed signal starts to decrease as the spins begin to dephase (Figure 4-15). The decay of the signal in the $x'-y'$ plane is faster than the decay of the magnetization along the z-axis.

This additional decay of the net magnetization in the $x'-y'$ plane is due to a loss of phase coherence of the microscopic components, which partly results from the slightly different Larmor frequencies induced by small differences in the static magnetic fields at different locations of the samples.

This process is characterized by T2, the *spin-spin* or *transverse relaxation*.

T2 is dependent on a number of parameters:

- resonance frequency (field strength), although for T2 this is less crucial than for T1;
- temperature;
- mobility of the observed spin (microviscocity);
- presence of large molecules, paramagnetic ions and molecules, or other outside interference.

In mobile fluids, T2 is nearly equal to T1, whereas in solids or in slowly tumbling systems (i.e., high-viscocity systems), static-field components induced by neighboring nuclei are operative and T2 becomes significantly shorter than T1. In solids, T2 is usually so short that the signal has died out within the first millisecond, whereas in fluids the magnetic resonance signal may last for several seconds.

To a large extent, this is the cause of the low or absent signal from solid structures such as compact bone or tendons in medical magnetic resonance imaging.

So, if we represent T1 and T2 versus the microscopic mobility of the spin system, we will obtain for T1 a curve passing through a minimum, corresponding to the Larmor frequency, and a continuously decreasing curve for T2 (Figure 4-16).

For pure water, the T2 value is approximately 3 seconds and the T1/T2 ratio is 1. The T1 value of tissues is usually under 1 second. Here, the T1/T2 ratio increases rapidly with values of 5-10 covering most tissue types. It is about 5 for muscle tissue at 0.1 T.

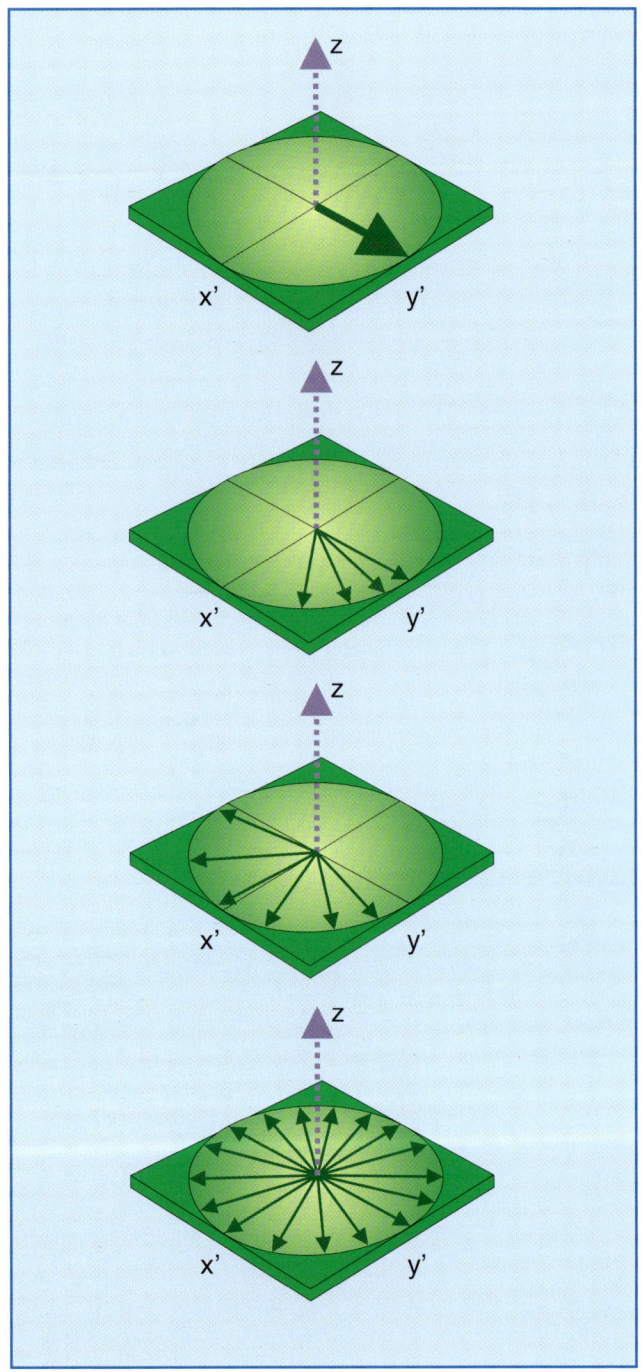

Figure 4-15:
Transverse relaxation phenomena induce an increase in dephasing of individual spins, so a progressive decrease of the macroscopic magnetization is observed.

In practice, it is observed that the same sample can show two different T2 relaxation times at the same field strength. This is because two phenomena contribute to the local inhomogeneity experienced by the nuclei:

- static and oscillating fields locally induced by neighboring magnetic moments (from other nuclei or unpaired electrons), and
- imperfections of the main static magnetic field B_0 (field inhomogeneities).

This leads to a decay of the observed signal which is faster than T2. It is called T2* (Figure 4-17).

We could calculate T1 of complex systems by adding the relaxation rates R1 (see the container example, Figure 4-6). In a similar way we can calculate T2* by adding the R2 relaxation rates. The observed decay rate R2* is thus related to the true spin-spin relaxation rate R2 and to that induced by the field inhomogeneities R2*inh* (or R2'):

$$R2* = R2 + R2inh.$$

To remove the effect of field inhomogeneities, a spin echo can be used, the amplitude of which depends on the time, TE, which has elapsed since the initial excitation.

This is done in one of the most common imaging sequences, the spin-echo pulse sequence, which has been the standard pulse sequence in magnetic resonance imaging and the mainstay of clinical diagnosis.

Even after the introduction of specialized pulse sequences for distinct diagnostic questions, SE remains the pulse sequence to use if any doubt exists.

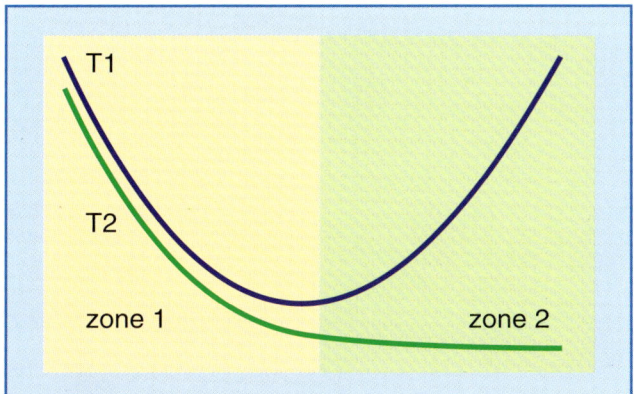

Figure 4-16:
Zone 1: high mobility with fast molecular motion; usually small molecules and 'free' water. Zone 2: low mobility with slow molecular motion; usually large molecules and 'bound' water.

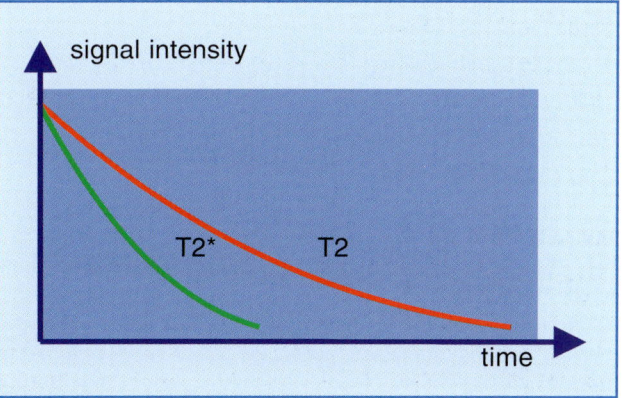

Figure 4-17:
T2 and T2*. The signal decay of T2* is faster than that of T2, because of field inhomogeneities and chemical shifts. However, the T2* can be made reappear by applying a second RF pulse.

Spin-Echo Pulse Sequence

After the system has been excited by a 90° pulse, the spins dephase in the x'-y' plane, i.e., they separate from each other and fan out, some moving faster, others moving slower.

If, after a time delay τ, the system is exposed to a 180° pulse, a refocusing is initialized. Now the faster spins lie behind the slower ones, but they catch up, which leads to an echo at time TE = 2τ (Figures 4-18 to 4-21). The creation of the echo is explained by an analogy in Figure 4-20.

The 180° pulse changes the phase of each spin by 180°; that is, it reverses its phase. The position of the spins has not changed, so they will continue to rotate in the same direction. However, the 180° pulse causes the spins to return towards their starting point (alignment), rather than rotating further away from it.

This 90°-180° pulse sequence is called *spin-echo* (*SE*) sequence. If several 180° pulses are transmitted, echoes of decreasing amplitude are created. This is described as a *multiple spin-echo* or *multiecho* sequence.

T2 is reflected by the envelope of the peaks of the echoes (Figure 4-21). At the center of the echo, the effects of inhomogeneities are cancelled out. Since the maximum amplitude of the echoes is not dependent on inhomogeneities and static gradients, echo amplitudes truly reflect the spin-spin relaxation of the sample.

Flow or diffusion irreversibly bring the spins from one location to another, and so lead to an attenuation of the echo.

The decay after the 90° pulse and on either side of the center of the SE is governed by T2* rather than T2. Therefore, the signal decays rapidly away from the echo center (Figure 4-21).

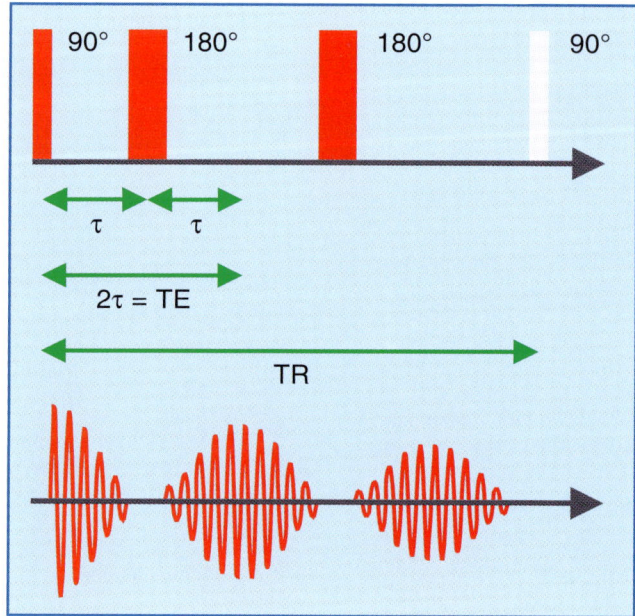

Figure 4-18:
Spin-echo pulse sequence. The spin system is excited by a 90° pulse. After a time delay (τ), one or several 180° pulses follow. This leads to the formation of an echo. The time between the 90° pulse and the peak of the echo is called *echo time* TE. TR is the repetition time between two complete pulse sequences.

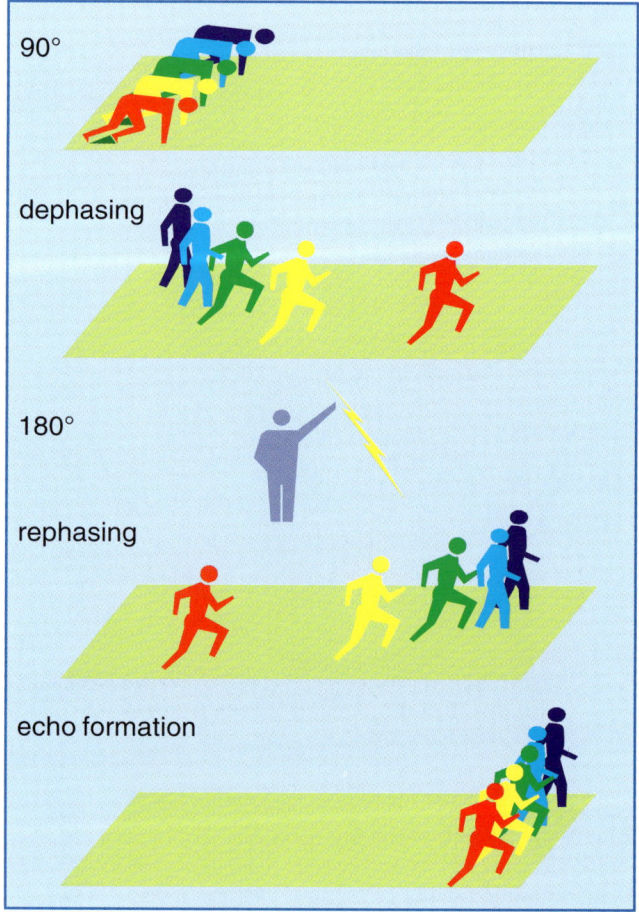

Figure 4-19:
The echo formation in a spin-echo pulse sequence can be compared with a race.
At the time of the 90° pulse, all runners are lined up at the starting line. After the 90° pulse, the faster runners separate from the slower runners (dephasing). At a certain time during the race, the runners are transposed (at the time τ when the 180° pulse is transmitted). Now the faster runners are behind the slower ones, but they catch up.
All reach the finishing line together (i.e., create an echo at the echo time 2τ = TE).

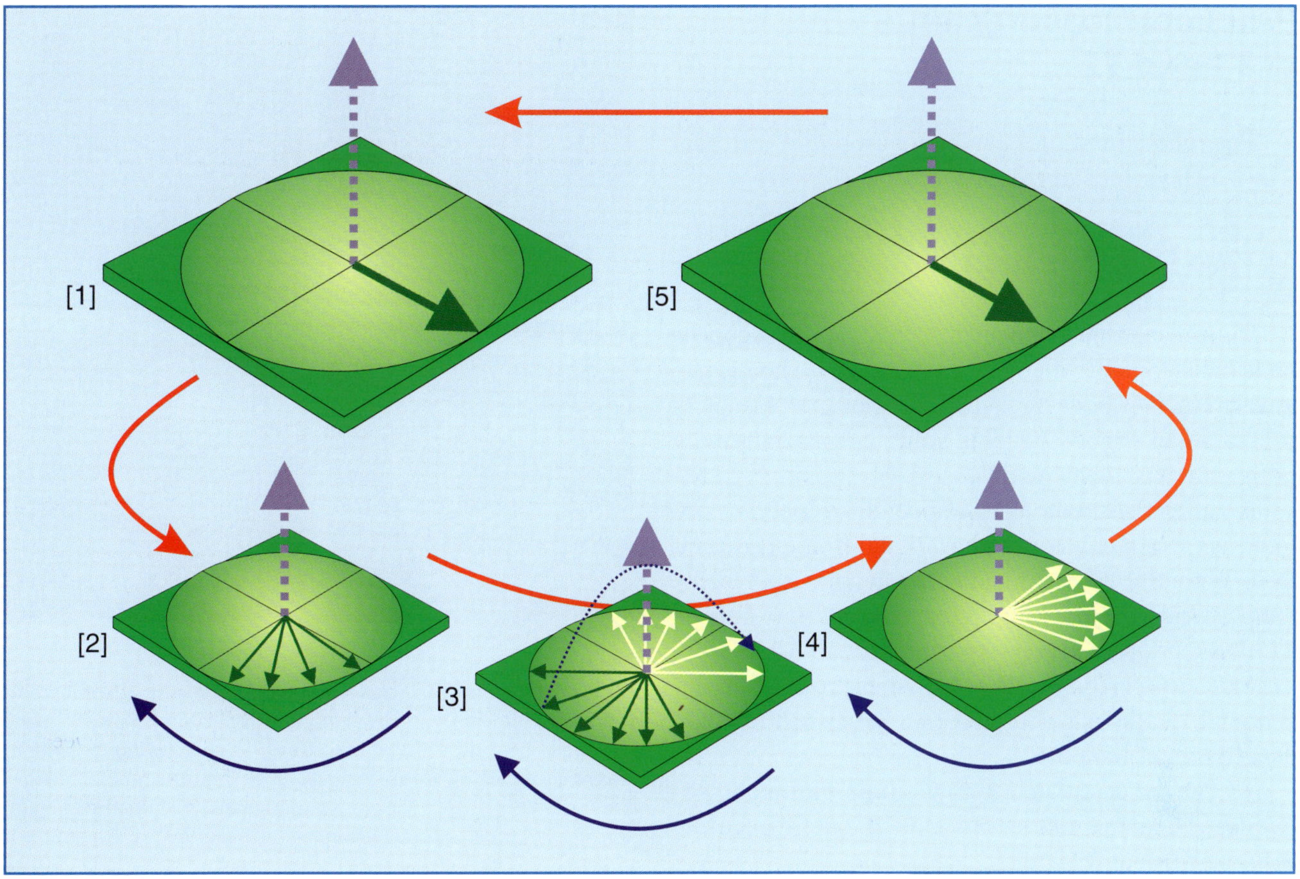

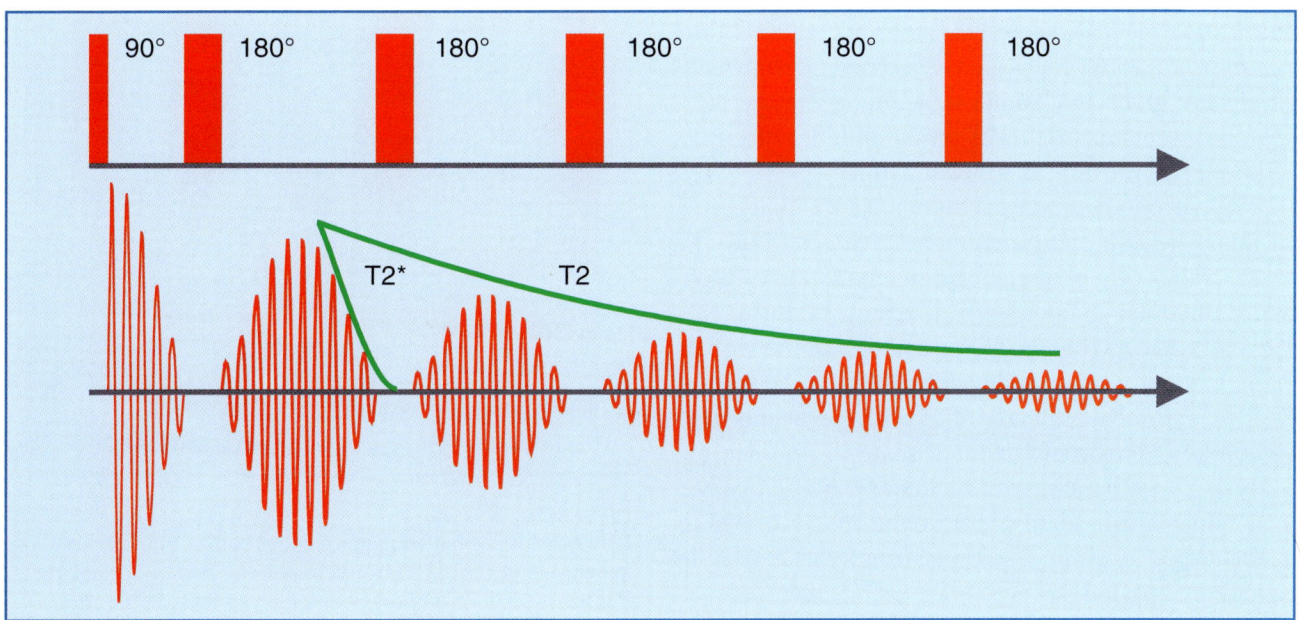

Figure 4-20:
After the system has been excited by a 90° pulse [1], the spins dephase [2]. When the system is exposed to a 180° pulse, the spins are refocused [3]. Now the faster spins are behind the slower ones [4], but they catch up with them, and create an echo at TE [5].

Figure 4-21:
The value of T2* can be obtained from the FID or single echoes while T2 is calculated from the peaaks of the echo amplitudes. Several 180° pulses create echoes of decreasing amplitude (multiecho sequence). The envelope curve drawn through their peaks is the T2 decay curve.

Practical Measurements of T1 and T2

The relaxation times can be measured in different ways with various degrees of accuracy.

In Vitro Determination

High-resolution magnetic resonance spectroscopists have measured T1 values for five decades. The *in vitro* measurement is done on a small sample, approximately 0.1-1.0 ml or slightly larger in volume, in an extremely homogeneous magnetic field.

A variety of methods have been developed to obtain maximal precision with minimal time consumption. Typically, 15 to 30 magnetization measurements are performed on the sample for different time delays, TI in inversion-recovery experiments or TR in partial saturation experiments. Based on these results, an observed T1 value is calculated, and the error limits are usually better than 5%.

T2 can be calculated with a single multiecho sequence. The more echoes one uses, the more accurate the measurment will be.

In Vivo Determination: Localized Measurements and Images

For magnet systems with larger bores, it is possible to examine whole organisms, animals, and people, which is more physiological than examining excised organs or tissues. One of the major problems of *in vivo* relaxation time measurements is the localization of the volume to be observed. Details of such localization techniques are given in Chapter 6. Actual accuracy of *in vivo* measurements depends on the number of points acquired and the quality of localization.

The current method used to obtain a T1 image, i.e., an image whose picture elements represent pure T1 values, relies on a mathematical manipulation of separately obtained images with different T1 influence. Typically two to four images are used and the signals mathematically processed to calculate pure T1 values. Bearing in mind that *in vivo* relaxation can be multiexponential, it is somewhat inadequate to perform the analysis by such a limited fit to an exponential curve.

T2 images are calculated from the images of a multiecho series. In clinical settings, usually four or eight echoes are applied.

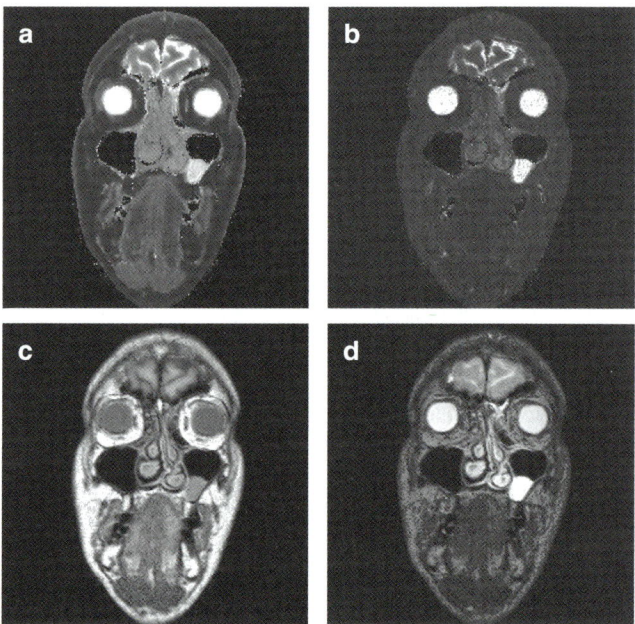

Figure 4-22:
(a) Calculated (pure) T1, and (b) calculated (pure) T2 images of a patient with a polyp in the left paranasal sinus. Figures (c) and (d) show T1- and T2-weighted images. There is far more contrast and detail on these images. Such multiparameter images are far more valuable for clinical diagnosis. The contrast behavior of T1- and T2-weighted images is described in Chapter 10.

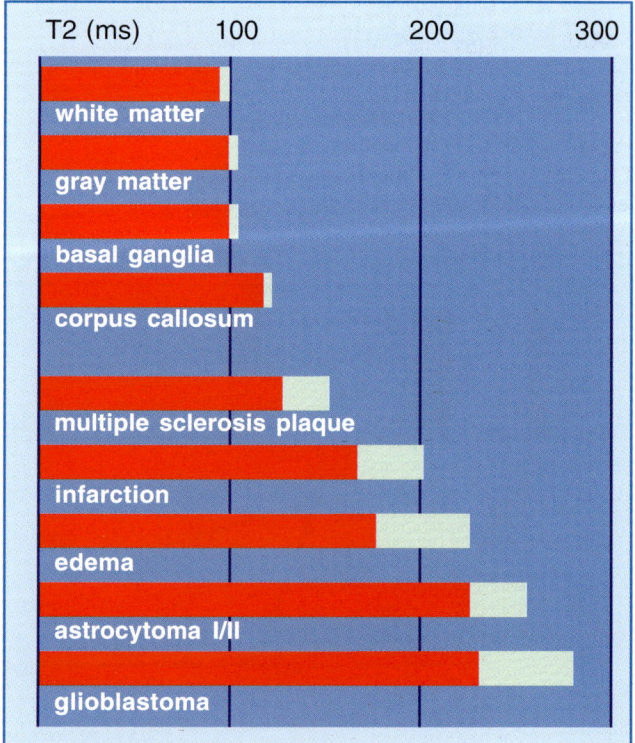

Table 4-3:
T2 values of normal and pathological human brain tissues. The standard deviation (SD) is given in green. SD of normal values is at least 20%, of pathological values at least 30% (after [12]).

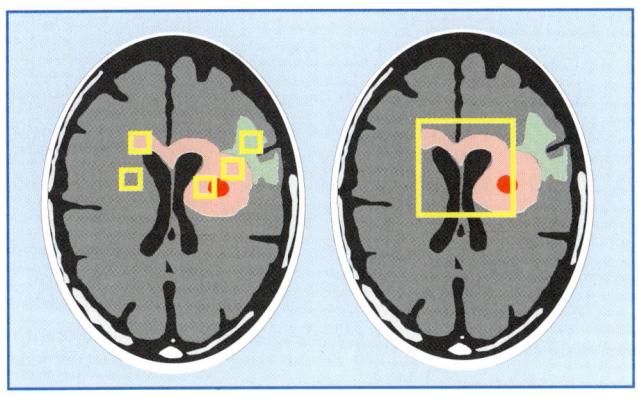

Figure 4-23:
Relaxation time measurements *in vivo* can be performed pixel by pixel and by regions of interest of different size. Left: small regions of interest covering edema (green), tumor (pink), necrosis (red), etc. Right: large region covering the entire tumor. Partial volume effects and other factors influence the measurements.

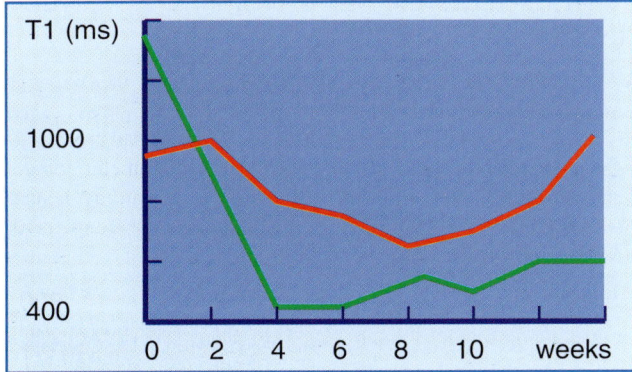

Figure 4-24:
T1 measurements. Follow-up of treatment of acute myeloblastic leukemia. Responder: green; non-responder: red [2].

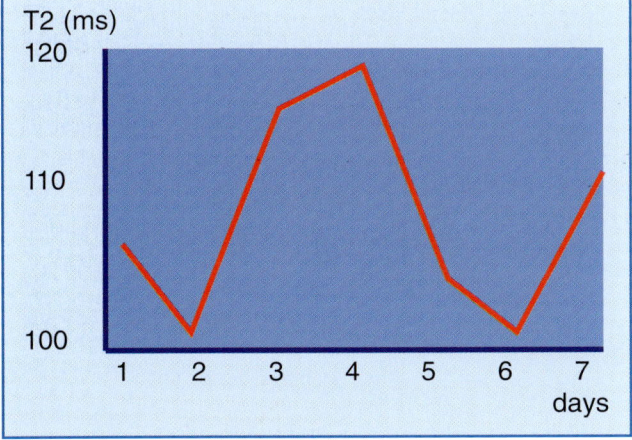

Figure 4-25:
Relaxation-time measurements of identical samples under identical measurement conditions can reveal great standard deviations, as shown in this example. Relying on *in vivo* measurements to evaluate the outcome of treatment is dubious. Only in some instances do massive changes allow a positive assessment.

All methods relying on slices through the examined object will have as additional error source partial volume effects from the edges of the slices. The only method which avoids this slice problem is the true 3D volume imaging method.

Recent progress has been made in using rapid imaging sequences to measure T1; such techniques may prove useful for estimating the concentration of paramagnetic contrast agents in an organ.

Figures 4-22a and b show examples of calculated pure T1 and T2 images.

T1 (T2) Images and T1- (T2-) Weighted Images

In clinical routine, people often talk about T1, T2 or proton-density images. The right terms should be *T1-weighted*, *T2-weighted*, and *proton density-weighted* (or better, *intermediately weighted*) images, because these images have only a certain T1, T2, or proton-density dependence. However, they are not calculated *pure* relaxation time or proton density images. Chapter 10 will explain this in detail. Figures 4-22 c and d are T1-weighted and T2-weighted images, to be compared with the pure T1 and T2 images of Figure 4-22a and b.

Relaxation Time Measurements in Medical Diagnosis

When magnetic resonance was first used for medical puposes, it was thought that relaxation times could help distinguish between tumors and normal tissue since most T1 (and in a similar way T2) values of pathologic tissue can differ markedly from the T1 of the similar normal tissue[5]. But the ability to discriminate, type, or even grade tumors using relaxation-time values has remained a dream, despite the sophisticated multi-point fits introduced over the years.

Table 4-3 shows that there are differences between T2 of normal and diseased tissues. Although values of T2 are more accurate than those of T1 because more points are used for their calculation, these differences are not significant between T2 values of, for instance, tumors and edema or infarction.

The overlapping is due to variations within the same lesion related to vascularity, necrosis, and cell behavior. Similar lesions may have more than a single exponential relaxation rate, e.g., brain tumors and multiple sclerosis plaques. This is not unexpected, considering the heterogeneous nature of tumors. Furthermore, matrix size and slice thickness, i.e., partial volume effects, are limiting factors in relaxation-time measurements *in vivo* because of the many different biological structures within one volume element. Standard deviation in fitting, artifacts, and variations in the selection of volume elements by the operators are all possible sources of error (Figure 4-23). System changes over time add to the uncertainty.

Every year, the literature announces new attempts to exploit relaxation-time measurements *in vivo*. There are some positive reports about its successful use. They all concern follow-up of therapy, with patients being their own reference. Publications include, for instance, the report that relaxation times from leukemic bone marrow can be used for the differential diagnosis of this disease (Figure 4-24)[8]. Similar results in high-grade gliomas have been published by another research group[2].

Another quite interesting relaxation-time study is the measurement of normal appearing white brain matter in multiple sclerosis (MS) patients. Pixel-by-pixel mapping suggests that there could be minute invisible changes in the white matter, which might explain brain function deficits that cannot be explained by the size and location of visible MS plaques[1, 9, 10].

Yet, the follow-up of treatment based upon relaxation-time values is difficult and in most instances dubious (Figure 4-25).

After absolute T1 and T2 values had been used unsuccessfully by researchers, combinations of T1 and T2, histogram techniques, and more sophisticated three-dimensional display techniques of factor representations were used[13].

Availibility of databases of *in vivo* relaxation-time measurements is very limited. The largest collection of data was published by Bottomley et al.[3, 4]. Unfortunately, these values are not very reliable. A comparison between *in vivo* and *in vitro* relaxation measurements is quite difficult because many T_1 relaxation-time values change rapidly after excision. Only brain tissues reveal a relatively stable relaxation behavior after they have been removed from the body[6, 7].

References

1. Barbosa S, Blumhardt LD, Roberts N, Lock T, Edwards RH. Magnetic resonance relaxation time mapping in multiple sclerosis: normal appearing white matter and the 'invisible' lesion load. Magn Reson Imaging 1994; 12: 33-42.
2. Boesiger P, Greiner R, Schoepflin RE, Kann R, Kuenzi U. Tissue characterization of brain tumors during and after pion radiation therapy. Magn Reson Imaging 1990; 8: 491-497.
3. Bottomley PA, Foster TH, Argersinger RE, Pfeifer LM. A review of normal tissue hydrogen NMR relaxation times and relaxation mechanisms from 1-100 MHz: dependence on tissue type, NMR frequency, temperature, species, excision, and age. Med Phys 1984; 11: 425-448.
4. Bottomley PA, Hardy CJ, Argersinger RE, Allen-Moore G. A review of 1H nuclear magnetic resonance relaxation in pathology: are T1 and T2 diagnostic? Med Phys 1987; 14: 1-37.
5. Damadian R: Tumor detection by nuclear magnetic resonance. Science 1971; 171: 1151-1153.
6. Fischer HW, Van Haverbeke Y, Rinck PA, Schmitz-Feuerhake I, Muller RN. The effect of aging and storage conditions on excised tissues as monitored by longitudinal relaxation dispersion profiles. Magn Reson Med 1989; 9: 315-324.
7. Fischer HW, Rinck PA, Van Haverbeke Y, Muller RN. Nuclear relaxation of human brain gray and white matter: analysis of field dependence and implications for MRI. Magn Reson Med 1990; 16: 317-334.
8. Jensen KE, Sorensen PG, Thomsen C, Christoffersen P, Henriksen O, Karle H. Prolonged T1 relaxation of the hemopoietic bone marrow in patients with chronic leukemia. Acta Radiol 1990; 31: 445-448.
9. Lacomis D, Osbakken M, Gross G. Spin-lattice relaxation (T1) times of cerebral white matter in multiple sclerosis. Magn Reson Med 1986; 3: 194-202.
10. Rinck PA, Appel B, Moens E. Relaxationszeitmessung der weissen und grauen Substanz bei Patienten mit multipler Sklerose. Fortschritte Röntgenstr 1987; 147: 661-663.
11. Rinck PA, Fischer HW, Vander Elst L, Van Haverbeke Y, Muller RN. Field-cycling relaxometry: medical applications. Radiology 1988; 168: 843-849.
12. Rinck PA, Meindl S, Higer HP, Bieler EU, Pfannenstiel P. Brain tumors: detection and typing by use of CPMG sequences and *in vivo* T2 measurements. Radiology 1985; 157: 103-106.
13. Skalej M, Higer HP, Meves M, Brückner A, Bielke G, Meindl S, Rinck P, Pfannenstiel P. T2-Analyse normaler und pathologischer Strukturen des Kopfes. Digit Bilddiagn 1985; 5: 112-119.

Interlude Two

Relaxation-Times Blues

Outstanding soft-tissue contrast is among the main characteristics of MR imaging that have enabled the technology to be developed so rapidly. This contrast is basically the result of the relaxation phenomena, T1 and T2.

Following the impulse given by a radiofrequency burst, the process of returning to a state of equilibrium from an excited state is called the *longitudinal* or *spin-lattice* relaxation process. It is characterized by the T1 relaxation time, which commonly lies in the range of several hundred milliseconds. The T2 relaxation time characterizes the dephasing of the spins (i.e., the separation of neighboring spins from each other), and therefore it is called the *spin-spin* or the *transverse* relaxation process. T2 times of tissues are much shorter than T1 times.

For example, at a magnetic-field strength of 0.5 T, human kidney tissue has a T1 relaxation time of approximately 500 ms and a T2 relaxation time of approximately 80 ms.

Although other factors contribute to contrast on an MR image, the three dominant factors are T1 and T2 times and proton density, the latter reflecting the water content.

Peter Mansfield of the University of Nottingham stated in 1980[3] that

>»NMR imaging of anatomical detail is feasible based purely on the measurement of water content.«

However, he also pointed out that images could reflect a combination of water content and relaxation times.

Proton density does not change much between different tissues. For instance, its difference between gray and white brain matter in an adult is approximately 10%, and the difference between brain pathologies and surrounding uninvolved brain tissue may be even less. Thus, proton density or water-content imaging of the human body is not particularly useful.

Today, magnetic resonance pictures dubbed as proton density-weighted images always depict a combination of water content and the two relaxation times; nobody uses pure water-content pictures for medical diagnostics. Usually, T1- or T2-weighted images are acquired in MR imaging because the two main relaxation processes govern the contrast in medical MR imaging.

Tumors, as well as other brain pathologies such as multiple sclerosis (MS) or brain infarctions, are barely visible on water-content images. This was demonstrated in the early days of MR imaging when, in a number of cases, already known brain lesions could not be discovered. The introduction of T2-weighted spin-echo pulse sequences changed this.

On these images, many pathologies are seen easily. The importance of T2-influenced pictures was demonstrated at a magnetic resonance conference in San Francisco in 1983[5].

Later, all manufacturers started offering this feature with their machines, and now it is part of any MR examination.

The use of relaxation times for medical applications was first proposed, attempted, and patented in 1974 by Raymond Damadian and his collaborators[1]. At that time, Damadian was a medical doctor at the State University of New York at Brooklyn.

Originally, he did not intend to use the relaxation times for imaging but for tissue characterization. The method, for which he gained a U.S. patent, was aimed at screening humans for cancer cells.

Since then, this idea has occupied the minds of many researchers because the ultimate goals of diagnostic medicine are noninvasive tissue characterization and the external identification of malignant cells within the human body, without even touching the body. Damadian's claim that relaxation-time changes highlight cancer cells seemed to be the pivotal step in medical progress. Thus, it is understandable that relaxation has been described as the Holy Grail of magnetic resonance.

Damadian was a colorful and controversial figure in magnetic resonance circles. He invested a lot in public relations and even sponsored two books written about him[2, 4]. Damadian had and has many opponents, not only because of his exuberant character and unrestrained behavior at conferences but also because of his scientific publications. Immediately after his first publication, his opponents showed that his claims were only founded on particular cases and not on any specific disease. However, this did not stop him continuing to propose his hypotheses.

In spite of Damadian's critics, nobody can deny that his description of relaxation-time changes in cancer tissue was one of the main motivations for the introduction of magnetic resonance into medicine. His assertion that this method can detect cancer has proved to be partly true, but in a different way: MR imaging with pictures influenced by relaxation times has become one of the main medical technologies applied in cancer diagnosis and follow-up.

However, the basic idea of obviating the need for hospital pathology departments and replacing them with MR imaging did not materialize.

In vivo relaxation-time measurements based on MR imaging have been tried out over the years by a large number of people, who have used relaxation-time values for tissue characterization in the brain, body, muscles, and bones. The task proved to be in vain because all efforts to characterize or even type tissue largely failed.

The reasons are manifold and include systematic measurement errors, inaccuracy of two-point plotting methods of relaxation curves, inherent variability of tissue composition, partial volume effects, and inter-observer variability. Researchers realized that it is futile to measure a point or a region of interest within a tumor because too many different components such as tumor and necrotic cells, small vessels, calcifications, and other structures can be found within a volume of interest. In addition, T1 and T2 values overlap with those of other pathologies and sometimes normal tissue: T1 and T2 of normal tissue change with age and hormonal cycles, breast tissue being a good example.

In 1985, it was realized that even carefully performed *in vivo* T2 measurements cannot be used as a diagnostic method in cancer detection, characterization, or typing[6].

After absolute T1 or T2 values had been used unsuccessfully by researchers, combinations of T1 and T2, histogram techniques, and more sophisticated 3-D display techniques of factor representations were applied. However, the heterogeneity of normal tissues as well as of pathological benign and malignant tissues did not allow the pathologist's view through the microscope to be replaced with MR techniques.

Damadian also claimed that T1 values of tumorous tissue are always higher than those of normal tissue. His dream of MR being the perfect screening method for cancer tissue in the human body was finally shattered when this claim was refuted. T1 values depend on the magnetic field strength (i.e., they increase with the magnetic field). Some tumor values can be lower than the values of normal tissue in certain fields while others are the same in certain fields, and therefore they cannot be distinguished.

Every year, the literature reports new attempts to change the relaxation-times blues into something more swinging. There are some positive stories about the successful use of relaxation-time measurements *in vivo*.

As mentioned in Chapter 4, a quite interesting relaxation-time study is the measurement of *uninvolved* white brain matter in MS patients. MS plaques in the brain have longer T2 relaxation times than surrounding tissue, which enables them to be visualized on T2-weighted spin-echo images. However, the inconspicuous-looking white matter in the rest of the brain is also changed by the disease. Relaxation-time measurements revealed longer T2 values than in normal subjects. This is not enough to diagnose MS, but it might be of use in follow-up therapy or in helping with the differential diagnosis.

References

1. Damadian R. Tumor detection by nuclear magnetic resonance. Science 1971; 171: 1151-1153; and Damadian R, Zaner K, Hor D, Dimaio T. Human tumors by NMR. Physiol Chem and Physics 1973; 5: 381-402.
2. Kleinfield S. A machine called indomitable. New York: Times Books. 1985.
3. Mansfield P, Morris PG, Ordidge RJ, Pykett IL, Bangert V, Coupland RE. Human whole body imaging and detection of breast tumours by NMR. Phil Trans R Soc Lond 1980; B289: 503-510.
4. Mattson J, Simon M. The pioneers of NMR and magnetic resonance in medicine. The story of MRI. Ramat Gan: Bar-Ilan University Press. 1997.
5. Rinck PA, Bielke G, Meves M. Modified spin-echo sequence in tumor diagnosis. Magn Reson Med 1984; 1: 237.
6. Rinck PA, Meindl S, Higer HP, Bieler EU, Pfannenstiel P. MR imaging of brain tumors: discrimination and attempt of typing by CPMG sequences and *in vivo* T2 measurements. Radiology 1985; 157: 103-106.

Chapter Five

Magnetic Resonance Spectroscopy

Chemical Shift

Figure 5-1:
Old wine bottles, full of spider nets — they must contain an excellent red wine: MR spectroscopy has many useful applications, for instance checking if wine (or orange juice) you want to drink has been adulterated or contaminated. This can be done with SNIF-NMR spectroscopy, which tells the specialist whether the juice (or wine) contains the sugars of a pure fruit juice or of added sugars. Unfortunately, we are more interested in medical applications of MR spectroscopy.

Until now we have assumed that everything resonates at the same resonance frequency in a given magnetic field. However, ^1H signals do not all come at the same frequency and so, for instance, the fat signal is generally shifted from its 'correct' position.

Why does ^1H in H_2O have a different resonance frequency to ^1H in fat ?

Even though both protons are within the very large, uniform external magnetic field, they actually experience slightly different magnetic fields due to their chemical environments. Each ^1H nucleus is surrounded by other nuclei and electrons, all of which have a small magnetic field associated with them.

In fact it is the electrons in chemical bonds which are most significant in affecting the magnetic field experienced by a nucleus. Thus, a ^1H nucleus in H_2O is mostly influenced by electrons in H-O bonds, whereas a ^1H nucleus in fat is mostly influenced by electrons in H-C bonds.

These differences in resonance frequency caused by the nuclei experiencing different chemical bonds are used for MR spectroscopy (Figure 5-1). The differences themselves are known as the *chemical shift*, δ (Figure 5-2).

Chemical shift is simply a difference in frequency and is measured in Hz. The difference in frequency varies with magnetic field so that the chemical shift between water and fat is about 350 Hz at 2.35 Tesla, but about 700 Hz at 4.7 Tesla.

Fortunately, the change in the frequency difference is directly proportional to the change in the external magnetic field. If the chemical shift in Hz is divided by the basic resonance frequency of the nucleus in Hz, one obtains a number for the chemical shift, e.g., between water and fat, which is identical regardless of the strength of the applied magnetic field.

Chemical shifts are typically in the range of tens to hundreds of Hz, whereas the resonance frequencies are typically in the range of tens to hundreds of MHz. This makes the values of the chemical shifts rather small, so the numbers are always multiplied by one million and expressed in a *parts-per-million* scale, or *ppm*.

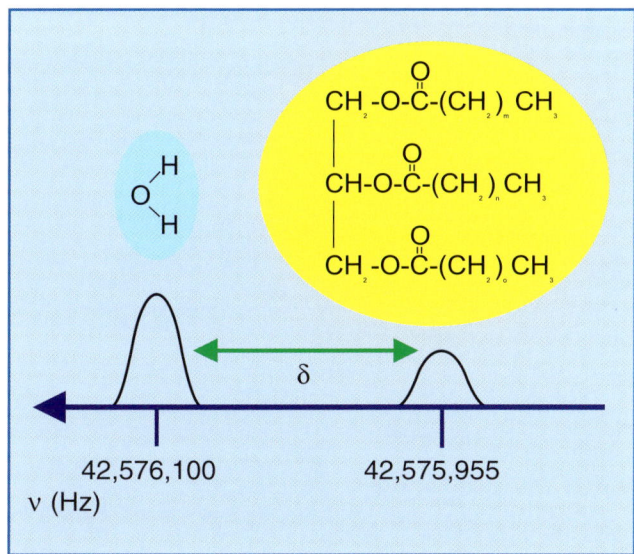

Figure 5-2:
Chemical shift (δ): a proton (^1H) spectrum of tissues often reveals two clearly distinct peaks. One is assigned to tissue water, the other one to protons in lipids (in this case triglyceride). Data given in Hertz for 1.0 Tesla.

Phosphorus Spectroscopy

Although it is very easy to see fat and water in a spectrum of tissue it is not very interesting from a medical point of view, and in fact the signals of the fat and water are so large that it can be very difficult to see the interesting metabolites.

Therefore, it is not too surprising to learn that early MRS studies tended to concentrate on other nuclei, and the most popular was ^{31}P.

Figure 5-3 shows that at a given magnetic field, the resonance frequency of ^{31}P will be approximately 0.405 that of ^{1}H, so that at 2.35 Tesla ^{1}H has a resonance frequency of 100 MHz and ^{31}P has a resonance frequency of 40.5 MHz. It is obvious then that a ^{31}P transmitter or receiver coil will have to be tuned to a very different frequency range than a ^{1}H transmitter or receiver coil, so that there is no possibility of detecting the wrong nucleus.

The same figure shows that ^{31}P has a wide chemical shift range due to the wide range of chemical bonds which can be formed to phosphorus. However, the chemicals which are of interest in biological studies fall within a much smaller range of chemical shift (about

ADP	adenosine diphosphate
ATP	adenosine triphosphate
Cho	choline
Cr	creatine
CSI	chemical-shift imaging
Gln	glutamine
Glu	glutamate
Glx	glutamate/glutamine multiplet
Lac	lactate
ml	myoinositol
NAA	N-acetyl-aspartate
P_i	inorganic phosphate
P C r	phosphocreatine
P D E	phosphodiester
P M E	phosphomonoester
p p m	parts-per-million
S V S	single voxel spectroscopy

Table 5-1:
List of abbreviations frequently used in MR spectroscopy.

Figure 5-3:
(a) An overview of the resonance frequencies of a number of nuclei relative to ^{1}H.
(b) An expansion of the frequency region where phosphorus nuclei give rise to signals.
(c) An expansion of the frequency region showing where phosphorus-containing molecules of biological interest give rise to signals.

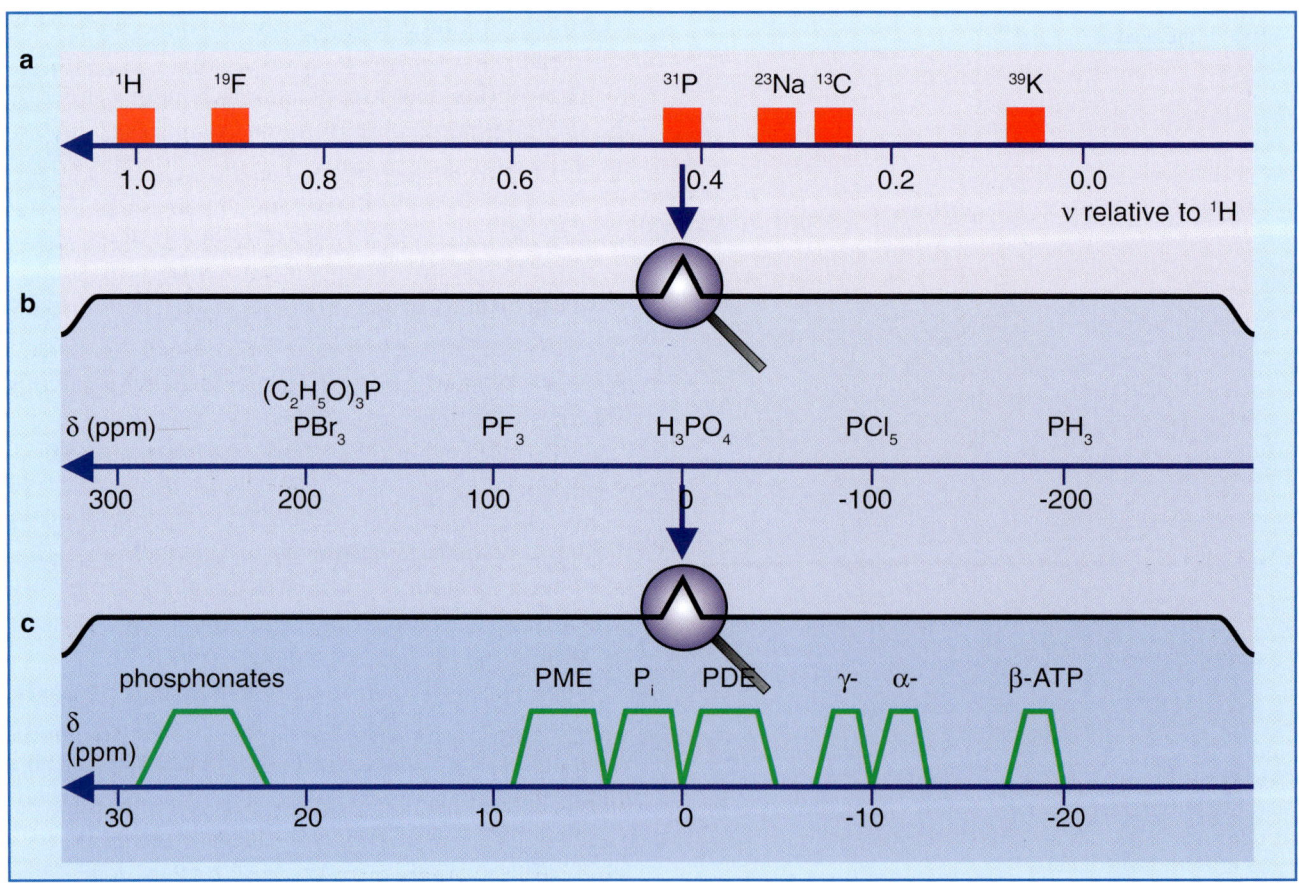

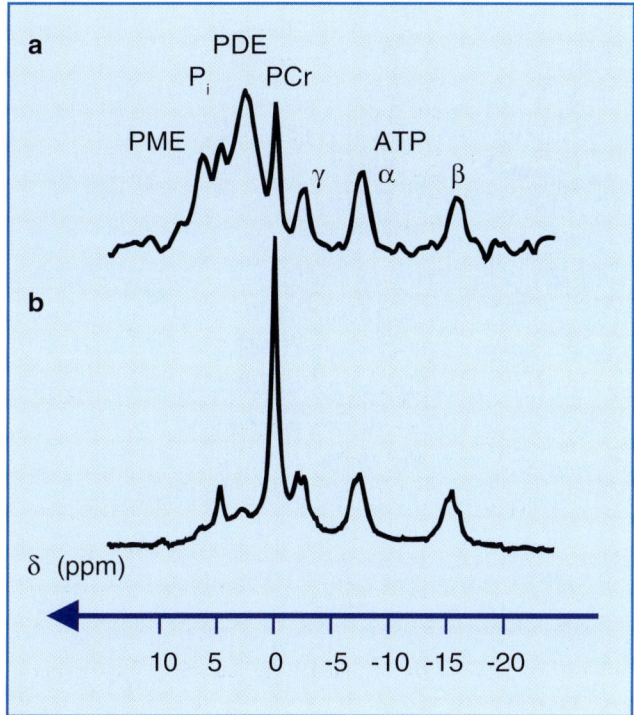

Figure 5-4:
(a) ^{31}P spectrum of a human brain; (b) ^{31}P spectrum of a human leg muscle.
The different concentrations of phosphorus metabolites in the two tissues affect the relative peak area, and the different physical environments of the metabolites in the tissues affect the widths of the peaks.

25 ppm). The metabolites which can typically be seen in *in vivo* spectra are phosphocreatine (PCr, important in energy metabolism), inorganic phosphate (P_i), phosphomonoesters (PME, including sugar phosphates like glucose-6-phosphate), phosphodiesters (PDE, which can include lipid precursors like phosphoryl-choline and phosphoryl-ethanolamine, and lipids themselves) and adenosine-triphosphate (ATP, which has a central role in energy metabolism) (see also Table 5-1).

Apart from ATP, all of these metabolites just have one phosphorus atom per molecule so that there is only one resonance or peak per molecule. This makes the spectra relatively simple to interpret, but it can also be a drawback if the cells have a number of chemically similar phosphorus-containing molecules. Thus, it is very difficult to distinguish between glucose-6-phosphate, fructose-6-phosphate, and other hexose-6-phosphates since all peaks tend to overlap.

This is a general problem in spectroscopy; the ability to distinguish between peaks is called *resolution*, and if two signals have a big enough difference in chemical shift so that they can be seen as two peaks, then they are said to be *resolved*.

ATP is different from the other molecules mentioned so far because it has three phosphorus atoms in each molecule. The α-phosphate has an adenosine group on one side and a phosphate group on the other; the β-phosphate has a phosphate group on each side; and the γ-phosphate has a phosphate group on one side of it. Thus, the three phosphorus atoms are chemically distinct and we see three lines in the ^{31}P spectrum of ATP.

Figure 5-4 shows the ^{31}P spectra of a human leg muscle and a human brain. The two spectra look very different, although most of the signals are present in both spectra. One difference is that the peak line widths are relatively narrow in the leg spectrum, whereas they are relatively broad in the head spectrum. This is due to the different physical environments experienced by the molecules in brain and muscle tissue.

However, the concentration of a metabolite detected in an NMR experiment is proportional to the area under a peak and not just the peak height, so a tall, narrow peak may not necessarily represent a higher concentration than a short, broad peak.

Another difference between the spectra in this figure is the relative signal strengths of the phosphorus metabolites. The muscle spectrum is dominated by the PCr peak and the ATP peaks, whereas the brain spectrum is dominated by the very broad PDE peak, with quite strong signals from PCr and ATP as well.

A good selection of articles about ^{31}P spectroscopy can be found in Matson's and Weiner's review [7], and information about applications can be found in Rubaek Danielsen's and Ross' book [12].

The spectra in Figure 5-4 reflect 'snapshots' of the metabolic state of a tissue, but magnetic resonance spectroscopy can also be used to follow changes in metabolism over a period as short as a few minutes or as long as several weeks or even months.

Figure 5-5 contains a series of ^{31}P spectra of a human calf muscle taken every 90 seconds. During the first three minutes, the muscle was at rest, then the muscle was exercised for 7.5 minutes during which the data acquisition was continued. Finally, spectra were collected for a further twelve minutes as the muscle was recovering. The spectra were plotted to emphasize changes in PCr and P_i. They demonstrate very nicely how PCr is used in the muscle during exercise to maintain the ATP at a constant level; the PCr peak declined, but there was no change in the ATP peaks. The increased P_i level following the hydrolysis of PCr to Cr and P_i during exercise can also be seen in the spectra. With such good time resolution, it would be possible to measure the rate of change of the metabolites as well as the size of the changes.

Spectroscopy can also be used to follow changes over periods of days or weeks, e.g., monitoring the response of a tumor to therapy [15].

The *advantages* of ^{31}P MRS are that phosphorus containing metabolites play an important role in energy metabolism, and they occur in reasonably high concentrations, particularly in muscle tissue, but also in brain and liver, for example.

One *disadvantage* of ^{31}P MRS is that it gives a relatively poor discrimination of metabolites *in vivo*. There might be in the order of 10 different metabolites contributing to the PME peak of an *in vivo* spectrum but the peaks all overlap each other so that the individual metabolites cannot be distinguished.

Another problem with ^{31}P is that outside the major energy metabolites the signals are either too broad to detect, for example phospholipids, or too weak to detect, for example the phosphorylated intermediates of glycolysis, or there are no phosphorus containing metabolites to detect, for example Krebs'-cycle intermediates, amino acids, lipids, sugars, etc.

This means that other nuclei need to be used to study these non-phosphorus containing metabolites.

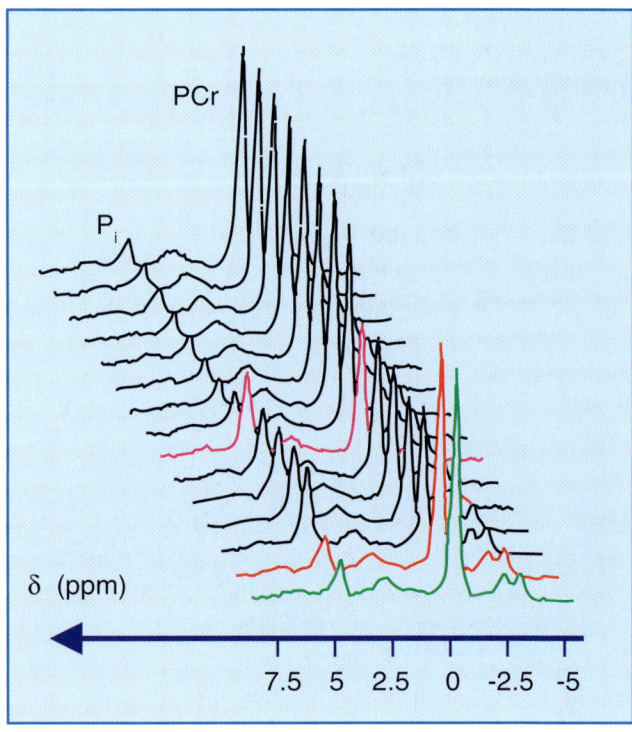

Figure 5-5:
A time-series of ^{31}P spectra of a human calf muscle showing how exercise and recovery affect the phosphorus metabolites. Each 90 seconds a new spectrum was acquired.[16]
Green = rest; red = beginning of exercise; magenta = beginning of recovery.

Nucleus	Spin Quantum Number	Natural Abundance	Relative/Absolute Sensitivity	
^{1}H	1/2	99.98	100	100
^{13}C	1/2	1.11	1.6	0.018
^{31}P	1/2	100	6.6	6.6
^{19}F	1/2	100	83	83
^{23}Na	3/2	100	9.3	9.3
^{39}K	3/2	93.1	0.58	0.047

Table 5-2:
Important NMR properties of selected nuclei used for *in vivo* MR spectroscopy.

	Advantages	Disadvantages
^{1}H	Strong signal; occurs in all biomolecules.	Large H_2O signal; overlapping peaks.
^{13}C	Well resolved peaks; occurs in all biomolecules.	Weak signal; ^{1}H decoupling necessary.
^{31}P	Strong signal; central role in energy metabolism.	Low concentrations; poor *in vivo* discrimination.
^{19}F	Strong signal; no naturally occurring signal.	High drug concentrations required.
^{23}Na	Strong signal; important role in ion balance.	No chemical shift range.
^{39}K	Important role in ion balance; high intracellular concentration.	Weak signal; no chemical shift range.

Table 5-3:
Advantages and disadvantages of selected nuclei for MR spectroscopy.

Spectroscopy of Other Nuclei

The description of spectroscopy has so far concentrated on ^{31}P since it is a good nucleus for describing the basic features of *in vivo* MRS and because historically it was the most commonly examined nucleus. However, interest in ^{1}H MRS has increased rapidly, and it is now more common than ^{31}P MR spectroscopy. Other nuclei like ^{13}C and ^{19}F are becoming more readily accessible by standard equipment. Table 5-2 shows some important properties of selected nuclei which are of interest in biological studies.

The spin quantum number, n, is a fundamental property of the atomic nucleus. Among other things, it is known that the nuclear spins can occupy 2n+1 energy levels, so nuclei with a spin of 1/2 have two possible energy states, whereas nuclei with a spin of 3/2 have 4 possibilities.

Nuclei with spins larger than 1/2 are said to be *quadrupolar*. An important practical feature of quadrupolar nuclei is that their relaxation is sensitive to fluctuating electric fields, as well as fluctuating magnetic fields, so that their T1 and T2 times are much shorter than for nuclei with a spin of 1/2.

Since we want as much signal as possible, it is desirable for a nucleus to have a high sensitivity, although the natural abundance is also important. The relative sensitivities of ^{31}P and ^{13}C differ by about a factor of 4, but because ^{31}P is 100% abundant and ^{13}C is only 1.1% abundant (i.e., about 98.9% of carbon nuclei are the nonmagnetic ^{12}C isotope), the absolute sensitivities differ by about 400 (Tables 5-2 and 5-3).

Alternatively, ^{39}K nuclei are about 31 times less sensitive than ^{13}C nuclei, but because ^{39}K is 93% abundant and ^{13}C is 1.1% abundant, a sample containing potassium would produce a stronger signal than a carbon sample at a similar concentration.

Proton Spectroscopy

^{1}H studies have become increasingly popular as technical difficulties have been overcome and as interest has switched to areas of metabolism which lack phosphorylated metabolites. ^{1}H possesses the strongest response of all the atomic nuclei, and it is found in all biochemicals. Thus, it is a good nucleus for monitoring metabolism[3, 7, 8].

However, there are some major technical problems. Probably the biggest problem with ^{1}H MRS is the large signal from water in tissue. If we assume a rather conservative figure for tissue water content of about 65%, then the molar concentration of the water would be about 36 M. Since there are two ^{1}H nuclei in each molecule of water, this gives a ^{1}H concentration of over 70 M. The metabolites which we wish to observe have a maximum concentration of 10 mM or less, which is at least 7,000 times smaller than the water signal. Therefore, special methods are required for reducing the size of the water signal to a level where it is comparable to that of the metabolites. The simplest method is to use a long selective frequency saturating pulse, but whereas this is very effective *in vitro*, it can lead to unacceptable heating of tissue *in vivo*.

Multiple pulse sequences, such as the binomial pulse sequences, can be used to reduce the signal from water. If the ^{1}H nuclei in water are not excited, then they cannot give rise to a signal. Another method of water suppression exploits the characteristics of a T1 relaxation curve. A selective 180° pulse is used to invert the water magnetization. At first the magnetization will be large and negative, but relaxation processes will start to take it back to its equilibrium values. After 0.69 × T1 of water, also known as the *null point* or the *zero-crossing* time, the water magnetization will be approximately zero. At this point a 90° excitation pulse would produce a relatively strong signal from the metabolites and very little signal from the water. If we were using a spin-echo pulse sequence in association with a particular volume selection method then we could arrange for the echo to be formed at the appropriate time after the selective inversion pulse so as to produce a minimum water signal.

Another problem with ^{1}H MRS is the narrow dispersion of the ^{1}H peaks. They lie in quite a narrow frequency range, thus there is a lot of overlapping. The problem can be helped by working at higher magnetic field strengths, but although 4 Tesla whole-body systems do exist, the majority of MR spectroscopy is performed at 1.5-2.0 Tesla on equipment for which MR imaging rather than MR spectroscopy considerations dictate the field strength.

The main clinical applications of proton MRS of the human brain include epilepsy, space-occupying lesions, multiple sclerosis, degenerative diseases such as Alzheimer's, Parkinson's, and Huntington's, hypoxia, and some metabolic diseases. There are numerous body diseases for which proton MR spectroscopy has been studied, particularly diseases of the muscular system.

Figure 5-6 shows an example of a proton spectrum of a normal human brain.

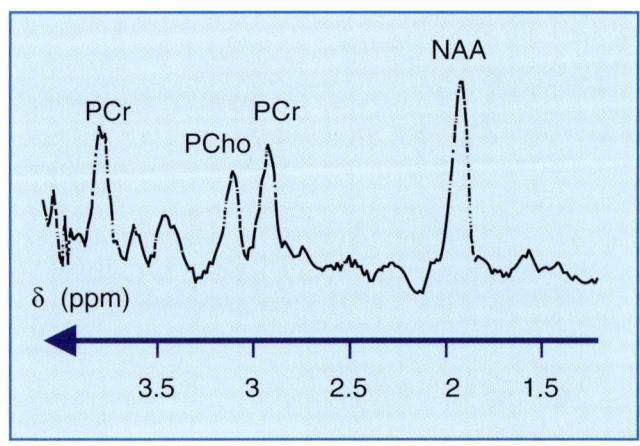

Figure 5-6:
Proton spectrum of a normal human brain. PCr: (phospho) creatine; PCho: (phospho) choline; NAA: N-acetylaspartate.

Carbon Spectroscopy

Unlike ^{1}H and ^{31}P, the magnetically active isotope of carbon, ^{13}C, is not the most abundant form of the nucleus. [13]C occurs in all biochemicals and the signals have a wide dispersion, i.e., they occur over quite a large frequency range, which reduces the likelihood of overlapping peaks. The biggest drawback of ^{13}C is the exceedingly weak signal and problems with [13]C-^{1}H coupling.

In a [1]H-coupled ^{13}C spectrum, most of the peaks will be split into two or more smaller peaks, which complicates the spectrum and reduces the signal-to-noise ratio, but the coupling effect can be removed by decoupling techniques. Direct irradiation at the ^{1}H resonance frequency is the simplest option, although this can lead to heating of tissue *in vivo*.

Multiple pulse techniques are available which are just as effective as direct irradiation but use a fraction of the power. The need for decoupling means that for ^{13}C MR spectroscopy, the instrument must be capable of operating two channels simultaneously, which increases complexity and also cost.

One *advantage* of ^{13}C MR spectroscopy is that we can do labelling studies. By giving ^{13}C-labelled compounds to an animal or patient, we can follow the large and distinctive signals from the labelled compound and try to work out how it is metabolized in the body. Because each carbon position in a molecule will produce a characteristic signal, the labelling experiment can be used to follow not only which molecules end up with the label but also the exact position in which they end up. This information is extremely useful in working out which biochemical pathways were used in the conversion of one molecule to another.

A major *disadvantage* of ^{13}C MR spectroscopy labelling is its costliness.

Similar labelling studies are not possible with ^{1}H and ^{31}P because they are already 100% abundant. ^{13}C MRS can detect signals from sugars, lipids, and glycogen in the liver and in muscle. It can obtain information about the carbon balance of energy metabolism, which is complementary to the information obtainable by ^{31}P MR spectroscopy about energy metabolism [7, 14].

A promising application for ^{13}C spectroscopy is the analysis of body fluids such as blood and urine. This can be done routinely with very high field analytical NMR spectrometers.

Proton spectroscopy can also be used to analyze such samples.

Fluorine Spectroscopy

^{19}F has a strong NMR signal and is 100% abundant. [19]F studies have been performed on the metabolism of fluorine-containing drugs; since there are no naturally occurring ^{19}F signals in the body, all ^{19}F signals must come from the drug or its metabolites. The drawback of ^{19}F MRS is that despite the strong signal, we still need drug concentrations in the order of 1-10 mM in the tissue, which is a rather high concentration for many drugs. The resonance frequency of ^{19}F is quite close to that of ^{1}H at the same magnetic field, and it is often possible to perform ^{19}F studies on the ^{1}H channel of MRS equipment without the need for major modifications [7, 10].

Sodium and Potassium Spectroscopy

[23]Na and [39]K differ from the other nuclei mentioned because they do not have spins of 1/2. They both have a spin of 3/2 and are therefore quadrupolar nuclei. Both are high-abundance isotopes ([23]Na is 100% abundant and [39]K is 93.1% abundant).

[23]Na has quite high extracellular concentrations, whereas [39]K has quite high intracellular concentrations, and both have an important role to play in ion balance. A big difference between the two is that [23]Na has a reasonably strong signal, comparable to ^{31}P, whereas [39]K has a weak signal.

The absolute sensitivity of [39]K is about three times that of ^{13}C, but the [39]K signal is much broader than the ^{13}C signal because the T2 of [39]K is very short. This is due to the effect of quadrupolar relaxation, which reduces the signal-to-noise ratio of the peaks.

[39]K also has a very low resonance frequency, which adds to the technical difficulties of the experiment. Animal studies with [39]K have been performed at 4.7 Tesla, but, to our knowledge, the nucleus has not been studied at 1.5 Tesla.

Although [23]Na also has a broad signal due to quadrupolar effects, the much greater signal strength means that reasonable spectra can be obtained. Unfortunately, [23]Na and [39]K have no natural chemical-shift dispersion. In other words, all the signal from an *in vivo* sample comes at the same frequency. Methods do exist for separating the chemical shift of intracellular [23]Na and [39]K from extracellular [23]Na and [39]K using chemical-shift reagents (rather like the relaxation contrast agents used in imaging), but studies are currently restricted to cells and animals [6, 7, 11].

Localized in vivo Spectroscopy

In traditional chemical/analytical NMR spectroscopy, the sample is placed inside the coil and the signal from the whole sample is observed. This is fine as long as we have a homogeneous sample.

However, if we want to obtain signals from living animals or humans, this approach would lead to signal from a mixture of the different tissues within the detecting region of the coil. Therefore, to obtain *in vivo* spectra from a particular type of tissue, it is necessary to limit or localize the volume from which we actually detect the signal.

The simplest localization technique is to use a surface coil. The size of the volume from which the signal is received is determined by the size and configuration of the surface coil. Whereas this technique is very simple and gives a good signal-to-noise ratio, it has a number of disadvantages, including contamination from surface tissue, a decrease in signal from tissues at increasing depth, and large variations of the flip angle across the excited volume.

The latter problem can be partly reduced if a standard RF coil is used for exciting the whole sample and a surface coil is used to detect the signal from the localized volume, or if specialized (adiabatic) RF pulses are applied.

The other techniques discussed here allow a volume to be defined on previously acquired positioning images, thus reducing the problem of contamination from other tissues. The quality of the magnetic field within the volume of interest is optimized by means of localized shimming. Smaller volumes allow for better shimming, but as the signal in the spectrum is proportional to the volume size, very small volumes are impractical except for ¹H spectroscopy.

For localized spectroscopy of nuclei other than ¹H we must be able to obtain a localized ¹H signal from the same volume; this is required for localized shimming since the sensitivity of the other nuclei is too poor. A number of localization schemes have been proposed. Four of the most widely used schemes will be discussed here.

An extended discussion of spectroscopy techniques can be found in the papers by Matson and Weiner[7] and Sauter and collaborators.[13]

Details on pulse sequences and, e.g., phase encoding are described in the following chapters (mostly in Chapter 6).

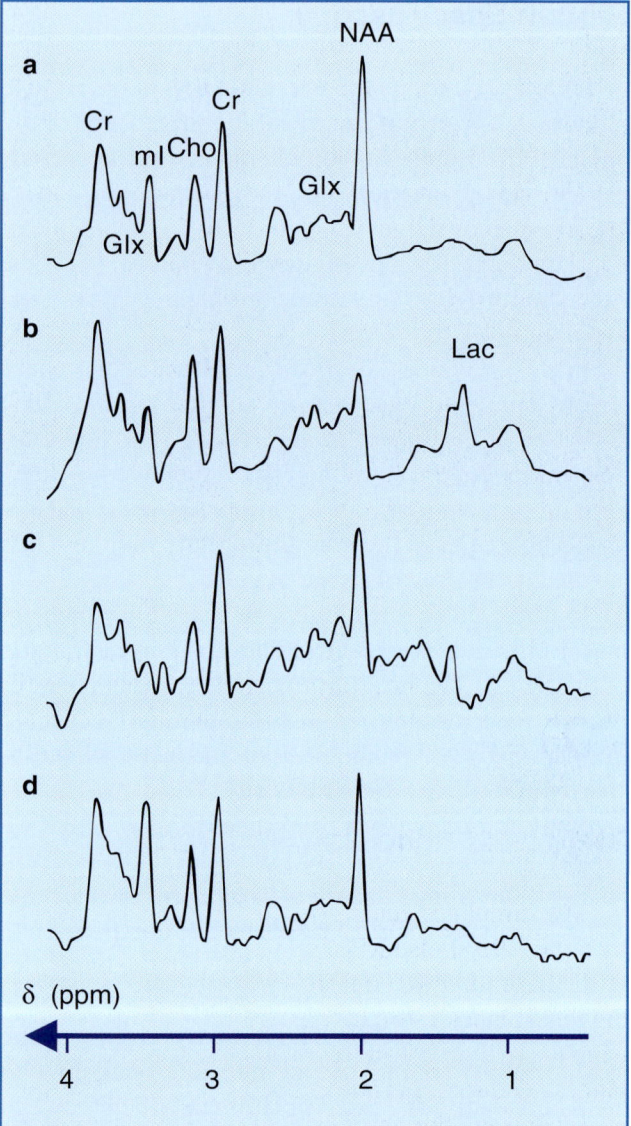

Figure 5-7:
STEAM proton spectra of (a) a normal human brain; (b) hypoxia; (c) hepatic encephalopathy; (d) Alzheimer's disease. The abbreviations are explained in Table 5-1.

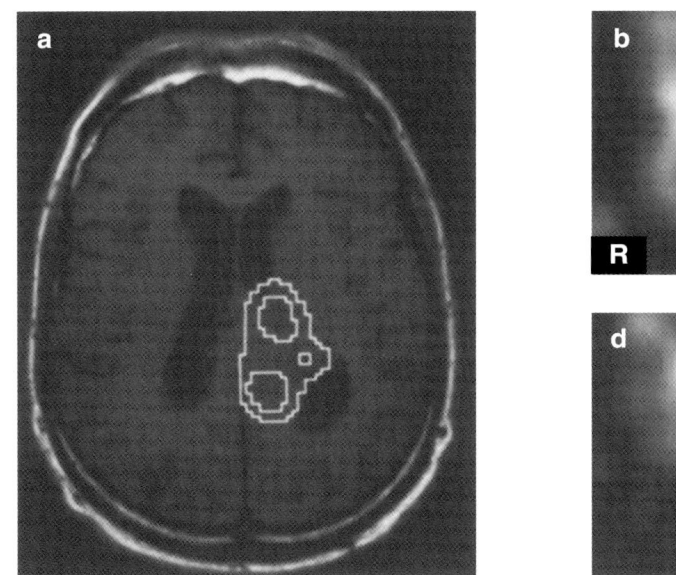
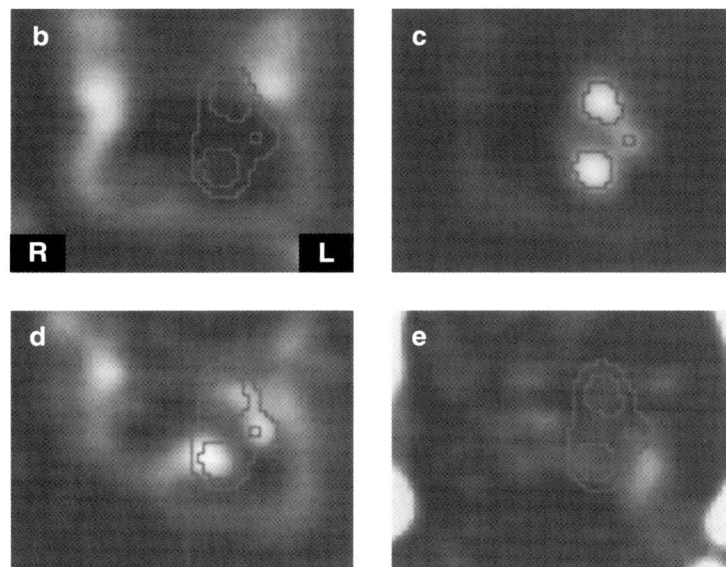

Figure 5-8:
Proton magnetic resonance spectroscopic imaging (MRSI) of a patient suffering from a high-grade glioma. (a) conventional T1-weighted spin echo MR image with overlayed choline contour. Metabolite maps of (b) N-acetylaspartate, (c) choline, (d) creatine, and (e) lactate. N-acetylaspartate is decreased in the tumor and the ventricles, whereas the choline level is elevated in the tumor. Regions with increased choline levels do not overlap regions showing lactate, which is an indication of the location of the active parts of the tumor. Images courtesy of Cologne University Hospital, Dept of Radiology.

Stimulated Echo Spectroscopy

The stimulated echo is one of five echoes generated by a three 90° pulse RF sequence.

If these three RF pulses are made selective in orthogonal planes a stimulated echo will only be obtained from the volume located at the intersection of the three planes. The signal is attenuated by T1 decay between the second and third pulse and by T2 decay in the other two intervals. The problem of T2 decay makes the scheme best suited to ^{1}H, rather than ^{31}P spectroscopy, due to the short T2 values of phosphorus.

A number of acronyms are used to describe pulse sequences of this type, including STEAM[2], STEV[5], and VOSY (Figure 5-7). A similar scheme could be implemented using a spin-echo signal from the same three pulse sequences.

Point-Resolved Spectroscopy

The point resolved spectroscopy localization technique (PRESS) consists of a spin-echo train with two 180° pulses. Each pulse is applied in the presence of an orthogonal gradient which allows the selection of a volume element within the sample to be examined. PRESS needs relatively long echo times.

Image-Selected *In Vivo* Spectroscopy

The ISIS (Image Selected *In Vivo* Spectroscopy) sequence[9] consists of eight steps. In each step, a sequence of selective inversion pulses is applied, followed by a nonselective excitation pulse. The combination of inversion pulses and receiver phases is unique for each of the steps. By combining the resulting eight signals in the correct manner, the signals from the volume of interest will add up, while those from the rest of the sample will cancel each other. Since the FID rather than an echo is observed, there is no T2 decay. This makes the sequence well-suited to ^{31}P spectroscopy.

For ^{1}H spectroscopy, the sequence should be modified so that the very large water signal from the outer volume (i.e., the rest of the sample) is suppressed to avoid dynamic range problems. This is done in an OSIRIS sequence, a modified form of the ISIS sequence in which the signal from the outer volume is suppressed with a noise pulse. Because of the eight steps involved, the sequence is rather susceptible to motion artifacts, since any movement will lead to incomplete cancellation in the outer volume. For the ISIS scheme to work efficiently, the magnetization must come close to full recovery before the next pulse is applied, so rather long TR values are required.

Chemical-Shift Imaging

Chemical-shift imaging (CSI)[1, 4] uses gradients on each of the three axes to encode the spatial position on to the signal using a technique known as phase encoding, which is dealt with in more detail in Chapter 6. The signal is then observed in the absence of any gradient.

In CSI, the whole volume is divided into a number of voxels. If we divide the volume into 512 voxels ($8\times8\times8$), then each phase-encoding gradient is stepped through eight values. Since they must be independently stepped, we require a total of 512 steps. A three-dimensional Fourier transformation then yields the spectrum for each of the voxels. The main drawbacks of this technique are that it generates very large amounts of data and is rather susceptible to motion artifacts. To reduce these problems, CSI can be combined with other localization techniques so that the phase encoding is applied to only one or two dimensions.

The main advantage of CSI is that it is a Fourier technique, and so the signal at each acquisition contributes to the signal in all of the voxels, making it very efficient in terms of signal-to-noise ratio and acquisition times.

CSI is a promising technique because it allows the depiction of metabolite concentrations in cross-sectional images similar to those depicting anatomy, but with much lower spatial resolution.

Usually in magnetic resonance spectroscopic imaging, the size of a volume element is at least 1,000 times larger than in MR imaging (Figure 5-8). However, compared to plain spectroscopy, the efficiency of MRSI is much increased because data of many volume elements are obtained at the same time and the regions of interest do not have to be identified before the examination.

Still, it remains to be seen whether this *in vivo* spectroscopic method will become an established clinical technique or remain purely a research tool.

References

1. Brady TJ, Wismer GL, Buxton R, Stark DD, Rosen BR. Magnetic resonance chemical shift imaging. Magn Reson Ann 1986; 55-79.
2. Frahm J, Bruhn H, Gyngell ML, Merboldt KD, Hänicke W, Sauter R. Localized high resolution proton NMR spectroscopy using stimulated echoes: initial applications to human brain *in vivo*. Magn Reson Med 1988; 9: 79-93.
3. Gadian DG. Proton NMR studies of brain metabolism. Phil Trans R Soc Lond A 1990; 333: 561-570.
4. Hugg JW, Matson GB, Twieg DB, Maudsley AA, Sappey-Marinier D, Weiner MW. Phosphorus-31 MR spectroscopic imaging (MRSI) of normal and pathological human brains. Magn Reson Imag 1992; 10: 227-243.
5. Kimmich R, Hoepfel D. Volume selective multipulse spin-echo spectroscopy. J Magn Reson 1987; 72: 379-387.
6. Kohler SJ, Perry SB, Stewart LC, Atkinson DE, Clarke K, Ingwall JS. Analysis of ^{23}Na NMR spectra from isolated perfused hearts. Magn Reson Med 1991; 18: 15-27.
7. Matson GB, Weiner MW. Chapter 15: Spectroscopy. In: Stark DD, Bradley WG (eds.). Magnetic resonance imaging. 3rd edition. Vol. 1. St. Louis (USA): Mosby Year Book Inc 1999. 181-214.
8. Miller BL. A review of chemical issues in ^{1}H NMR spectroscopy: N-acetyl-L-aspartate, creatine and choline. NMR Biomed 1991; 4: 47-52.
9. Ordidge RJ, Connelly A, Lohman JAB. Image selected *in vivo* spectroscopy (ISIS). A new technique for spatially selective NMR spectroscopy. J Magn Reson 1986; 66: 283-294.
10. Pouremad R, Wyrwicz AM. Cerebral metabolism of fluorodesoxyglucose measured with ^{19}F NMR spectroscopy. NMR Biomed 1991; 4: 161-166.
11. Rashid SA, Adam WR, Craik DJ, Shehan BP, Wellard RM. Factors affecting ^{39}K NMR detectability in rat tissue. Magn Reson Med 1991; 17: 213-224.
12. Rubaek Danielsen E, Ross B. Magnetic resonance spectroscopy diagnosis of neurological diseases. New York: Marcel Dekker 1999.
13. Sauter R, Schneider M, Wicklow K, Kolem H. Current status of clinically relevant techniques in magnetic resonance spectroscopy. Electromedica 1992; 60: 32-54.
14. Shulman GI, Rothman DL, Shulman RG. ^{13}C NMR studies of glucose disposal in normal and non-insulin-dependent diabetic humans. Phil Trans R Soc Lond A 1990; 333: 525-529.
15. Southon TE, Gribbestad IS, Nilsen G, Nordlid K, Svarliaunet AJ, Unsgaard G, Rinck PA. 31P magnetic resonance spectroscopy in the follow-up of therapy of patients with astrocytomas. Radiol Diagn 1993; 34: 11-18.
16. Timm G, Rinck PA, Southon TE, Aasly J, Michler RP. 31P nuclear magnetic resonance spectroscopy in patients with fibromyalgia versus normals and patients with different muscular diseases. Quarterly Magn Res in Biol and Med 1996; 3: 89-92.

The Slow Life of Clinical Spectroscopy

Magnetic resonance imaging has taken off like a rocket and become the diagnostic runner of the last twenty years, but MR spectroscopy has stayed in the back rooms of the researchers. There are two main reasons for this development: there are not many clinical applications for MR spectroscopy, and there is no reimbursement for such examinations in most countries. This makes the method unattractive for physicians, hospitals, and in particular for private practices.

One of the first papers on medical MR spectroscopy applications was published in the *New England Journal of Medicine* in 1981 by Ross and his collaborators. They described spectroscopic changes of phosphorus in McArdle's syndrome [3].

McArdle's syndrome is not a major global disease, nor are other muscular diseases in which MR spectroscopy has shown changes of phosphorus or proton spectra.

Thus, it is understandable that both the clinical users and the manufacturers of MR machines have reduced or even ceased to use whole-body MR spectroscopy machines. In 1990, a spokesman for one of the major manufacturers of whole-body MRI/MRS equipment stated that there are no clinically efficient applications for MR spectroscopy. Therefore, his company and other producers of high-field equipment have limited their investments in whole-body MR machines below 2 Tesla although in recent years some higher field machines have reached the market.

However, this trend is not reflected by the research output — MR spectroscopy research is thriving. In 1982, at the first meeting of the Society of Magnetic Resonance in Medicine in Boston only two papers dealt with MRS. In 1983 less than 100 papers were published about MRS, in 1991 MedLine counted 500, and in 1999 700 publications. As well as there being more papers, there was also an increase in complexity.

The following statement is typical of many articles dealing with MR spectroscopy and its applications:

> »It is hoped that the new information provided by (in this case) multidimensional spectroscopic imaging of metabolites *in vivo* will further enhance the clinical and scientific value of this technology.« [1]

The overwhelming majority of publications about MRS either focus on anecdotal clinical cases, in which some changes in spectra were (or were not) seen, or they discuss improvements of MRS technology. It is always 'hoped' that one day MRS will enhance the horizons of medicine.

MR spectroscopists sometimes claim that whole-body MRS is not accepted by clinicians because the latter cannot read and interpret the spectra. They postulate that:

»The arrogance of the ignorants hinders the development of spectroscopy,«

This might be partly true because radiologists are not trained in biochemistry or in reading and recalculating spectra.

However, the ball is played back into the spectroscopists' court by the physicians. The latter underline that spectroscopists, with a background in chemistry or physics, have no idea of the possible medical relevance of spectra and are, by and large, only interested in playing scientific games. They also claim that spectroscopists create a sea of irrelevant data in which potentially useful information is drowned.

Another important argument is that spectroscopy is insensitive. Phosphorus spectroscopy is sometimes dubbed 'the Twin Peaks of MR', although in reality there are three main peaks in *in vivo* phosphorus spectra.

The technique of phosphorus spectroscopy suffers because of the large volumes (50-100 cm^3) that are necessary to acquire decent spectra within the time period a patient can remain motionless in the magnet. However, tissue volumes of 50-100 cm^3 are of no relevance to clinical diagnosis. When examining brain tumors, an MRS examination volume usually includes vital tumor tissue, the necrotic tumor center, edema, perhaps hemorrhage, and also normal non-involved tissue. This type of volume is too inhomogeneous to clarify or even grade such a tumor. Follow-up examinations may reveal whether a tumor responds to therapy, but even this is doubtful.

However, proton spectroscopy has a greater sensitivity and possesses a wider range of metabolic information than phosphorus MRS. It saves between a half and two-thirds of the time necessary to acquire a similar phosphorus spectrum at 1.5 T.

Spectroscopic data usually require spectral analysis to indicate the metabolite concentration, ratios, and tissue pH. These data give a momentary picture of macroscopic local metabolism and the distribution of metabolites. To date, both time and space resolution are restricting factors of MRS, and therefore MRS examinations cannot compete directly with single photon emission computed tomography (SPECT) or positron emission tomography (PET). However, MRI can now begin to compete with these radioisotope technologies.

It is also possible to convert the spectroscopic result into metabolic maps. Thus, images can be created that reflect the concentrations of certain metabolites on an anatomical background.

Maps of phosphates or other metabolites can deliver spectroscopic information as pictures that can be more easily understood by radiologists. Proton spectra might become the solution for creating such maps because numerous metabolites such as creatine, choline, and lactate can be depicted.

The interpretation of such maps still requires considerable knowledge of diagnostic biochemistry. Because today's radiologists are not trained in this field, this is a job for skilled spectroscopists. Worldwide, there are few scientists with such knowledge, and training is limited because of financial restrictions.

The question remains as to how MRS can be accepted by clinicians using whole-body MR machines.

First, relevant clinical and diagnostic applications have to be found. These applications must be better than competing techniques, and if possible, the MRS examinations must become faster and cheaper than comparable diagnostic methods. In addition, for its implementation in clinical routine, there should be a therapy for the patient's disease. MRS must be able to exclude certain differential diagnoses better than other diagnostic techniques, and/or MRS must be superior to other diagnostic methods in the follow-up period.

Secondly, MR spectroscopy must be easy-to-use and accepted by radiologists, otherwise it will stay a research tool.

On the other hand, there is no doubt that MRS has already contributed greatly to the furtherance of medical knowledge and the understanding of certain aspects of human physiology and pathophysiology. MRS examinations of muscle metabolism, tumors, tissue damage caused by ischemia and infarction, and transplant rejection have added to the understanding of these diseases.

Still, to date, most examinations have not proved useful for daily medical routine. And, what makes it even more difficult for medical spectroscopy, functional and dynamic magnetic resonance imaging have become possible during the last few years. Functional imaging allows users to depict some of the working mechanisms in the body such as the response of the visual cortex of the brain upon light, enabling almost direct assessment of neuronal function.

However, unrelated events can influence and boost medical techniques — such as diseases of presidents or monarchs or wars. MR spectroscopy of the brain, for instance, hit the frontpages of newspapers when a research group was able to show brain abnormalities in veteran military personnel after the Gulf War [2].

References

1.   Bottomley PA, Charles HC, Roemer PB, Flamig D, Engeseth H, Edelstein WA, Mueller OM. Human in vivo phosphate metabolite imaging with 31P NMR. Magn Reson Med 1988; 7: 319-336.

2.   Haley RW, Marshall WW, McDonald GG, Daugherty MA, Petty F, Fleckenstein JL. Brain abnormalities in Gulf War syndrome: evaluation with MR spectroscopy. Radiology 2000; 215: 807-817.

3.   Ross BD, Radda GK, Gadian DG, Rocker G, Esiri M, Falconer-Smith J. Examination of a case of suspected McArdle's syndrome by 31-P nuclear magnetic resonance. N Engl J Med 1981; 304: 1338-1342.

Chapter Six

Image Formation

Composition of MR Images

The manner in which the spatial information is obtained in magnetic resonance imaging is referred to as the *reconstruction technique* (Figure 6-1).

Images can be produced point-by-point, line-by-line, in slices, or in slices calculated from a whole volume (Figure 6-2). Nearly all MR imaging techniques currently in use are either planar (slice) or volume techniques. In the former case, the MR experiment is restricted to a slice through the sample and is often referred to as a *two-dimensional (2D) experiment* since only two spatial dimensions have to be encoded.

Volume techniques which spatially encode the whole volume are hence referred to as *3D techniques*.

The formation of an image involves the following procedures:
- localization of the spins of interest;
- excitation of selected spins;
- spatial encoding of their signal; and
- signal detection and reconstruction.

Each of these procedures, as well as their incorporation into a complete image formation experiment, will be discussed in detail over the following pages.

In the preceding chapters we have discussed the magnetic resonance phenomenon as such, the relaxation times, and the application of magnetic resonance to chemical analysis. However, the most important medical application of magnetic resonance is imaging.

To create an image from a patient, the magnetic resonance signal from the nuclei has to contain information about where the nuclei are positioned in the patient. The MR equipment, as we have described it so far, does not provide us with any such information.

In MR spectroscopy experiments a sample is placed in a magnetic field which is shimmed to make it as uniform as possible. Now a particular molecule will give a signal of the same frequency at any point in the sample. Thus, any frequency changes observed in the Fourier-transformed signal reflect chemical shifts within the sample which can be used to create analytical spectra.

Figure 6-1:
The first three-dimensional picture of the heart - but where do such pictures come from ?

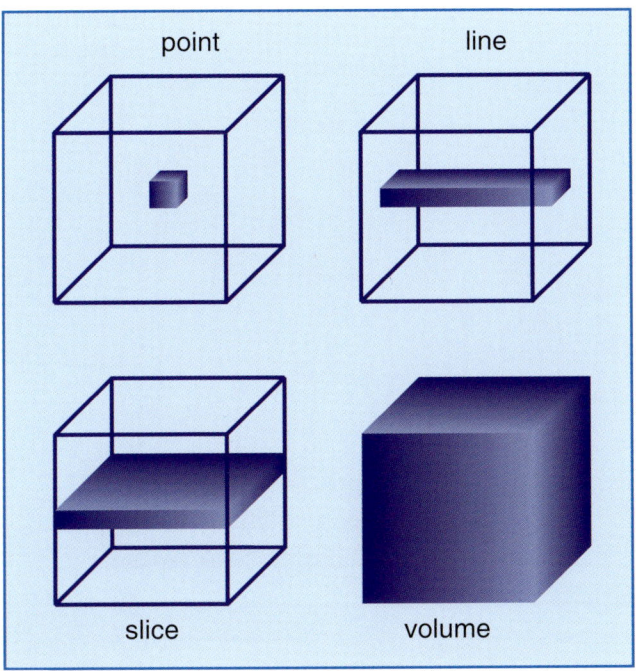

Figure 6-2:
Excitation volumes: point, line, slice, and volume excitation.

Localization of Spins with Field Gradients

In imaging experiments we are not concerned with chemical shift information, rather we require positional information — we want to know from where within our sample a certain signal originates.

As stated the Larmor frequency is proportional to the magnetic field strength. If one varies the frequency of the signal by changing the magnetic field linearly across the sample, the frequencies at different locations will also vary linearly. This technique is known as *imposing a field gradient*. Today, all MR imaging methods utilize such magnetic field gradients (Figure 6-3) for *spatial encoding*.

In Figure 6-4 we examine three small samples of water placed at different positions along the x-axis. Without the magnetic field gradient, applying an RF pulse produces a signal consisting of a single frequency; Fourier-transforming this signal gives a *spectrum* with a single peak. If a magnetic field gradient is present when we measure the signal, the signal consists of three different frequencies corresponding to the three different positions.

Fourier-transforming the signal gives a spectrum of three peaks corresponding to the three different sample positions. The frequency differences between the samples depend on their physical separation and the strength of the magnetic field gradient. At the center of the magnet, the resonance frequency is unchanged since the gradient has no effect at the center. On either side, the resonance frequency will be either higher or lower depending on the polarity of the gradient (Figure 6-5).

The (magnetic) field gradients are generated by a set of coils positioned within the magnet. They can produce fields which vary uniformly along each of the three main axes (x, y, z). These linear field gradients have a strength of up to 30 milli-Tesla per meter (mT/m) in standard clinical systems, but much stronger gradients can be obtained by using smaller gradient coils or in specialized imaging methods.

Although the frequency variations produced by the gradients are very small compared to the resonance frequency, the range of resonance frequencies created is sufficient for high-resolution MR imaging. For example, to produce a frequency distribution of 25 kHz over a distance of 30 cm requires a gradient of only 2 mT/m.

The principle of the generation of a field gradient and the shape of gradient coils in a whole-body imaging system have been explained in Figures 1-4 and 3-7. Figure 6-6 shows how a pulsed magnetic field gra-

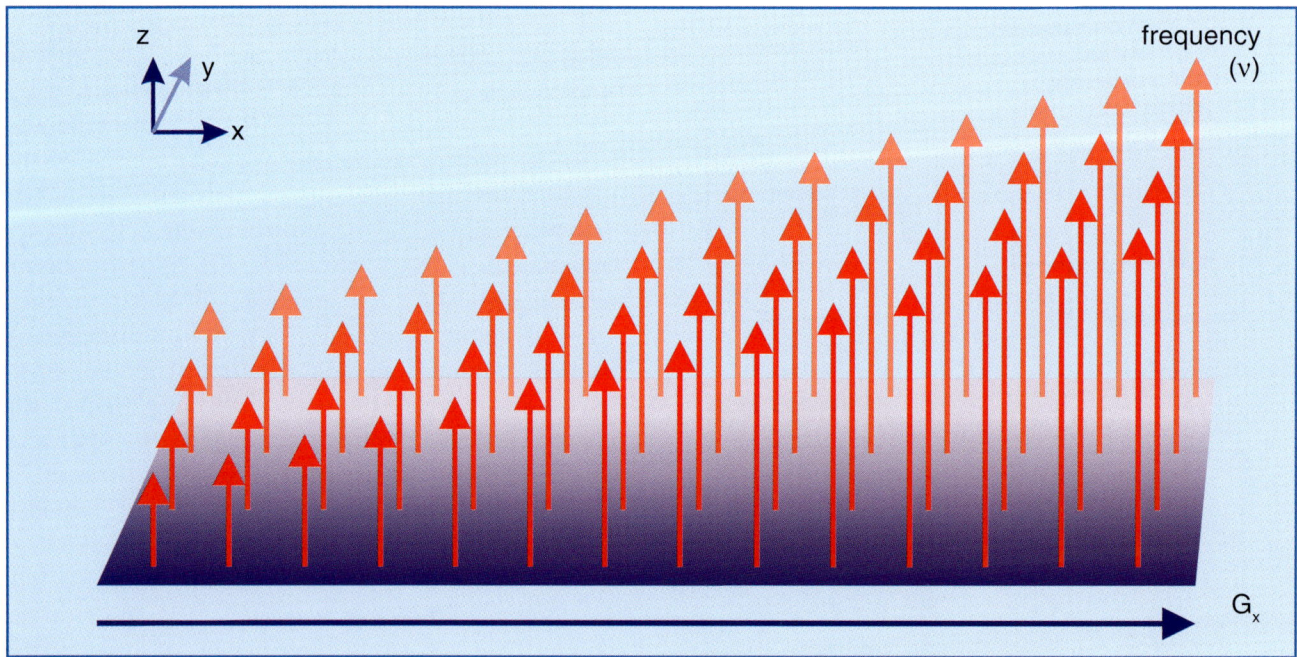

Figure 6-3:
Effect of field gradients: the frequency range is spread out. In this case, the gradient follows the x-direction. The frequency in the center is the 'exact' resonance frequency.

dient is usually depicted in pulse sequence diagrams. In this case the gradient pulse is positive. Because there are three gradient directions x, y, and z, one can find gradients depicted in three 'electronic channels' in pulse sequence diagrams. Gradient pulses can consist of several different components.

The amplitude of the gradient is determined by the current flowing in the gradient coils. Shortening the rise time requires a faster rate of change of the voltage in the gradient coils.

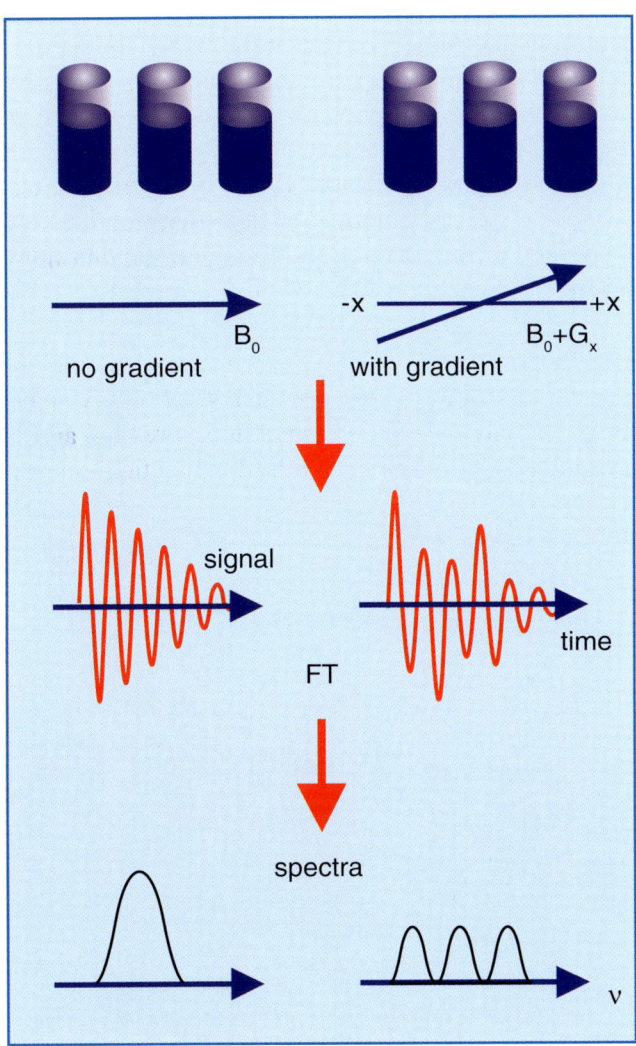

Figure 6-4:
The signals and spectra from three water samples at different positions along the x-axis with and without the application of a magnetic field gradient along the x-axis. In the presence of the gradient, the signals from the three samples are resolved, with the separation of the signals being dependent on their spatial separation in the x-direction and on the strength of the magnetic field gradient (FT = Fourier transformation).

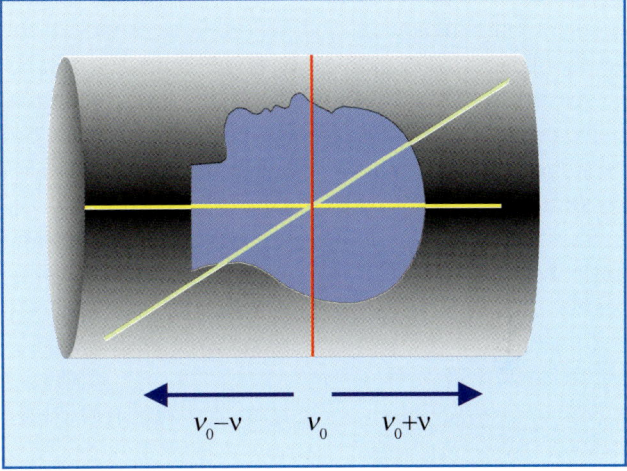

Figure 6-5:
Magnetic gradient fields are superimposed upon the static magnetic field. Different parts of the sample experience different magnetic field strengths. Only in the center of the sample is there no change of the static magnetic field, and thus of the resonance frequency ν_0.

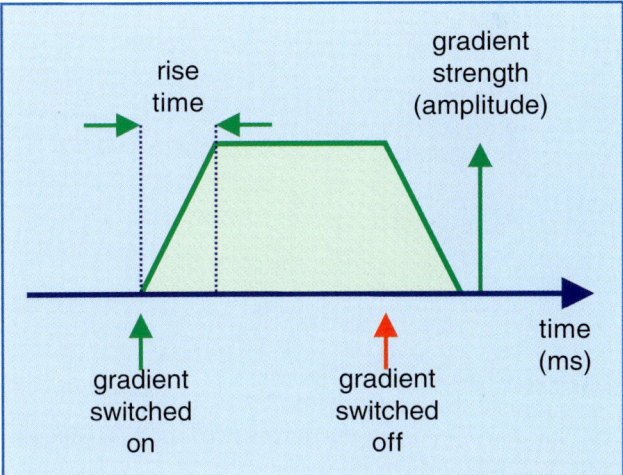

Figure 6-6:
Schematic representation of a pulsed field gradient used for pulse sequence diagrams.

Excitation of Selected Spins

With the implementation of gradients, we can locate the nuclei in our sample, but we also have added a problem. If we switch on a gradient after the RF pulse the mere act of switching on the gradient significantly reduces the magnitude of the magnetic resonance signal from a sample (Figure 6-7). Ideally, the signal will remain aligned along the y'-axis and decay at a rate determined by T2. However, even small imperfections in the magnetic field cause a spreading out (dephasing) of the magnetization. The dephasing of the spins is amplified by the field gradients we want to use to localize them. Thus, by the time we have a stable gradient and can measure the signal in the presence of the gradient, we will have little or no net signal.

To avoid this problem, we have to reform the signal in the presence of the magnetic field gradient. This can be achieved using either a spin-echo or a gradient-echo pulse sequence, thus restoring the original signal in the presence of the gradient, allowing its detection and spatial encoding.

The Spin-Echo Imaging Experiment

A spin echo is formed by applying a 180° pulse at a time τ after a 90° pulse. Following the 90° pulse, the magnetization vectors spread out because of the variations in resonance frequency caused by field inhomogeneities (ΔB_0). Applying the 180° pulse reverses the dephasing, so that at a time τ after the 180° pulse, the effects of ΔB_0 are cancelled out and an echo is formed. Only T2 decay reduces the intensity of the echo.

Complete rephasing only occurs at the center of the spin echo. With increasing distance from the center the effects of field inhomogeneities grow. The spin-echo sequence also refocuses chemical-shift effects at the center of the spin echo. Hence the water and fat signals will be in phase at the echo center.

In a non-imaging spin-echo experiment, the dephasing effects in the two halves of the experiment before and after the 180° pulse are equal. In an imaging experiment, we can control both the duration and the amplitude of the imaging field gradients and arrange for the gradient areas to be equal (Figure 6-8).

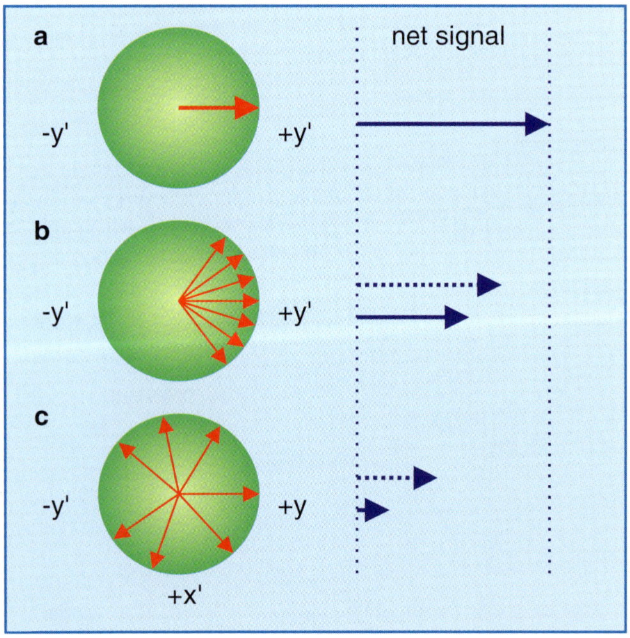

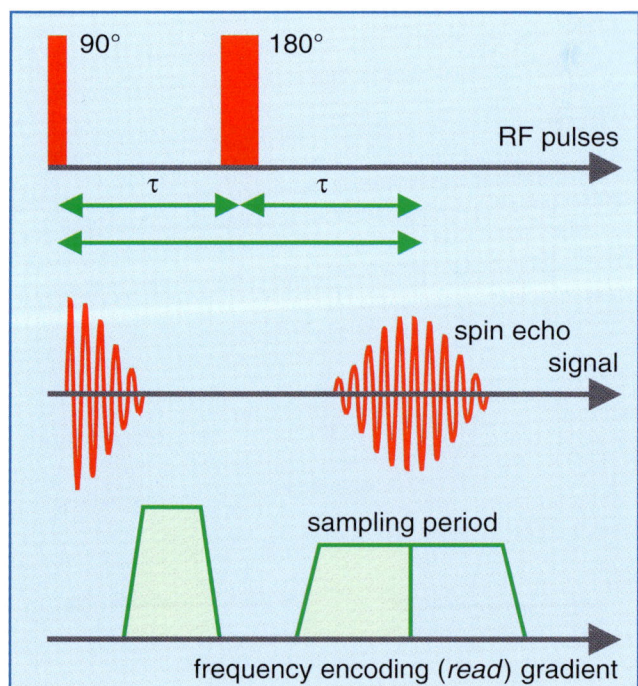

Figure 6-7:
Dephasing produced by the application of a gradient. (a) The signal is initially aligned along the y'-axis and decays at a rate determined by T2. (b) If we turn on a magnetic field gradient, we observe an accelerated spreading of the spins. (c) Increasing the time delay increases the fanning out of the signal. The solid arrows on the left of the figure represent the net signal. The dotted lines in (b) and (c) are the signal that would be measured in the absence of the gradient.

Figure 6-8:
Spin-echo experiment with balanced gradients during the sequence. The (green) gradient pulse between the 90° and 180° pulses equals the green area of the gradient after the 180° pulse. Since the 180° pulse induces a phase reversal, the effects of the two gradients cancel at the center of the spin echo. Therefore, a spin echo forms in the presence of a gradient.

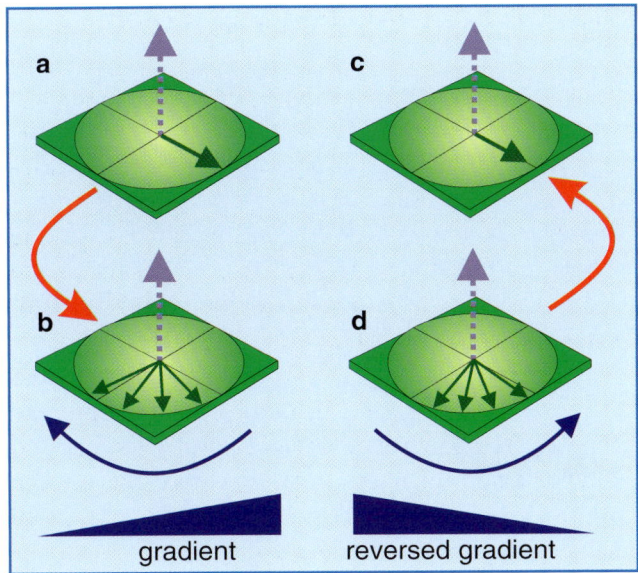

Figure 6-9:
Formation of a gradient echo in the absence of local field inhomogeneities. From counter clockwise, (a) immediately after the RF pulse the transverse magnetization is strong; the spins are in phase; (b) the spins begin to dephase; the application of a field gradient accelerates this process. The net magnetization vanishes; (c) the gradient is switched to the opposite polarization and the spins begin to rephase, until (d) a gradient echo is formed.

Gradient-Echo Imaging Experiment

To create an echo, we do not necessarily need a 180° pulse; we can also use the field gradients to build up an echo. This leads to *gradient echoes*, which today are widely used in *rapid* (or *fast*) *imaging sequences*.

Following an RF pulse, the signal decays due to the combined action of T2 decay and local field inhomogeneities. This effect is described by T2*.

By altering the polarity of the gradient, we change the direction of the induced precession, the spins start rephasing, and at the echo time TE grow into a *gradient echo* (Figures 6-9, 6-10, and 6-11). To create such an echo, the areas of the gradients with different polarities must be equal[2]. A gradient-echo (GRE) experiment measures a delayed reformed version of the FID. This is necessary due to the gradient switching.

An important point to remember with gradient echoes is that, unlike spin echoes, the effects of field inhomogeneities are not cancelled out. The signal decays faster; therefore, GRE experiments require a relatively short echo time.

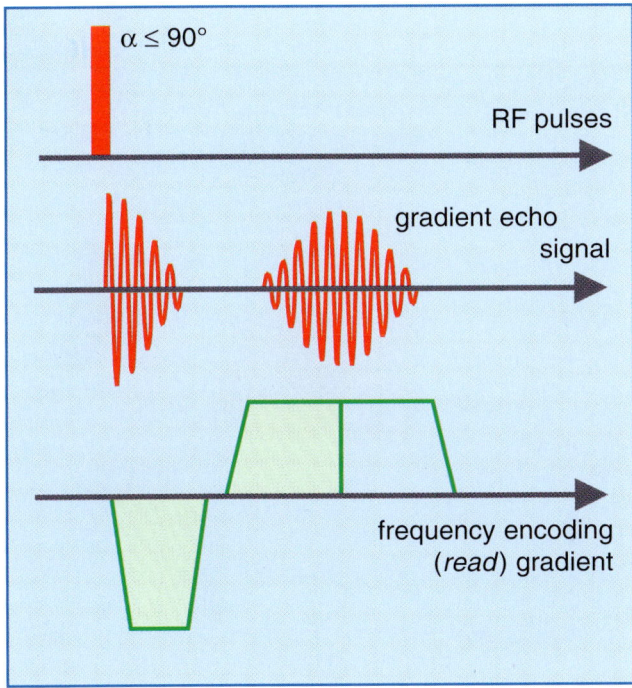

Figure 6-10:
Formation of a gradient echo. Instead of the 180° pulse, a gradient pulse (-G) is used followed by a second gradient pulse of opposite polarity (+G). For spin echoes, the signal decay is determined by T2, since the effects of local field inhomogeneities are cancelled out. With gradient echoes the signal decay is determined by T2*, which is always less than T2.

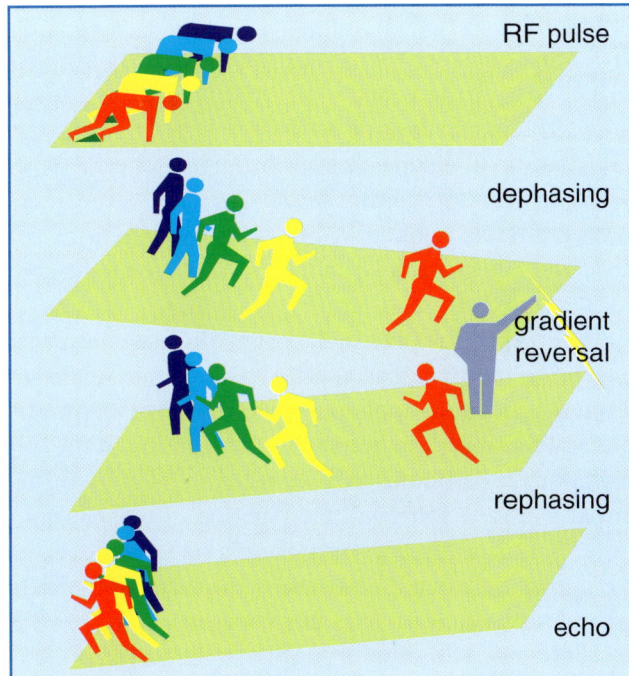

Figure 6-11:
Formation of a gradient echo using the example of the runners. All participants start together; they begin to separate from each other, accelerated by the gradient. By the reversal of the gradient, they are recalled: they turn around at their present position and run back to the starting line. Unlike in the spin-echo experiment, they return on their own track to form a gradient echo.

Spatial Encoding

Until now we have focused entirely upon the methods with which we can *excite* the region of interest. In order to obtain an image of this region defining spatial distribution and other characteristics, we need to discuss the different methods for spatial encoding. These methods can be divided into two groups: *frequency encoding* and *phase encoding*.

Frequency Encoding

Frequency encoding is conceptually the simplest of the two encoding methods.

If a sample is placed in a homogeneous field, no information concerning the spatial location is present in the magnetic resonance signal, since all regions of the sample have the same Larmor frequency. However, when a gradient is imposed on the sample, the magnetic resonance signal contains information about the spatial location of the resonating spins. If we rotate the gradient in all three dimensions, we can both distinguish different samples or different tissue contents in the body, as well as locate them in all three spatial dimensions.

Let us consider two water samples. They are displaced in both the *x*- and *y*-directions. When we apply only a gradient in the *x*-direction, we do not gain enough information to deduce their position in space.

However, if we perform the experiment three times with gradients applied in the *x*- and *y*-directions respectively, then we have enough information to be able to estimate their positions (Figure 6-12).

The projection obtained through the use of a gradient can be thought of as being a shadowgram resulting from a light beam applied perpendicular to the gradient. Each projection thus shows the position of the object along a particular axis. In order to be able to calculate both the form and the position of one or more objects, it is necessary to measure a set of projections for angles between 0° and 180° with steps of 1° or 2° between projections. It is not necessary to cover the full 360° since this would simply duplicate the projections obtained over 180°. The projections obtained in this way can then be mathematically processed to produce an image of the object. This approach is conceptually similar to that used in x-ray CT and was the method used for the first demonstration of MRI[6].

The method is generally referred to as *projection reconstruction* or *backprojection* (Figure 6-13).

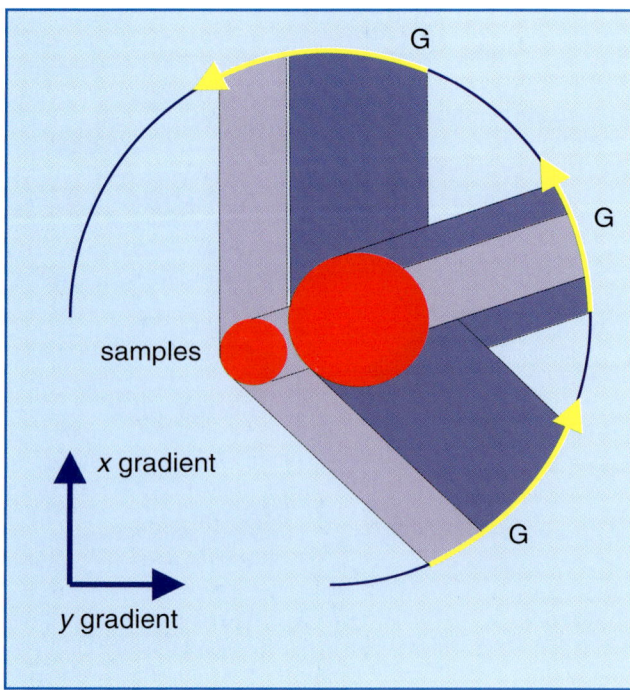

Figure 6-12:
By using field gradients along the *x*- and *y*-axes in three different experiments, the positions of two samples of water with displacement in both the *x*- and *y*-directions can be estimated.

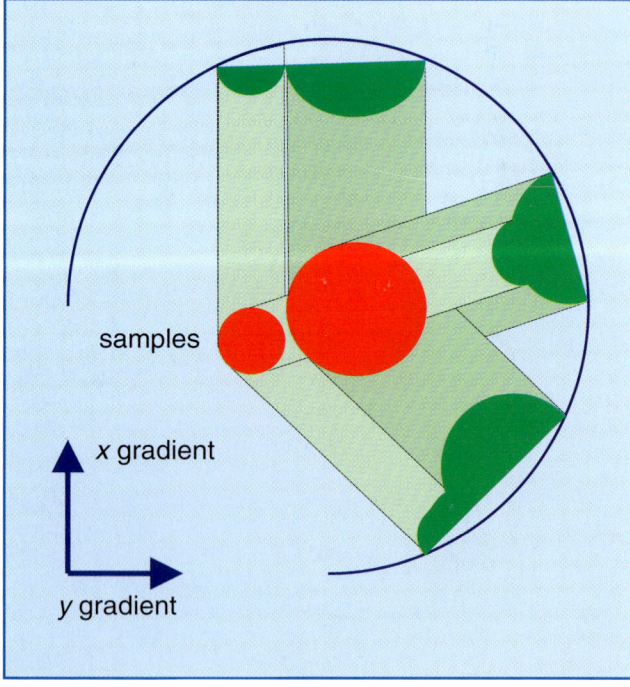

Figure 6-13:
Backprojection method: three projections of a sample allow the location of the objects and creation of a crude image. Additional projections will lead to a more exact definition of the shape of the objects.

Projection reconstruction is currently little used in clinical MRI, mainly because it is very sensitive to the effects of inhomogeneities in the main magnetic field. However, recent improvements in magnet technology have reduced the level of these imperfections, which has led to renewed interest in these techniques. One of the main potential applications for such techniques is in diffusion imaging, where the errors produced by motion seriously degrade the images acquired using other reconstruction schemes [4].

Phase Encoding

In frequency encoding, the system is excited while no gradient is present; then the signal is recorded in the presence of a gradient. The phase encoding of the signal takes place *before* the signal is recorded, but again in the presence of a gradient.

Immediately after the excitation, the spins are coherent, no difference in phase has yet developed, and if we wait, natural T2 processes (and field inhomogeneities) will start acting on our sample, and some dephasing will start. However, if we suddenly switch on a gradient, then the spins will start dephasing. The rate of dephasing will depend on the location of the individual spin and the strength of the gradient.

The phase of the spins contains spatial information. Phase encoding compares the phase with an MR signal of the same frequency. The information can be recovered by proper use of the Fourier transformation.

The effect of applying different gradients to samples at different positions is shown in Figure 6-14.

To obtain a resolution of n pixels in the *y*-direction, we have to repeat the experiment n times. The phase gradient is stepped (changed) at each repetition with a certain increment. The changes in the phase gradient can be obtained by changing either the duration or the amplitude of the phase gradient. The former method was the first to be suggested [5], but has the disadvantage of giving a different T2- (or T2*-) weighting to the different phase steps. Thus, the method of varying the amplitude of the phase gradient is preferable.[1]

Frequency and phase encoding are in fact very closely related. The main difference between the two methods is that the phase encoding is completed before we start to measure the signal, whereas frequency encoding is applied during the measurement process. In frequency encoding, we can use the evolution of signal with time to give us the desired number of points; in the phase-encoding direction, we do not have this option and hence have to repeat the experiment.

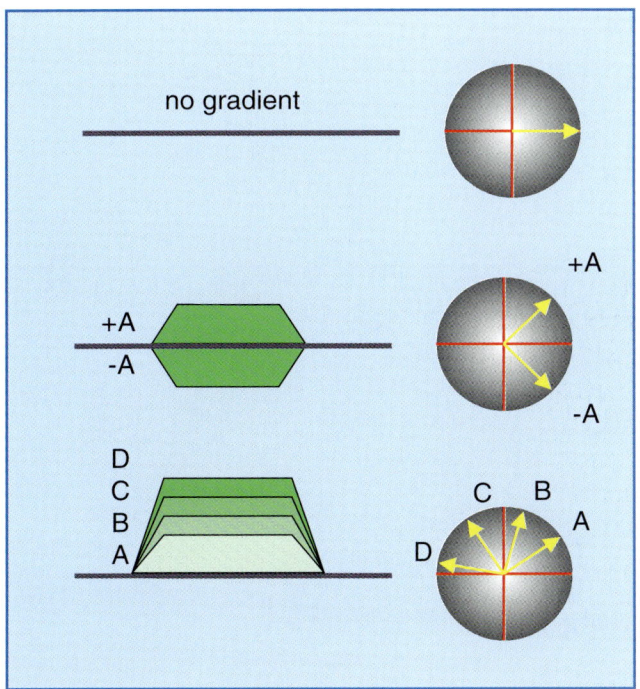

Figure 6-14:
The phase angle induced by a gradient depends on both the amplitude and the duration of the gradient. When using a constant duration for the gradient, we can control the phase angle by varying the amplitude (A-D) and polarity of the gradient (+A, -A).

Slice Definition and Selection

In an imaging experiment, definition and selection of a slice are of great importance. They are determined by characteristics of the excitation pulse. One distinguishes between shaped and hard pulses (compare Figures 1-9 and 1-10).

Slice Definition. In analytical NMR studies, the maximal RF power is applied for long enough to give the desired pulse angle (hard pulse). In more complicated experiments, it is necessary to adjust the pulse amplitude with time so as to give a better defined frequency content (shaped pulse). The pulse shape is used to give an approximately rectangular slice profile for the slices in the imaging experiments (Gaussian and sinc pulses; see Chapter 1) and can heavily influence image contrast in magnetic resonance imaging.

The phase of the RF pulse is also determined at this stage, with many imagers only allowing phases of $0°$, $90°$, $180°$ or $270°$ to be selected. The resulting excitation pulses can be as short as 10 ms for nonselective hard pulses, and typically a few milliseconds for the frequency-selective shaped pulses used in magnetic resonance imaging with peak-to-peak amplitudes of up to several hundred volts.

Slice Selection. We can express the gradient strength in either $mT \times m^{-1}$ or in $Hz \times m^{-1}$. Since the pulse has a fixed bandwidth (provided that the pulse duration is held constant), raising the gradient strength increases the number of $Hz \times m^{-1}$; this results in a decrease in slice thickness (Figure 6-15).

For example, for a sinc pulse with a bandwidth of 2 kHz, increasing the slice gradient from $4 \ mT \times m^{-1}$ (1.7 kHz/cm) to $8 \ mT \times m^{-1}$ (3.4 kHz/cm) reduces the slice thickness from 11.8 mm to 5.9 mm.

Applying an RF pulse in the absence of any field gradients will excite the whole sample. If a field gradient is applied at the same time as the pulse, the magnetic field, and hence the resonance frequency, will change with position within the sample.

For an RF pulse at the resonance frequency excitation will occur at the magnet center where the gradient has no effect (compare with Figure 6-5). Off-center, the nuclei cannot be excited by RF pulses at the Larmor frequency.

The distance (or slice thickness) over which the nuclei in the center resonate is determined by the range of frequencies (bandwidth) contained in the excitation

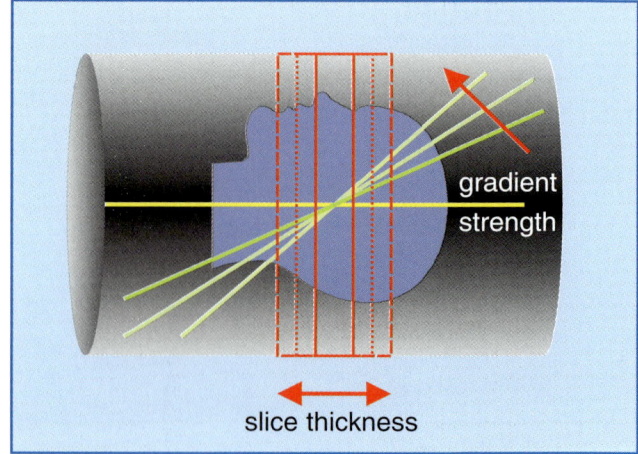

Figure 6-15:
Slice thickness: moving the gradient in the direction of the arrow increases the number of Hz/m, and thus gradient strength. It decreases slice thickness.

pulse and the strength of the field gradient. If the RF pulse contains only a well defined band of frequencies, then excitation will occur for a well defined range of positions. This excitation corresponds to the selection of a slice in the sample.

The length of the RF pulse, and thus also its bandwidth, is the second factor influencing the slice thickness. The longer the pulse, the thinner the slice will be (Figure 6-16). The trade-off for thinner slices is the prolongation of the echo time (TE). Because TE is measured from the center of the pulse, longer pulses for thinner slices mean a longer initial TE, which, in turn, influences imaging time, image artifacts, and contrast.

Changing the frequency of the RF pulse corresponds to moving the position of the nuclei on resonance from the center of the sample. In this way we can move the slice to any desired location along the axis (Figure 6-17). For a transverse slice, the slice gradient is applied along the z-axis; for a coronal slice, the slice gradient is applied along the y-axis; and for a sagittal slice, it is applied along the x-axis.

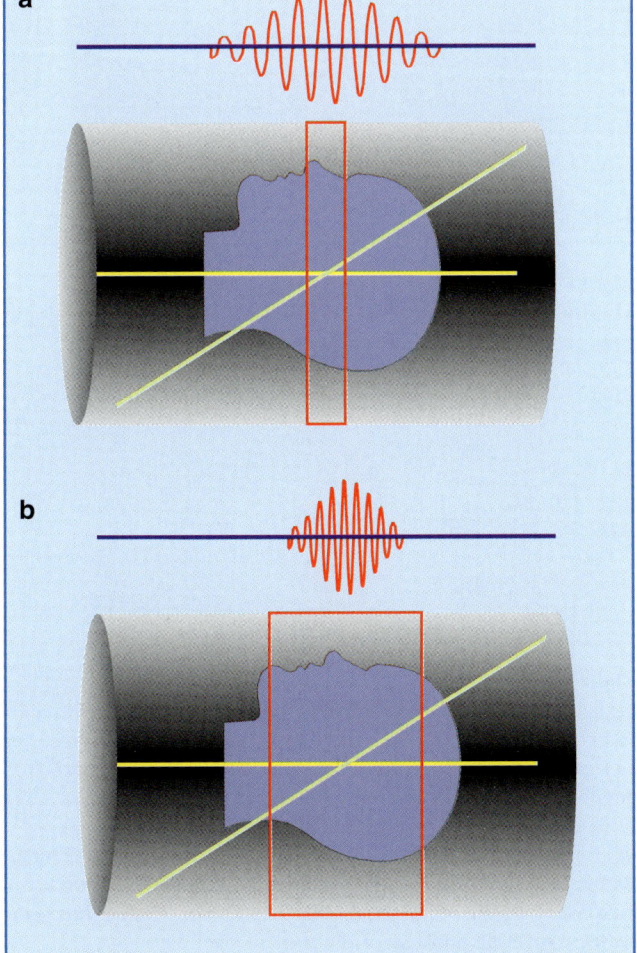

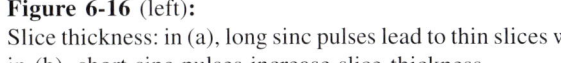

Figure 6-16 (left):
Slice thickness: in (a), long sinc pulses lead to thin slices whereas in (b), short sinc pulses increase slice thickness.

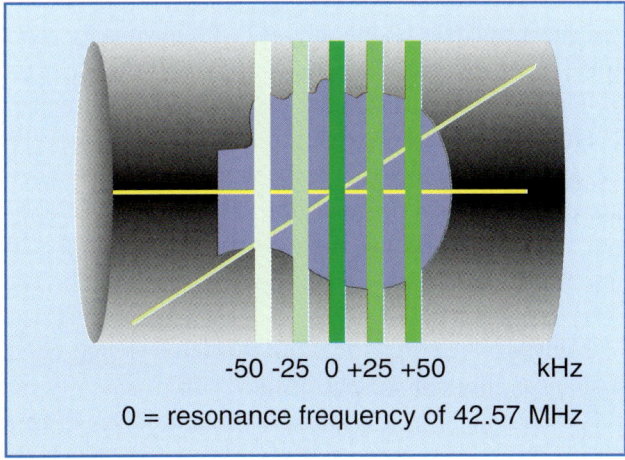

Figure 6-17:
Moving the slice position: at 1.0 T, the resonance frequency in the center of the sample corresponds to 42.57 MHz. Changing the pulse frequency by several kHz moves the slice off-center.

Multiple Slices

A number of possible slice configurations in multiple slice imaging is depicted in Figure 6-18. Provided there is no overlap between the slices, then each slice should be totally independent of the other slices.

If there is an overlap, the RF pulses used for excitation of the different slices interfere with each other and interrupt, for instance, the T1 recovery in adjacent slices, leading to reduced signal-to-noise ratio and influencing signal intensities.

Commonly, a small gap is retained between slices to avoid this kind of 'crosstalk', or the excitation pulses are distributed over two packages; the first one excites the odd, the second one the even slices.

In many pulse sequences, there is a quite long delay between each excitation (repetition time = TR) of a particular slice while the magnetization recovers. Because of the relatively long T1 relaxation time of tissues, a delay of up to three seconds may be necessary before repeating the excitation. To make the most efficient use of this time, we can excite a number of parallel slices in each interval, which is achieved by changing the frequency of the RF pulse.

This procedure can be repeated to produce a series of slices (Figure 6-19). The number of slices obtainable can be calculated by dividing the repetition time TR by the time required for each slice. For example, if TR = 400 ms and TE = 50 ms, the theoretically possible number of slices is eight (in practice seven, since each slice requires slightly more than TE).

If the repetition time is long enough, not only several slices but also several images with increasing echo times per slice can be created.

This is known as *multi-slice multi-echo* sequence. Commonly in an investigation of the brain 15 or 16 parallel slices in the transverse view with two echoes are acquired. The repetition time (TR) is between 2,000 and 3,000 ms, the times of the two echoes (TE) are 20 and 80 ms.

Multiple slice imaging is not restricted to spin-echo sequences, but can be performed with practically all pulse sequences. You can add inversion pulses to create inversion-recovery images with a relatively long inversion time (TI); during this time, additional slices can be inverted. However, usually the number of slices is limited in IR sequences.

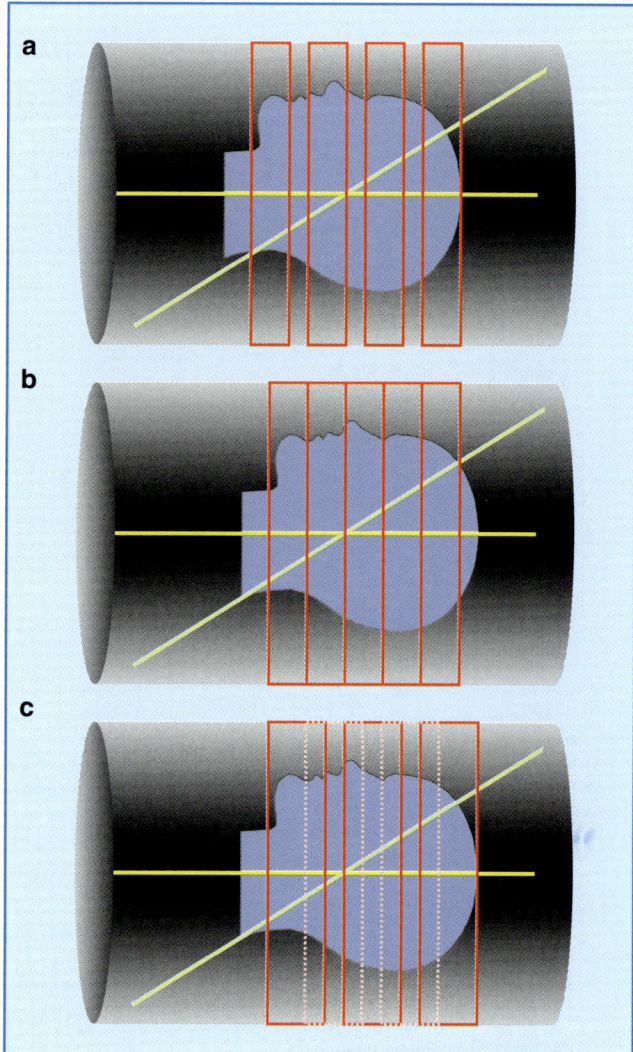

Figure 6-18:
Multiple slices: (a) multiple slices with wide gaps; (b) multiple contiguous slices; and (c) multiple overlapping slices.

Figure 6-19:
(a) Excitation of multiple slices within one TR cycle (spin-echo pulse sequence: multi-slice single echo). In this case, five slices can be excited within one repetition cycle. The excitation frequency of the individual pulses is slightly changed so that only selected nuclei, and thus slices, are excited (Figure 6-17).
(b) Excitation of multiple slices within one TR cycle (SE pulse sequence). In each slice, one echo has been added to create a multi-slice double-echo sequence. The number of slices has been reduced to three because of the time restriction.

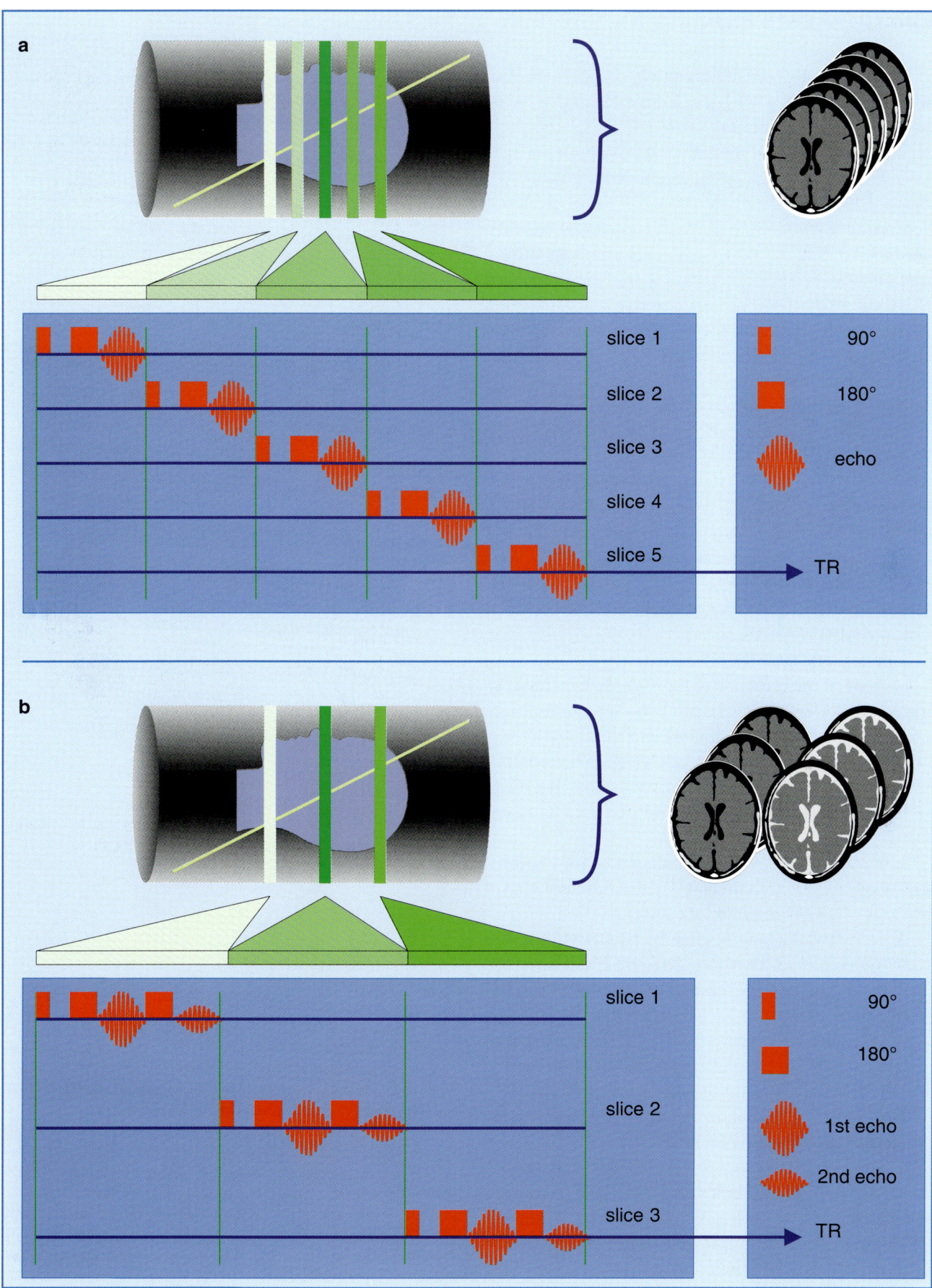

The Complete Imaging Experiment

In two-dimensional MR imaging, a slice is excited, for example, by applying a selective RF-pulse in the presence of the z-gradient.

There are two different methods of obtaining spatial information for the remaining two dimensions:

Frequency-encoding only. The remaining two gradients (x and y) are now combined with a resulting gradient of a certain strength and spatial direction. The FID is recorded in the presence of this gradient. Then the combined gradient is rotated a certain angle. Once again the FID is recorded, and this process continues until enough information is collected.

Based on the frequency spectra, an image can be reconstructed using the backprojection method, as depicted in Figure 6-13.

Two-dimensional FT Method. This method is a combination of phase- and frequency-encoding. Today it is the standard image formation method.

One gradient, for example the y-gradient, is switched on and the spins are allowed to dephase in the presence of this gradient. After a certain time, the y-gradient is switched off and the FID, or alternatively the spin echo, is recorded in the presence of the x-gradient. Thus, the y-gradient serves as the phase-encoding gradient (often called *preparation* gradient) and the x-gradient is encoding the frequency information (*readout* gradient).

The system is re-excited, but with either the duration or, more commonly, the strength of the y-gradient changed.

The whole process is repeated n times for a resolution of n pixels in the y direction, with a different phase-encoding gradient applied for each excitation.

Since the field inhomogeneity effects will be the same for each repetition, they have no effect upon the final image; rather, they represent a baseline. This represents one of the main advantages of the 2D Fourier-transform technique. The resulting 2D raw data matrix is processed using a 2D Fourier-transform to produce a 2D image (Figure 6-20).

The combination of frequency- and amplitude-stepped phase-encoding techniques is called *2D spin warp imaging* [1].

Figure 6-21 summarizes an entire 2DFT imaging experiment, in this case using a spin-echo sequence, although one can use nearly any other pulse sequence applied in clinical MR imaging.

Figure 6-20:
2DFT owes more to MR spectroscopy than to CT reconstruction algorithms because both amplitude and phase information are acquired to spatially encode the signal. The first step generates a one-dimensional projection, but here a phase-encoding gradient is applied just before the original gradient is turned on. The phase-encoding gradient is applied in right angles to the original gradient and its duration or amplitude is successively increased. The corresponding points from each projection are Fourier-transformed a second time to generate the final image [8].

Figure 6-21:
Complete 2DFT spin-echo imaging experiment. The procedure consists of the selection of the 90° and 180° RF pulses, a transverse slice through a brain (z-direction), phase-encoding (y-direction), and frequency-encoding (x-direction). The phase-encoding gradients change the phase in the respective row of the transverse slice; the frequency-encoding gradients allocate a specific frequency to each column. Combining both phase and frequency information allows the creation of a grid in which each pixel possesses a distinct combination of phase and frequency codes. The entire procedure is usually repeated 256 times, with changing phase-encoding gradients to produce a 256×256 image.

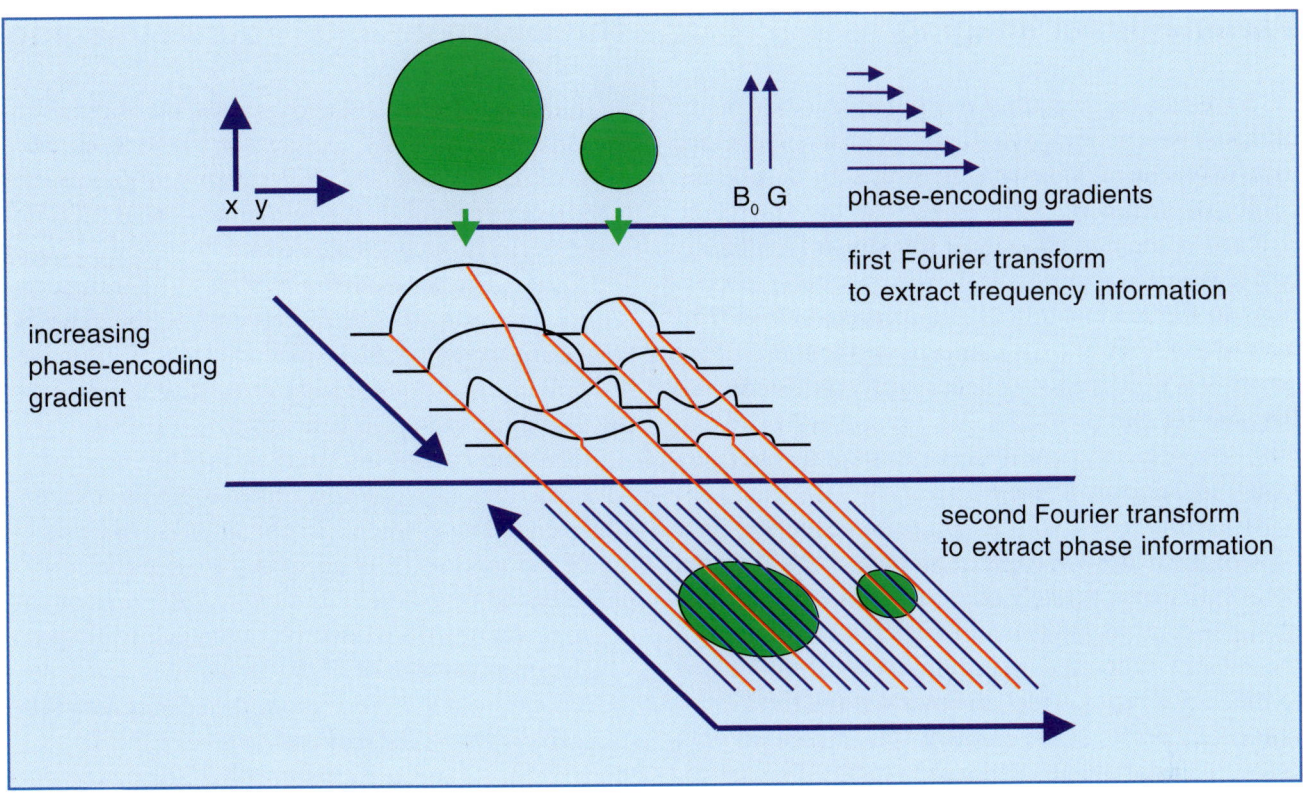

x y

B₀ G   phase-encoding gradients

first Fourier transform
to extract frequency information

increasing
phase-encoding
gradient

second Fourier transform
to extract phase information

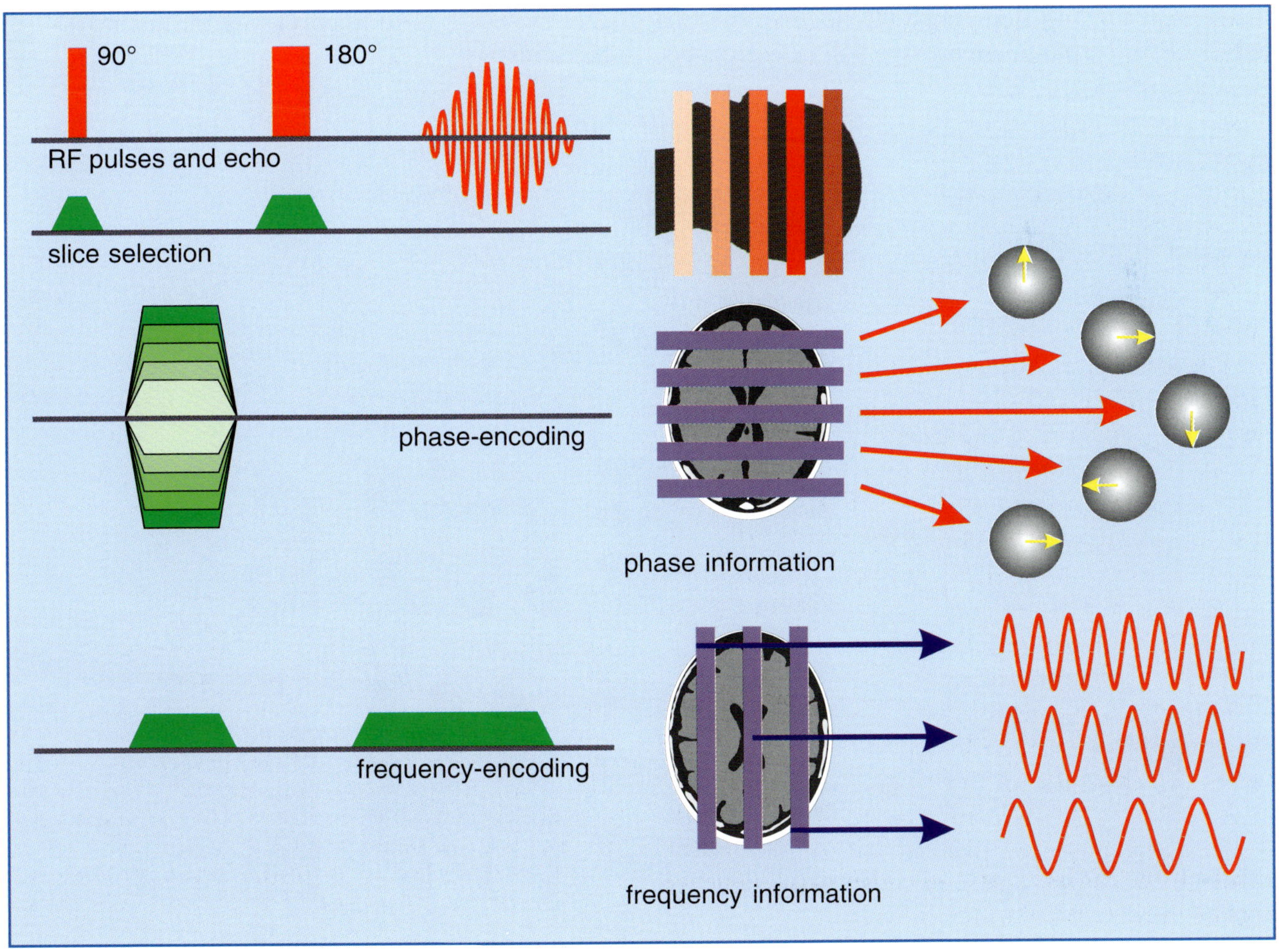

90°   180°

RF pulses and echo

slice selection

phase-encoding

phase information

frequency-encoding

frequency information

Partial Fourier Imaging

To reduce the scan time, it is possible to omit some of the phase-encoding steps and use the acquired data to estimate the missing data, given that the data set has conjugate symmetry [7]. The image can then be reconstructed in the normal manner. Ideally, it would only be necessary to acquire half the data set, but imperfections in the magnetic field and effects such as flow lead to phase errors. To compensate for these phase errors, it is necessary to collect slightly more than half the data set and then calculate a phase correction. If 70% or more of the raw data is acquired, it is not necessary to calculate a phase correction.

If used together with a spin-echo sequence, only 55% of the phase-encoding steps have to be acquired before we can accurately reconstruct the image. With gradient-echo techniques, a rather larger fraction of the data set is required for accurate reconstruction due to the larger phase errors arising from the field inhomogeneity errors in the gradient-echo signal. In both cases, the final images will have a reduced signal-to-noise ratio compared with images acquired using all the phase-encoding steps since the noise in the two halves of the partial Fourier image will be correlated.

Three-Dimensional Fourier Imaging

In the late 1970s and early 1980s, all original images acquired at Paul C. Lauterbur's laboratory were three dimensional (3D). Other research groups and manufacturers introduced two-dimensional (2D) imaging because such images could be acquired faster and easier. The 3D imaging methods available today are all based on 3D-Fourier reconstruction. The 2D spin-warp sequence can be extended to a 3D sequence by applying the RF pulse without a slice gradient, causing the whole sample to be excited.

To obtain spatial information in what was previously the slice direction, we have to apply a second phase-encoding gradient. To obtain full spatial encoding of the volume, it is necessary to step the second phase-encoding gradient through its full range of values for each step of the first phase-encoding gradient.

The disadvantage of 3DFT imaging is that unless the repetition time is very short, the scan time will be excessively long. The main advantage of the 3D technique is that it has a signal-to-noise advantage over 2D techniques (if the voxel size is kept constant, the signal-to-noise ratio improves by the square root of the number of slices).

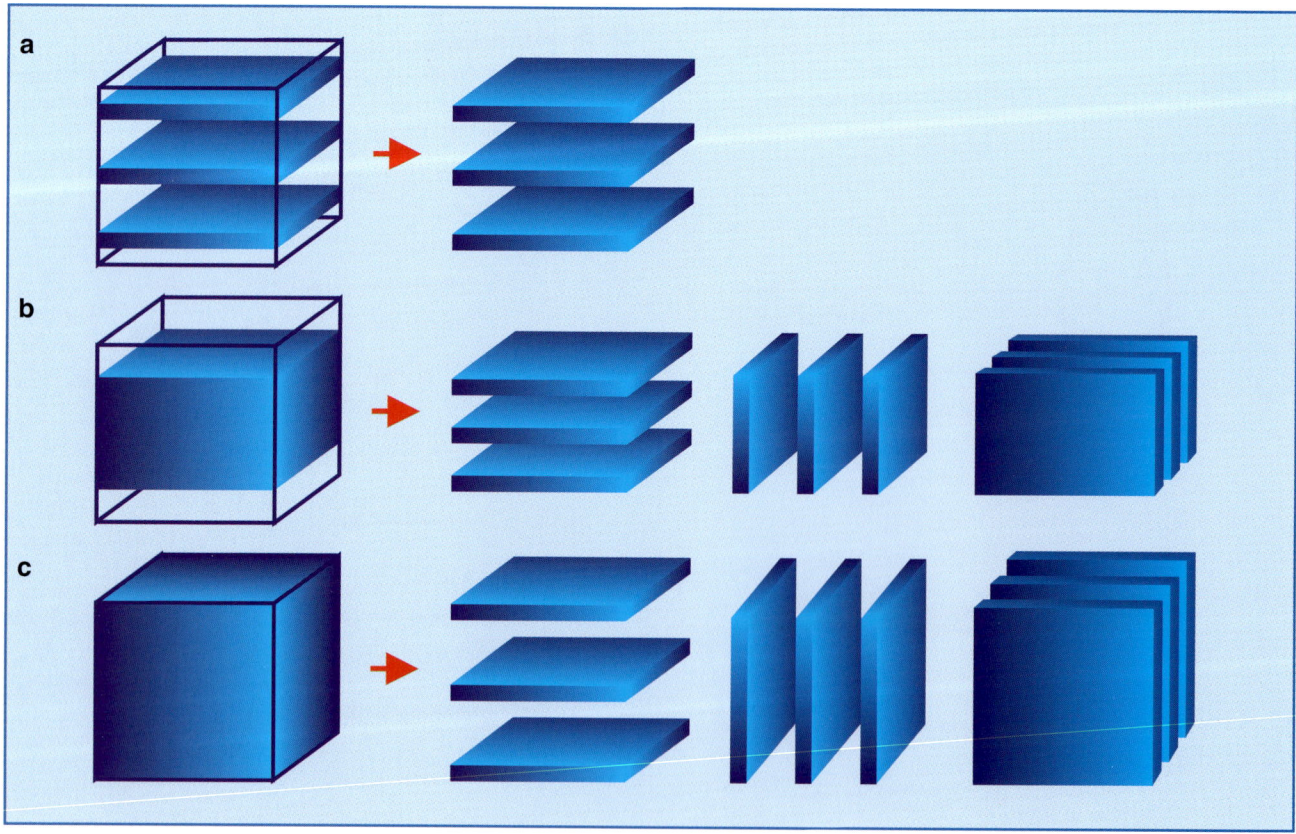

Other advantages are that the slices are contiguous (which is not the case with multiple-slice techniques), that any desired slice orientation can be reconstructed from the data set, that very thin slices can be obtained, and that the slice profile of the 3D set is rectangular.

Among the additional problems with 3D imaging are that data processing requirements are greatly increased, that viewing 3D data sets with typically 64-128 images created during one data acquisition generally requires a separate workstation, and that for each 3D image set we have only one type of contrast.

A compromise between full 3D and 2D imaging is to excite a thick slice (slab), which is then sub-encoded into slices using a 3D sequence (Figure 6-22)[3]. This allows the 3D region to be accurately defined, but if used with only a small number of phase encodings in the third dimensions, significant ringing artifacts can occur. The slabs must have a good slice profile; otherwise some of the slices must be used to encode the edges of the slabs to avoid ringing artifacts[3].

Three-dimensional imaging usually does not use spin-echo sequences because of the long scan time, but rather gradient-echo (GRE) or rapid spin-echo (RSE/FSE/TSE) sequences.

References

1.   Edelstein WA, Hutchison JMS, Johnson G, Redpath TW. Spin warp NMR imaging. Phys Med Biol 1980; 25: 751-756.
2.   Hutchison JMS, Edelstein WA, Johnson G. A whole body NMR imaging machine. J Phys E Sci Instrum 1980; 13: 947-955.
3.   Johnson G, Hutchison JMS, Redpath TW, Eastwood LM. Improvements in performance time for simultaneous three-dimensional NMR imaging. J Magn Reson 1983; 54: 374-384.
4.   Jung KJ, Cho ZH. Reduction of flow artifacts in NMR diffusion imaging using view-angle tilted line projection reconstruction. Magn Reson Med 1991; 19: 349-360.
5.   Kumar A, Welti D, Ernst RR. NMR Fourier zeugmatography. J Magn Reson 1975; 18: 69-83.
6.   Lauterbur PC. Image formation by induced local interactions: examples employing NMR. Nature 1973; 242: 190-191.
7.   MacFall JR, Pelc NJ, Vaurek RM. Correction of spatially dependent phase shifts for partial Fourier imaging. Magn Reson Imag 1988; 6: 143-155.
8.   Pykett IL. NMR imaging in medicine. Scientific American 1982; 246: 78-88.

Figure 6-22:
(a) 2D multiple slices; (b) 3D slab, which can be used to create 2D slices; (c) 3D volume, which can be used to calculate slices in any direction of the entire volume.

k-Space

Introduction

Its flexibility distinguishes MR imaging from all other medical imaging modalities. The ultimate reason for this is the unique handling of MR raw data in an abstract data collection matrix called 'k-space', where the data stay to be deciphered. This space consists of the raw data that have been collected during image acquisition but has not yet been converted into the final anatomical image.

The motto in the foreword to this book fits very nicely with this chapter. The easiest way to deal with k-space is seeing and believing; this, however, is not very helpful when one wants to understand how some imaging techniques function and what their pitfalls are (Figure 7-1).

A k-space is a mental concept. There is no hardware in an MR scanner corresponding to it. It is a platform to collect, store, and process complex data. These data represent thousands of sine and cosine waves which build the MR image.

The term 'k-space' is mathematical. The letter 'k' is used by mathematicians and physicists to describe spatial frequency, for instance, in the propagation of sound, light, or, in general, electromagnetic waves.

The Optical Equivalent

One way of understanding the concepts and mechanisms of k-space is looking at a different physical property which, perhaps, is simpler to imagine: the collecting and processing of light by a lens, as Mezrich explains in his excellent introduction to k-space.[4]

The processing of the incoming light to an image by the lens determines to a great extent its resolution, size, and contrast. The light passing through the lens is bent slightly in the center, increasingly towards the edges. In a perfect lens, the light will meet in one point, the focus, and then create an inverted image (Figure 7-2).

The processing of the light data by a lens is more complicated than generally thought: there is no point-to-point correspondence between points within the lens

Figure 7-1:
There is something wrong here: we do not talk about cat-space — or do we ?
The picture on the left was taken during daytime, the picture on the right at night. Look at the cat's eyes: the pupils are small, with a lot of light, but then they are wide with little light.
The central part of the retina displays extraordinary visual discrimination, thanks to the tiny size of the closely packed, light sensitive cones located there. This area with maximum resolution covers only 1° of the eye's field-of-view. At night, the periphery of the retina is used; it has an incredible sensitivity to light but a very poor ability to distinguish details.
k-Space behaves differently, but there are some similarities as explained in the text.

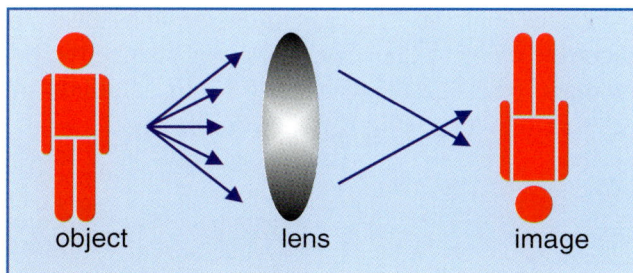

Figure 7-2:
Image processing by a lens.

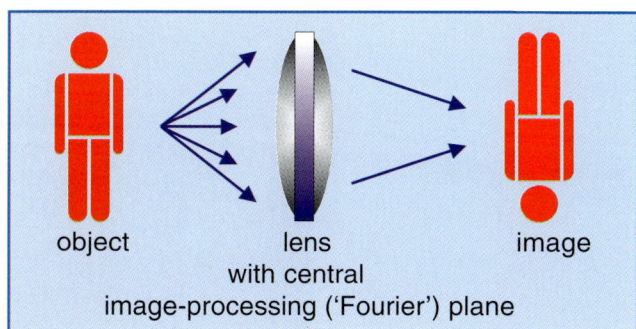

Figure 7-3:
Image processing by a lens with fictitious image-processing plane.

— or within a center plane in the middle of the lens — and the final image created by the lens. All points within the lens process data from all points of the original object. However, for our purposes we could imagine such a center plane as the location where processing takes place (Figure 7-3).

Visible light actually consists of different frequencies. As we have seen in Chapter 2, a prism can make a frequency analysis. A lens is more sophisticated. We can consider it as a special filter which, depending on its characteristics, lets some or all of these frequencies pass. It accepts signals, analyses them, processes them, and creates an image; basically, it performs a Fourier transform. We have assumed that the Fourier transform is accomplished in a fictitious central plane of the lens. In front of the lens, we can set instruments performing optical functions, for instance an iris, or we can change the size of a lens (Figure 7-4).

Changing the size of the lens or an iris influences the size of our processing plane. The steeper the angle the light makes within the lens, the sharper the focus will be; the larger the lens, the better image resolution will be. The sharpness of the final image is determined by the outer parts of our 'Fourier' plane. Points in the outer regions of the plane contribute more to image resolution than points close to the center because they allow higher spatial frequencies to pass through.

Lower spatial frequencies are closer to the center. Their main responsibility is the distribution of brightness and darkness. This means that they are responsible for image contrast.

MR Imaging and k-Space

What we have said about optical lenses holds, in a similar way, for k-space in MR imaging (Figure 7-5). As the lens, k-space collects image raw data for Fourier transform. One of the main differences is the shape. Lenses are round, k-space is rectangular.

In k-space, the iris of the camera is replaced by gradient strength, in one direction for frequency-encoding, in the other direction for phase-encoding (Figure 7-6).

In MR imaging, k is divided into three dimensions (k_x, k_y, and k_z) which define a domain or a space. Only two of them are commonly included, k_x and k_y. The third, k_z, is the slice-selecting gradient which is mostly disregarded in k-space.

The points at the center of this raw data matrix represent small gradients; increasing the offset from the center corresponds to increasing gradient strength.[3, 7]

Again, in an MR image the low spatial frequencies determine the gross signal levels (and hence contrast), while the higher spatial frequencies principally determine the edge definition (sharpness), as shown in Figure 7-7.

The definition of small objects is an integral part of the contrast and requires high spatial frequencies; thus, in this situation the high spatial frequencies also contribute to contrast. The maximum signal intensity is recorded close to the center of k-space since the net read and phase gradients applied for these points are relatively small, resulting in less dephasing.

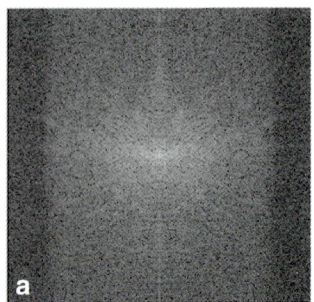
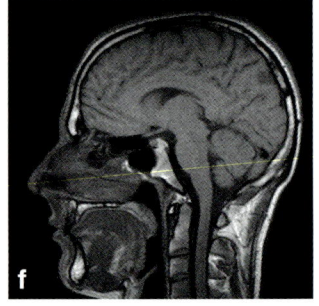

Figure 7-4:
Increasing the size of a lens with the same focus improves image resolution because the individual image points are smaller. The same holds for k-space: larger k-space with the same field-of-view means better spatial resolution of the image.

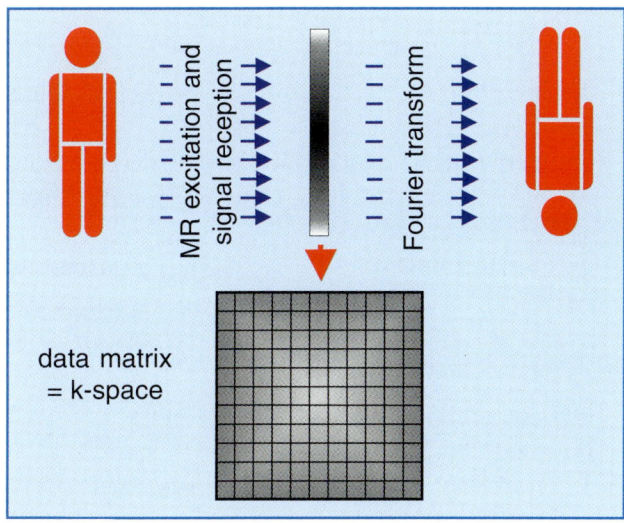
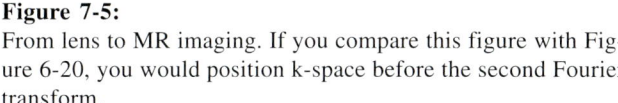
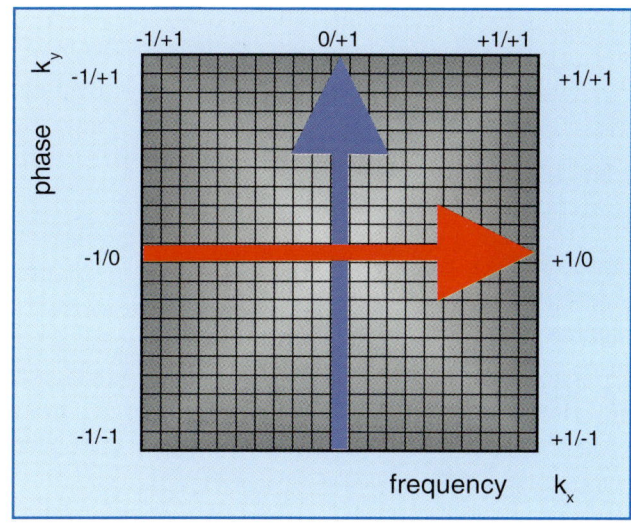

Figure 7-5:
From lens to MR imaging. If you compare this figure with Figure 6-20, you would position k-space before the second Fourier transform.

Figure 7-6:
The k-space raw data matrix consists of an area to be filled with the information needed to form an MR image. The coordinates of k-space are called 'spatial frequencies' (measured in cycles per millimeter).
They are filled depending on gradient strength of the frequency-encoding gradient (readout gradient: red arrow; x-direction) and phase-encoding gradient (preparation gradient: blue arrow; y-direction), moving from low gradient strength (-1) to zero gradient strength in the center (0) and high gradient strength (+1).

Figure 7-7:
k-Space with spatial frequency filtering. On the left side (figures a and f), we see regular k-space with image reconstruction. Figures b and g show the same k-space with filtering of the high frequencies; the reconstructed image has lost sharpness, it looks blurred; however, image contrast has hardly been affected. Figures c and h show spatial frequency filtering of the low frequencies; the reconstructed image has lost image contrast, but image details have hardly been affected. Figures d and i show low pass filtering in the readout direction, and Figures e and j high pass filtering in the preparation direction with similar results to b and c.
The signal amplitude (or magnitude) corresponds with the absolute brightness on the image because k-space is the representation of the amplitudes of the sampled echoes. Therefore, the highest intensity is in the center.

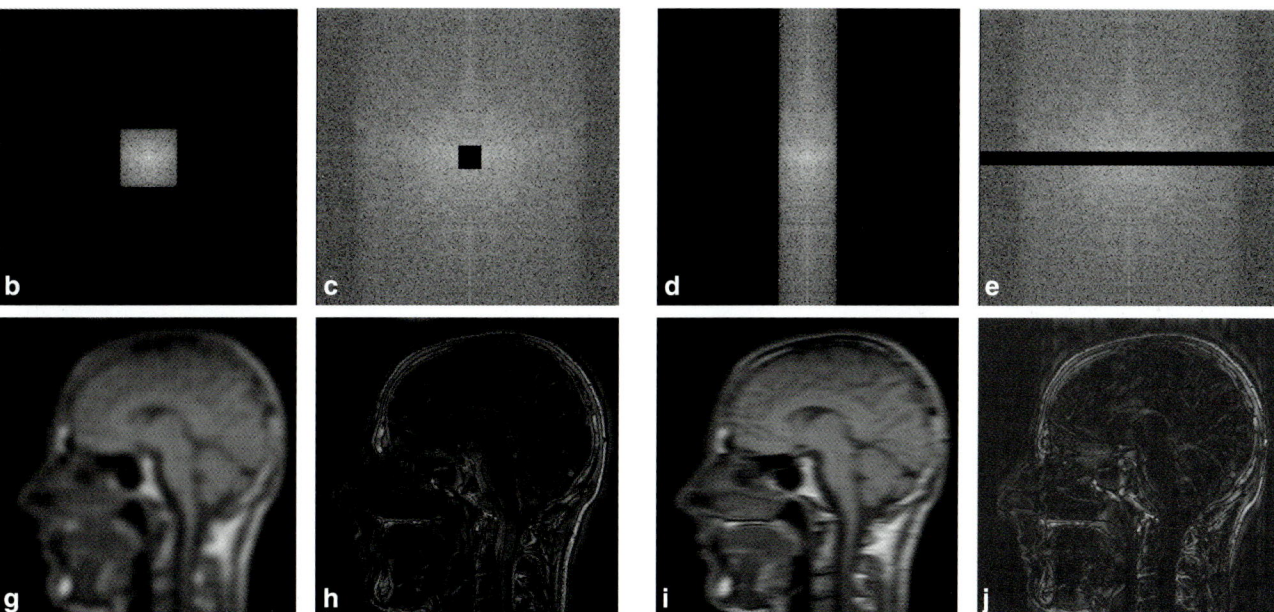

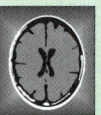

You can acquire interactively experience of k-space filling with **MR Image Expert**.

k-Space

MR Image Expert allows you to get an impression of what k-space looks like and what happens when you manipulate it.

- Select *File/Open* or the *Brain* icon in the **MR Image Expert** menu bar. Double-click on the file **1.5 T: Normal Brain, transversal slice**.

- Now select *Options/k-Space Filling*, leave the selected choices (*Direction / Frequency*, *Filter Type / Low Pass*, *100%*) and click OK.
  The foggy image that will appear is a k-space map; move it to the right; in the meantime the original brain image has been reversely Fourier-transformed, but the new image looks like the original.
  You do not recognize any anatomical structures in the k-space map you have created.

Let's try the same with another part of the human anatomy. Move the two brain images to the right side of the screen.

- Select *File/Open* or the *Body* icon in the **MR Image Expert** menu bar. Double-click on the file **1.5 T: Knee, sagittal slice**.

- Now select *Options/k-Space Filling*, leave the selected choices (*Direction / Frequency*, *Filter Type / Low Pass*, *100%*) and click OK.
  Another foggy k-space image will appear; move it to the right; again the original knee image has been reversely Fourier-transformed
  Both the brain and the knee k-space maps look similar — and there is no hint as to which kind of human anatomy is hidden behind them.

Let's now filter out some parts of k-space and see what happens. We will start with frequency filters.

- First close the two knee images and the k-space map of the brain image.

- Now select *Options/k-Space Filling*, click *Frequency*, *High Pass Filter*, choose 5%, and click OK.
  The k-space shows where the frequencies have been cut off. The reversely Fourier-transformed MR image displays the anatomical structures of the head and brain, but there is no contrast.

Close the k-space map and repeat the same experiment with the Low Pass Filter. Now the MR image has lost its sharpness, but image contrast is visible.

Let's continue with another example:

- Select *File/Open* or the *Spine* icon in the **MR Image Expert** menu bar. Double-click on the file **1.5 T: Lumbar Spine, sagittal slice**.

- Now select *Options/k-Space Filling*, click *Phase*, *High Pass Filter*, choose 5%, and click OK.
  The k-space shows where the frequencies have been cut off. Again, the reversely Fourier-transformed MR image displays the anatomical structures of the head and brain, but there is no contrast.

Close the k-space map and repeat the same experiment with the Low Pass Filter: Now the MR image has lost its sharpness, but image contrast is visible.

What happens when we use filters in both directions ?

- Select *Options/k-Space Filling*, click *Frequency + Phase*, *Low Pass Filter*, choose 25%, and click OK.
  The k-space shows where the frequencies have been cut off. The resulting image is similar to the one in Figure 7-7 b/g; it possesses contrast, but little sharpness.

Repeat the same procedure for the High Pass Filter with 5%.

See for yourself what happens with different percentages of filtering in different directions.
You will observe that with 80% Low Pass Filtering the final image quality is sufficient. This kind of image acquisition is often used to reduce scanning time.
More about these methods can be found in the following chapters.

For artifacts created by k-space manipulation read Chapter 17 and see the **MR Image Expert** Tutorial on page 206.

Filling k-Space with Data and Reconstructing an Image

In most MR imaging sequences applied in clinical routine today, raw data are placed in a rectangular k-space grid [1].

In a standard spin-echo sequence, each 90° pulse creates a new line (Figure 7-8). The length of the line is determined by the strength of the frequency-encoding gradient and the sampling time, its position by the strength of the phase-encoding gradient.

The position of the line is determined as follows. After the initial 90° excitation pulse, the spins evolve in the direction given by the phase-encoding gradient G_y and the frequency-encoding gradient G_x (yellow arrow in Figure 7-9a). They are then turned around by the 180° pulse (red/magenta arrow). Then the frequency encoding gradient is switched on again and sampling starts. This is repeated for different amplitudes of the phase-encoding gradient until k-space is filled (Figure 7-9b).

The time needed for such an imaging experiment is the number of phase-encoding steps (NG_y) multiplied by the repetition time (TR) and the number of excitations (NEX): $NG_y \times TR \times NEX$.

Now we have filled the data matrix with each row containing information from one echo.

Each data point is then Fourier-transformed in the x-direction, which leads to a new data matrix where every point in each column contains information stemming from a certain frequency; the phase information differs point-by-point per row.

The second Fourier transform is performed in the y-direction to extract phase information. This again leads to a new data matrix containing combined phase and frequency information.

The output is a matrix showing a 'modulus' or 'magnitude' image which corresponds to the bulk of MR signals from each point. Phase correction might be necessary to correct for phase jumps between 0° and 360°.

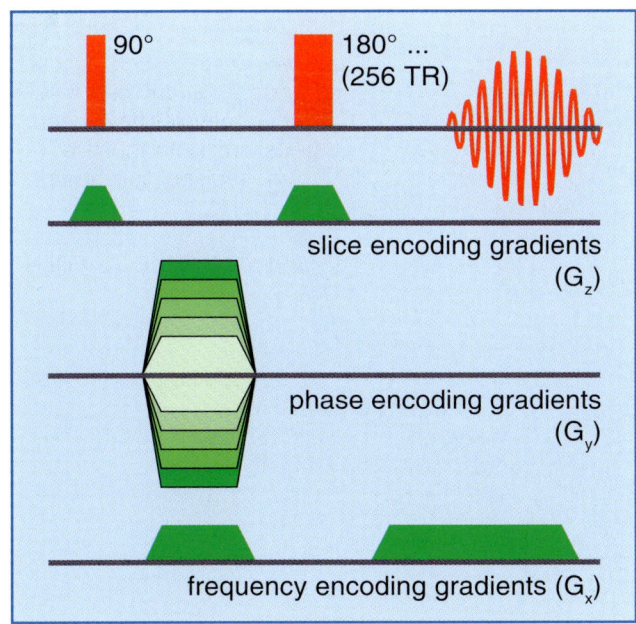

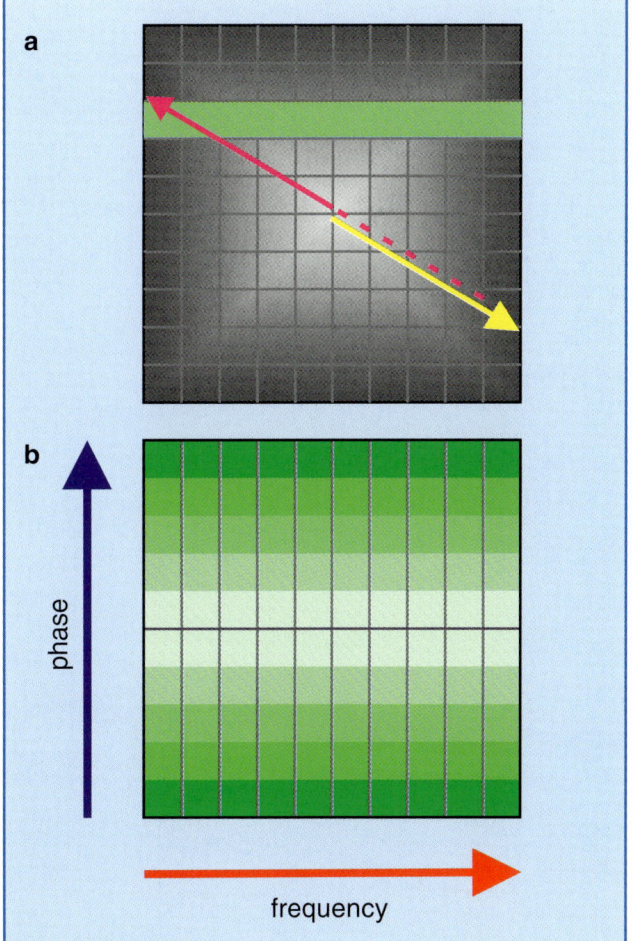

Figure 7-8 (top):
Graphic depiction of a spin-echo pulse sequence.

Figure 7-9 (right):
Mapping of k-space in a spin-echo pulse sequence.
(a) Positioning of a single line. (b) Filling of the entire k-space.
Phase direction: blue arrow, frequency direction: red arrow.
In conventional pulse sequences, such as the spin-echo sequence, one line of k-space is filled per repetition time (TR) cycle (commonly there are 256 cycles per imaging experiment, not only ten as in this figure).

Some general parameters influenced by k-space are listed in Table 7-1. You can find an in-depth discussion of k-space criteria and applications in reviews by Hennig[2], Mezrich[4], Pelc et al.[5], and Peters[6].

References

1.   Edelstein WA, Hutchison JMS, Johnson G and Redpath TW. Spin warp NMR imaging. Phys Med Biol 1980; 25: 751-756.
2.   Hennig J. K-space sampling strategies. Eur Radiol 1999; 9: 1020-1031.
3.   Ljunggren S. A simple graphical representation of Fourier-based imaging methods. J Magn Reson 1983; 54: 338-343.
4.   Mezrich R. A perspective on k-space. Radiology 1995; 195: 297-315.
5.   Pelc NJ, Glover GH. A stroll through k-space. In: Medical Physics Monograph no. 21: The Physics of MRI. American Institute of Physiscs 1993; 21: 771.
6.   Peters TM. An introduction to k-space. In: Medical Physics Monograph no. 21: The physics of MRI. American Institute of Physiscs 1993; 21: 754.
7.   Twieg DB. The k-trajectory formulation of the NMR imaging process with applications in analysis and synthesis of imaging methods. Med Phys 1983; 10: 610-621.

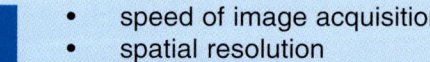

- speed of image acquisition
- spatial resolution
- field-of-view
- contrast
- artifacts

Table 7-1:
Some parameters that can be influenced by manipulating the data represented in k-space.

Chapter Eight

Rapid Imaging

Introduction

Conventional MR imaging is a slow imaging technique. All of the classical imaging sequences have long scan times; for instance, the acquisition of a single spin-echo image takes between four and twenty minutes.

The principal constraints of imaging times are the long relaxation times on the one hand, and the desired signal-to-noise ratio and spatial resolution on the other hand.

After $5 \times T1$ of a tissue, the spins have nearly completely recovered. To get the optimum signal, one would have to wait for this period of time between each excitation pulse. Thus, to acquire images in which the recovery of T1 during TR has little effect requires relatively long TR values, particularly at high fields where T1 of tissues is longer than one second.

In the preceding chapter, we have seen that the time t to acquire the data for one image can be calculated as follows:

$$t = NG_y \times TR \times NEX$$

where TR is the repetition time; NG_y the number of phase-encoding steps (or the number of lines in the image; usually 256 in routine imaging); and NEX the number of excitations (= number of data averages).

If TR = 2 s, NG_y = 256, and NEX = 2, the data acquisition time adds up to 1024 seconds (i.e., 17 minutes and 4 seconds).

This used to be the common imaging time for a classical T2-weighted spin-echo image. As Figure 8-2 shows, many parameters influence imaging time and changing one of them will immediately have an interdependent effect on the others.

From the early days of MR imaging, efforts were made to shorten the imaging time. Better signal-to-noise, for instance, was achieved by increasing the magnetic field strength of MR machines. Hardware, particularly coils, and software were also improved.

Multi-echo multi-slice techniques use the dead time during the long delays to create a large number of images, but until the early 1990s imaging within seconds or even real-time imaging seemed to be impossible.

Figure 8-1:
Some people like it slow, others like it fast. Both styles have their advantages and disadvantages. This Viennese cabman prefers slower procedures; we would like to get it going a little bit faster. The question is: is the quality of the output better — or, in the long run, who gets better results ?

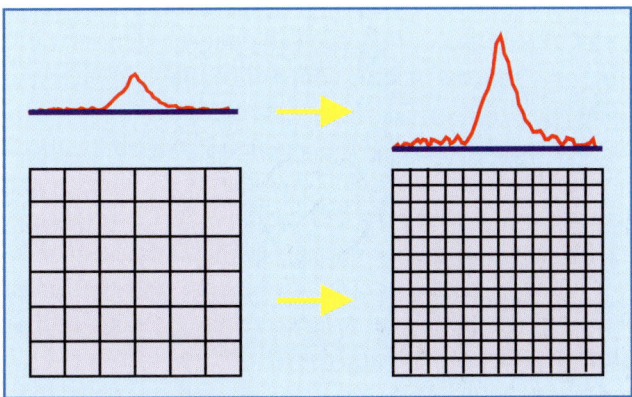

Figure 8-2:
Imaging time depends on a multitude of factors, among them signal-to-noise ratio and spatial resolution. Better spatial resolution with the same signal-to-noise, or better signal-to-noise with the same resolution, for instance, may require more averages and thus leads to a time penalty.

At the beginning of the 1980s, many physicists believed that rapid MR imaging would be impossible because of the limitations set by the relaxation times. Several seconds of recovery after every excitation was the main obstacle. Only in the mid-1980s did new ideas develop on how to accelerate imaging procedures.

To boost the understanding of rapid (or fast) sequences, let us recall some of the main factors of conventional MR imaging. Obtaining a signal which contains spatial information represents the first step in an MRI experiment. The next step is the manipulation of contrast in the images, which is achieved using a pulse sequence. Generally, the reconstruction technique and the pulse sequence are independent, so any pulse sequence can be combined with any reconstruction technique. For the purpose of understanding the basics of rapid imaging sequences, we will assume that the acquisition is performed with a 2D spin-warp sequence with selective excitation of one or more slices and the reconstruction with a 2DFT to produce a 2D image.

The form of the magnetic resonance signal is determined by a large number of factors, including spin density ρ, T1, T2, flow, and diffusion. By suitable preparation of the spins, we can emphasize the contribution of one or more of these factors.

The basic pulse sequences (spin-echo, inversion-recovery, partial-saturation) were outlined in Chapters 4 and 6. These pulse sequences have to be modified for an imaging experiment since the FID signal has to be reformed in the presence of the imaging gradients by using either a spin- or gradient-echo sequence. This also provides sufficient time for the other position-encoding gradients (slice- and phase-encoding gradients in a 2D spin-warp experiment) to be applied.

The spin-echo sequence modified in this way produces a single spin-echo image, with the signal levels being mainly determined by the repetition time (TR) and the echo time (TE). Since the sequence is generally used with a relatively long TR, multiple slices can be collected to improve the efficiency of the sequence.

By manipulating TR and TE, we can induce T1-weighted and T2-weighted contrast, respectively. This will be discussed in detail in the following chapters.

T1 contrast can also be generated by applying an inversion pulse (180°) and a delay (TI) prior to the excitation (90°) pulse. A 180° pulse always inverts the z-magnetization; in addition, when transverse magnetization is present, it also refocuses the transverse magnetization, leading to a spin echo.

To increase the amount of information available from a spin-echo sequence, we can apply a series of 180° pulses and create multiple echoes. Normally, each spin echo contributes one line of raw data in k-space to one image for each excitation. Thus for n spin echoes, n images are obtained. Multiple-echo sequences provide images with different contrast with increasing echo time. The various forms of the spin-echo sequence and their contrast behavior will be discussed in detail in the following chapters.

RARE

The RARE sequence (*R*apid *A*cquisition with *R*elaxation *E*nhancement; also called *R*apid *S*pin *E*cho, *RSE*, *F*ast *S*pin *E*cho, *FSE* or *T*urbo *S*pin *E*cho, *TSE* — cf. Table 8-1) was introduced by Jürgen Hennig in 1986 [9]. It is based on the multiple-echo sequence. Over the years, RARE has partly replaced the conventional multiple spin-echo pulse sequence, which was the most common sequence used in clinical imaging. Blurring in fine detail, however, will hinder a complete replacement.

Rather than using the same amount of phase-encoding for each echo and each echo as one line for an image associated with a particular TE, in RARE sequences different amounts of phase-encoding can be applied to each echo. This enables them to be used as different lines in a single image (Figures 8-3 and 8-4).

For example, a multiple echo sequence with 8 echoes, implemented as a RARE sequence can contribute 8 phase-encoding lines to a single image, or 4 phase-encoding lines to two images. This results in reductions of 8 or 4, respectively, in the number of excitations required to collect the full data set, while retaining the clinically useful contrast of the spin-echo sequence.

Considering a RARE sequence in which 8 echoes are used to provide 8 lines for a 128 ×1 28 image, each line has a different echo time and hence T2 weighting. This is an undesirable constraint of the RARE sequence, but it can be overcome by altering the assignation of the particular echoes to respective lines in k-space. For clinical use, probably the most effective version of the RARE sequence uses the first half of the echo train to provide the lines for a proton density image and the second half for a T2-weighted image. In this way, one retains the clinically useful double-echo sequence, while reducing the scan time by a factor of between 2 and 8, depending on the number of echoes used [13].

90°
180°...
(followed by additional 180° pulses)
1st echo (followed by additional echoes)
phase-encoding gradients (G_y)
frequency-encoding gradients (G_x)

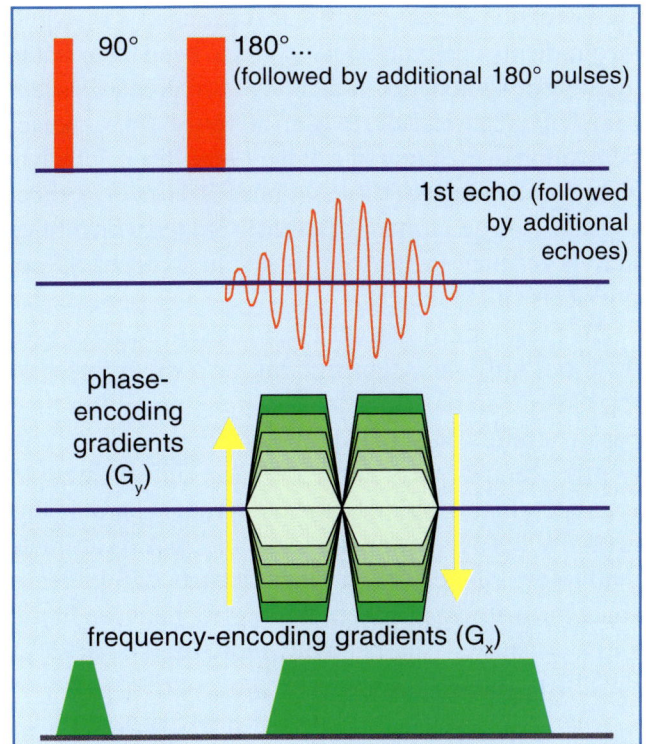

Figure 8-3:
The RARE pulse sequence. The sequence is a modified multiple spin-echo sequence. A train of echoes is created and each echo is individually phase-encoded. Usually eight to sixteen echoes are used. There are several different variations with different styles of gradient switching.

Figure 8-4:
A comparison of a multiple spin-echo (a) and a RARE sequence (b). Every echo in the SE sequence is used to create an individual image (one echo = one line per image), whereas in a RARE sequence several echoes contribute lines to a single image of the raw data matrix (k-space), as in this example, or to two images.

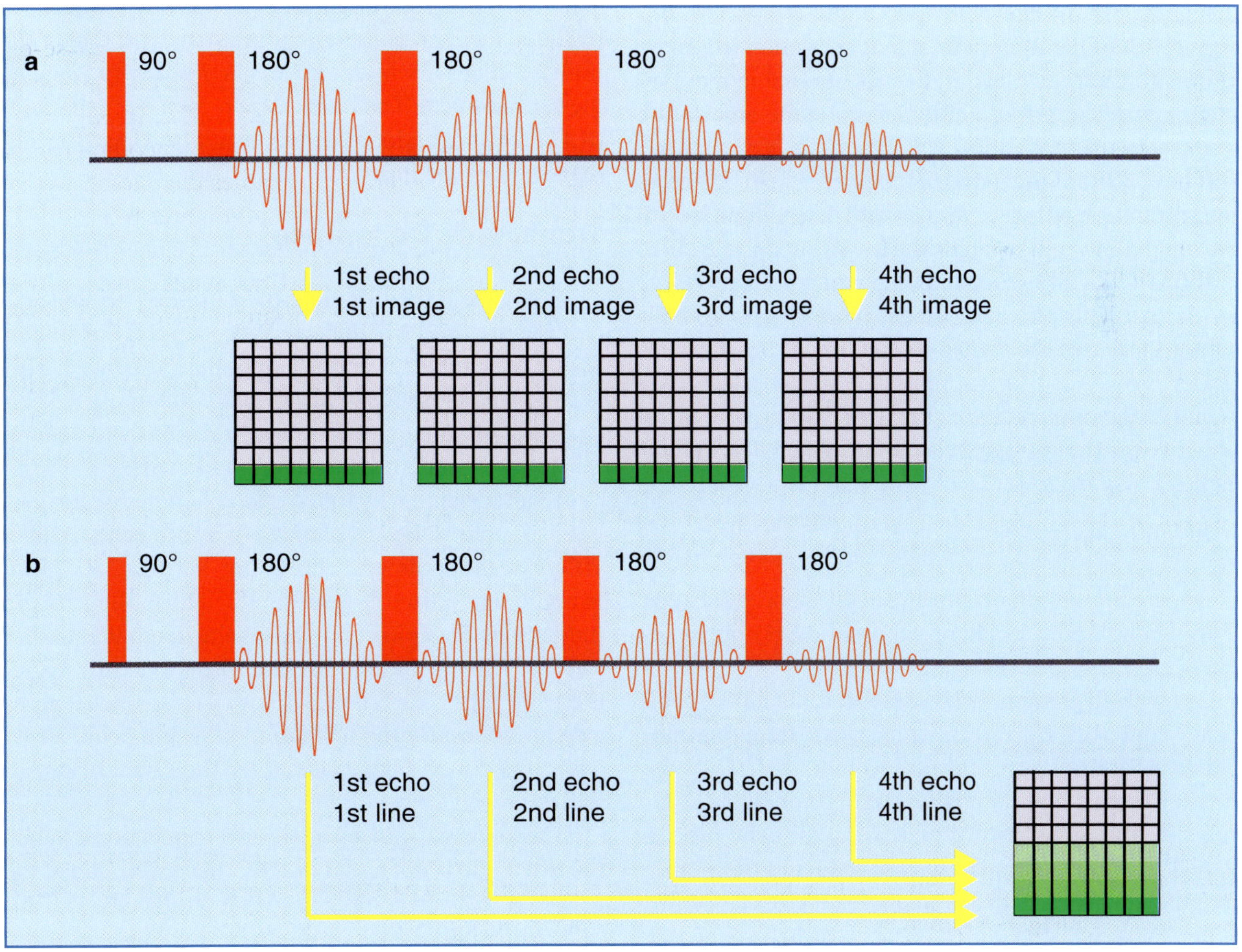

a
90°
180°
180°
180°
180°
1st echo
1st image
2nd echo
2nd image
3rd echo
3rd image
4th echo
4th image
b
90°
180°
180°
180°
180°
1st echo
1st line
2nd echo
2nd line
3rd echo
3rd line
4th echo
4th line

Gradient-Echo Sequences

A completely different approach to rapid imaging was used by the first pulse sequences, which shortened imaging time in routine clinical settings. The generic name of these sequences is *gradient-echo (GE) sequences* or, better, *gradient-recalled (GRE) sequences,* and they come in a plethora of different acronyms. An overview of the different rapid pulse sequences, their names and acronyms can be found in Table 8-1. A basic description of GRE has been given in Chapter 6.

The first sequence in this group was presented in 1986 by Axel Haase and collaborators and dubbed *FLASH* [6]. The FLASH (*Fast Low Angle Shot*) sequence is a saturation recovery sequence with a short repetition time (TR < 200 ms), a low flip angle (< 90°), and a gradient echo for refocusing.

The application of flip angles different from 90° and 180° brought an end to the ideology of long waiting times, which was based upon the belief that T1 is the limiting time factor of MR imaging.

The reason for using a low flip angle is illustrated in Figure 8-5. When a 90° flip angle is applied, we convert all of the longitudinal magnetization (in the *z*-axis) into transverse magnetization (signal in the x'-y' plane), while, e.g., for a 30° flip angle the amount of transverse magnetization is halved (sin 30°), but we still have 87% of the *z*-magnetization (cos 30°). The *z*-magnetization will recover at a rate determined by T1 during the interpulse interval. However, since the TR is short in FLASH sequences, the *z*-magnetization left by the previous pulse becomes dominant and significantly increases the signal obtained after the next RF pulse.

For a given repetition time, the flip angle which will give maximum signal can be calculated. It is known as the *Ernst angle* [3]:

Ernst angle = cos⁻¹[exp(-TR/T1)].

Figure 8-6 summarizes the main differences between a spin-echo and a gradient-echo (FLASH) pulse sequence.

As with all gradient-echo sequences, but unlike spin-echo sequences, the effects of magnetic field inhomogeneities are not compensated so that short TE must be used if high-quality images are to be obtained. This rules out the possibility of increasing the echo time to give T2 contrast. Another way of reducing field inhomogeneity effects is to use small voxel sizes since this limits the dephasing which occurs within a voxel.

To reduce the echo times, it is necessary to switch the gradients relatively quickly and to keep them stable after being switched. Gradient-switching requires less energy to create an echo than a 180° pulse. Thus, power deposition in the body of a patient is reduced, which is a major advantage of these sequences. However, there are also a large number of disadvantages which have not yet resulted in FLASH replacing standard SE sequences in all instances.

Because of the shorter TR, FLASH sequences reduce not only the scan time but also the number of slices that can be acquired. Optimum repetition time has to be adjusted to the number of slices required and to other factors such as the duration of a breathhold for abdominal imaging or the heart rate in cardiac imaging. When decreasing the scan time, motion artifacts tend to be reduced, while flow artifacts will increase since the difference in signal intensity between blood and stationary tissue becomes more marked at short repetition times.

The feature can be exploited in FLASH-based cine-MR imaging where 8-32 lines of the same slice are acquired during one cardiac cycle, then the sequence is repeated for each phase-encoding step to produce 8-32 images, each of which represents a different stage of the cardiac cycle. The images are presented in the form of a closed movie loop, which depicts the function and dynamics of the heart.

Transverse Coherence

When the repetition time for a FLASH sequence is reduced to a level where the repetition time TR is shorter than T2, the relaxation behavior is also influenced. This is due to the presence of transverse coherences [5]. Their exploitation or suppression forms the basis of several fast imaging schemes based upon FLASH.

To understand why transverse coherence occurs, we have to modify the simple idea of a spin echo. After a 90° pulse the spins start dephasing. When a 180° pulse is applied at a time τ after the 90° pulse, the rotation induced by the spin echo causes the magnetization to start refocusing and a spin echo forms at a time τ after the 180° pulse. This model is very useful since it gives a clear picture for the formation of a spin echo. However, it is not so easy to visualize the effect of pulses which are < 180°.

These < 180°-pulses also form spin echoes. When the flip angle is not equal to 180°, the amplitude of the echoes is reduced compared to that produced by a 180° refocusing pulse. In addition to the evolution of the *z*-

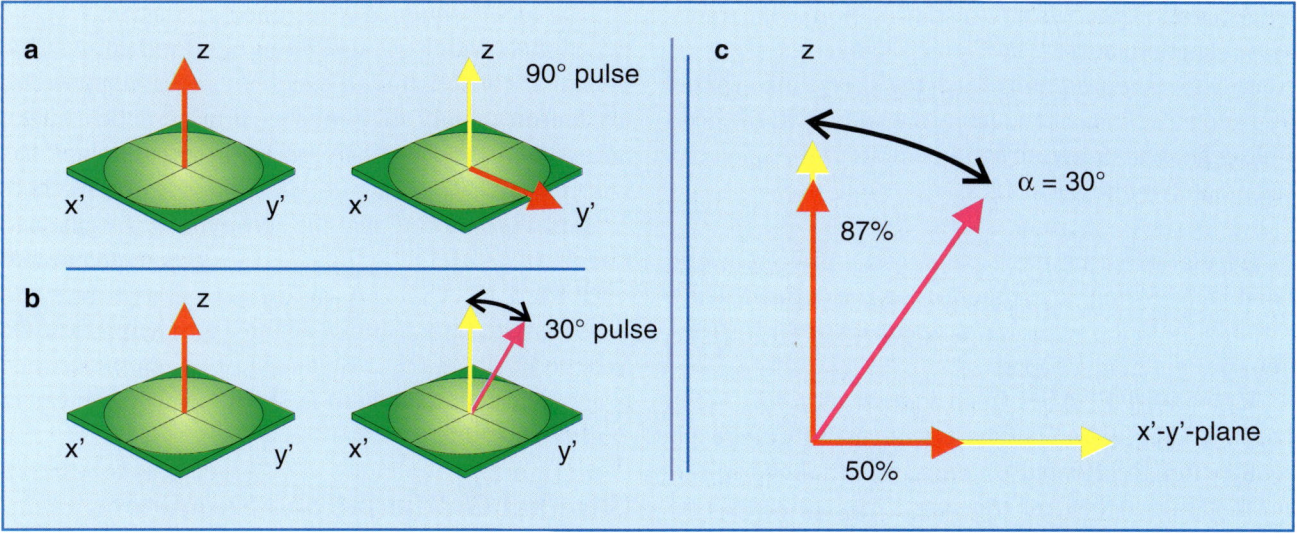

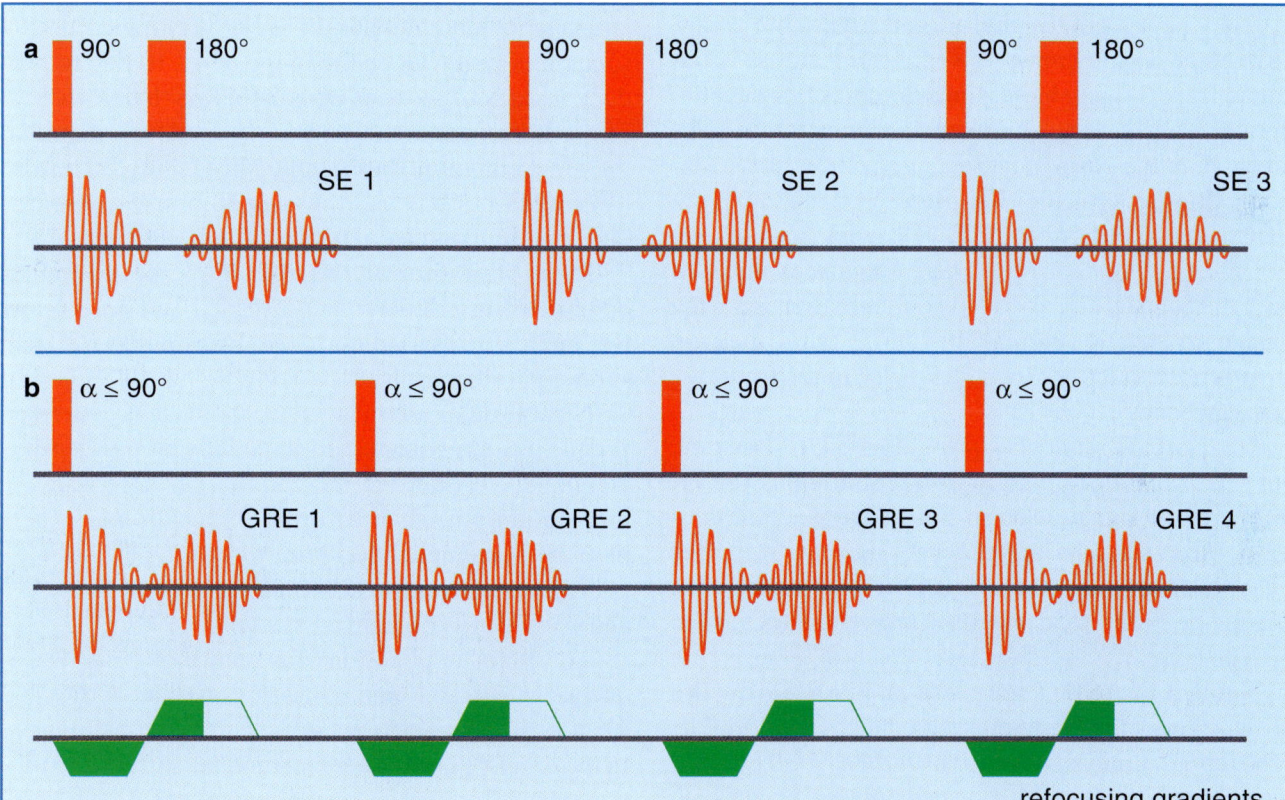

Figure 8-5 (top):
Principle of a standard pulse sequence (a) compared to a rapid imaging sequence of the FLASH type (b).

In both cases, the net magnetization during equilibrium is aligned with the z-axis. In the standard sequence, a 90° pulse tilts the magnetization into the x'-y' plane. No longitudinal component remains. In the FLASH-type sequence, a flip angle $\alpha < 90°$ is applied. Such a pulse divides the magnetization in transverse and longitudinal components.

In our example (c) α equals 30°. This results in a reduction of the longitudinal magnetization to 87%, whereas the transverse magnetization is 50% of the available longitudinal magnetization.

Figure 8-6 (bottom):
Principle of a standard spin-echo pulse sequence (a) compared to a rapid imaging sequence of the FLASH type (b).

(a): in the spin-echo pulse sequence, the echo is created by a 180° pulse. This involves relatively long time delays and high power deposition in the examined sample. Because of the dependence of TR on T1, TR has to be relatively long.

(b): in the FLASH sequence, any pulse angle can be used instead of the initial 90° pulse. The echo is formed by gradient switching. This can be done faster and with less power deposition (potentially less hazardous for the patient). Thus, TR (and TE) can be shortened.

SE = spin echo; GRE = gradient (recalled) echo.

magnetization, there is now also evolution of the transverse magnetization. The signal received consists of contributions representing fresh transverse magnetization and an *echo term*, which is the sum of all the possible echoes arising from combinations of the spin echoes created by pulses < 180°.

By manipulating these parameters, three types of rapid FLASH imaging sequences can be defined. They can be simulated with **MR Image Expert**. Instructions are given in Chapter 10.

Refocused FLASH (also known as *FFE*, *FISP*, *FAST*, *GRASS*, *ROAST*). These sequences measure the signal after the RF pulse, which corresponds to the combination of the fresh transverse magnetization and the echo term [4, 17].

They have a good signal-to-noise ratio, but generally rather poor contrast. A very strong signal is obtained from flowing blood since the spins flowing into the slice will have equilibrium magnetization (M_0) rather than the steady state magnetization of the stationary tissue (typically 10% of M_0).

Contrast-Enhanced (CE-) FLASH (also known as *CE-FFE*, *PSIF*, *SSFP*). These sequences measure only the echo term [8]. To avoid contamination from the fresh magnetization present after the RF pulse, the echo term is observed prior to the RF pulse in the form of a gradient echo.

CE-FLASH sequences provide good T2 contrast, but relatively poor signal-to-noise. Shortening the TR improves the signal-to-noise, but also reduces the contrast. Flow artifacts are generally absent from CE-FLASH scans since the blood flows out of the slice during the TR interval and thus cannot be refocused to give an echo.

Spoiled FLASH. This type of rapid pulse sequence observes only the fresh transverse magnetization. The echo term is removed (spoiled) by the use of either spoiling gradients or phase spoiling techniques. When a high flip angle is used, the spoiled FLASH sequence can give good T1 contrast.

Two other variants of FLASH sequences are the FADE sequence [16] and the FISP sequence [14].

The FADE sequence combines the refocused and CE-FLASH sequences into a single sequence in which the two resulting signals are observed in separate acquisition periods during a single interpulse interval. Therefore, the minimum TR is longer, but the sequence is more efficient because we obtain two images with different contrast.

The FISP sequence is designed to superimpose the two signals which are separately acquired in FADE to give a single signal with excellent signal-to-noise ratio. Unfortunately, the sequence is not practical since, unless the two images are perfectly aligned, artifacts will result [18].

It is worth noting that the acronym FISP is used to refer to two different sequences (i.e., this sequence and refocused FLASH). A modification of refocused FLASH with refocusing of all three gradients is known as True FISP (and also Balanced FFE). This sequence is about to become one of the most used sequences in cardiac imaging.

Ultrafast Gradient-Echo Sequences

A recent innovation is the use of very rapid FLASH sequences (with TR in the range of 4 - 10 ms) to produce images in seconds or even in less than a second. These sequences are, for instance, used for abdominal imaging, commonly as a single-slice method. They allow breath-holding and thus can eliminate ghost artifacts and blurring from respiratory motion.

In the basic *snapshot FLASH* sequence, no spoiler or refocusing gradients are included and a very low flip angle corresponding to the Ernst angle for such short TR values is used.

Now, little or no transverse coherence is generated and the resulting images are essentially proton-density-weighted. To improve the contrast in these scans, a preparation pulse can be used. Its function is to prepare the z-magnetization prior to starting the scan [7].

The scan times for 128^2 matrices vary between 0.5 and 1.0 seconds on clinical systems.

The main applications for snapshot FLASH sequences are in abdominal imaging, cardiac studies and functional (dynamic) imaging using contrast agents. In the first two cases, other techniques suffer from motion artifacts or long scan times when triggering is used. For dynamic imaging, the time resolution required (1-3 seconds) means that snapshot sequences have to be used if a reasonable (128^2) resolution is to be obtained. Standard 2D head examinations are better carried out using standard spin-echo or RARE sequences since motion artifacts are not usually a serious problem. Only when highly motion-sensitive sequences (such as diffusion sequences) are used is it necessary to revert to high-speed imaging.

The following third group of rapid imaging methods is completely different from those mentioned above.

Acronym	Spelled-out Name	Acronym	Spelled-out Name
Balanced FFE	balanced fast field echo	**GRECHO**	gradient-recalled echo
CE-FAST	contrast-enhanced Fourier-acquired steady state	**IR FGR**	inversion recovery fast gradient-recalled acquisition in the steady state
CE-FFE T1	contrast-enhanced fast field echo (T1-weighted)	*MESS*	*multi-echo single shot*
		MPGR	multiplanar gradient-recalled
CE-FFE T2	CE-FFE (T2-weighted)	**MP-RAGE**	magnetization-prepared rapid gradient echo
CSFSE	*continguous-slice fast-acquisition spin echo*	**PFI**	partial flip imaging
DE FGR	driven-equilibrium fast gradient-recalled acquisition in the steady state	**PS**	partial saturation
		PSIF	reverse fast imaging with steady precession
DEFAISE	*dual-echo fast-acquisition interleaved spin echo*	**QUEST**	quick echo-split imaging technique
DESS	double-echo stead- state (combination of FISP and PSIF)	**RAM-FAST**	rapidly acquired magnetization-prepared Fourier-acquired steady state
DFSE	*double-fast spin echo*	*RARE*	*rapid acquisition with relaxation enhancement*
E-SHORT	steady-state gradient echo with spin-echo sampling	**RF-FAST**	RF-spoiled Fourier-acquired steady state
FADE	FASE acquisition double echo	**RF spoiled**	RF-spoiled Fourier-acquired
FAME	*fast-acquisition multi-echo*	*RISE*	*rapid imaging spin echo*
FASE	*fast spin echo*	**ROAST**	resonant offset averaging in the steady state
FAST	Fourier-acquired steady state		
FATE	fast turbo echo (=FADE)	**RS**	rapid scan
FE	field echo	*RSE*	*rapid spin echo*
FEDIF	field echo with echo time set for water and fat signals in opposition	**SHORT**	short repetition technique
		SMASH	short minimum-angle shot
FEER	field even-echo by reversal	**SPGR**	spoiled gradient-recalled
FESUM	field echo with echo time set for water and fat signals in phase	**SSFP**	steady-state free precession
		STAGE	small tip angle gradient echo
FFE	fast field echo	**STAGE:T1W**	small tip angle gradient echo: T1-weighted
FGR	fast gradient-recalled acquisition in the steady state		
FISP	fast imaging with steady precession	*STEP*	*stimulated echo progreessive imaging*
FLARE	*fast low-angle recalled echo*	**STERF**	steady-state technique with refocused free induction decay
FLASH	fast low-angle shot		
FRE	field-reversal echo	**TFE**	'turbo'-field echo
FS	fast scan	**3D MP RAGE**	3D magnetization-prepared rapid gradient echo
FSE	*fast spin echo*		
F-SHORT	steady-state gradient echo based on free induction decay	**T1 FAST**	Fourier-acquired steady state (T1-weighted)
FSPGR	fast spoiled gradient-recalled	**T1 FFE**	contrast-enhanced fast field echo (T1-weighted)
GE	gradient echo		
GFE	gradient field echo	**T2 FFE**	contrast-enhanced fast field echo (T2-weighted)
GFEC	gradient field echo with contrast		
GRASE	*gradient and spin echo*	**TRUE FISP**	fast imaging with steady precession (heavily T2-weighted)
GRASS	gradient-recalled acquisition in the steady state		
		TSE	*'turbo' spin echo*
GRE	gradient echo (as 'generic' name); gradient-recalled echo	**Turbo-FE**	'turbo' field echo
		TurboFLASH	'turbo'-fast low angle shot
GREC	gradient field echo with contrast	*Turbo SE*	*'turbo' spin echo*
GRECO	gradient-recalled echo	**Turbo-SHORT**	'turbo' short repetition technique

Table 8-1:
Rapid imaging techniques.
Gradient-echo techniques are shown in normal font, spin-echo based techniques in italics.
Several companies use different acronyms to describe certain rapid, usually gradient-echo-based techniques. Some of the acronyms mentioned are registered trade marks. Note that some acronyms such as FAST, FISP, and SSFP are used by different companies for different pulse sequences. We have included only the most common usage. Table 10-7 tries to give an overview of the different families of gradient-echo sequences.

Echo-Planar Imaging

Echo-planar imaging (EPI) is the fastest imaging sequence currently available, and unlike the other sequences discussed in this chapter, it does not use the spin-warp technique.

However, in its modern implementation, it is conceptually very similar to spin warp. EPI was proposed by Peter Mansfield in 1977 [12]. It is based on the principle of a single excitation of the spins, followed by the rapid switching of a strong gradient to form a series of gradient echoes, each of which is given a different degree of phase-encoding and thus can be reconstructed to form an image. The phase gradient can be applied as a constant gradient, as in the original echoplanar scheme (Figure 8-7), or a series of small 'blips', each of which corresponds to one phase-encoding step [10]. The example of Figure 8-7 consists of 9 sampling periods. For a 64 × 128 image matrix, 64 sampling periods are necessary. During each sampling period, 128 points are sampled. The k-space trajectory is a single sawtooth-pattern path (Figure 8-8).

One of the main problems with the original echoplanar sequence is the T2* dephasing of the signal during the scan. We can reduce this effect by forming the EPI echo train about a spin echo, even though substantial T2* dephasing will remain at the start and end of the EPI sequence. To minimize such effects, very short scan times have to be achieved [2, 15].

However, as we reduce the sampling period, we also reduce the signal-to-noise ratio and increase the amplitude of the read gradient required to obtain a given resolution. For these reasons, single-shot EPI tends to be limited to a maximum of a 128×256 matrix.

Whereas the true snapshot capability of echo planar is very attractive, the problems associated with the technique mean that it cannot be used for a number of possible clinical applications.

New developments in gradient and switching techniques have, however, partly overcome these problems, and single-shot EPI is nowadays available on most new 1.5 T machines.

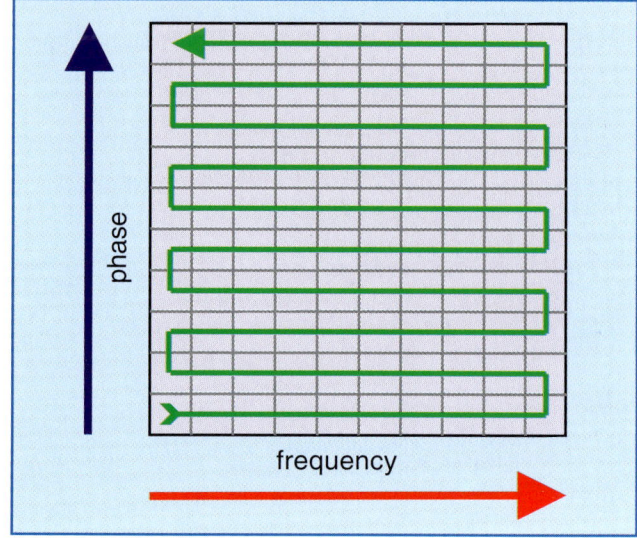

Figure 8-8 (top):
The k-space trajectory of an echo-planar imaging experiment.

Figure 8-7:
An EPI pulse sequence (FID-based MBEST sequence).

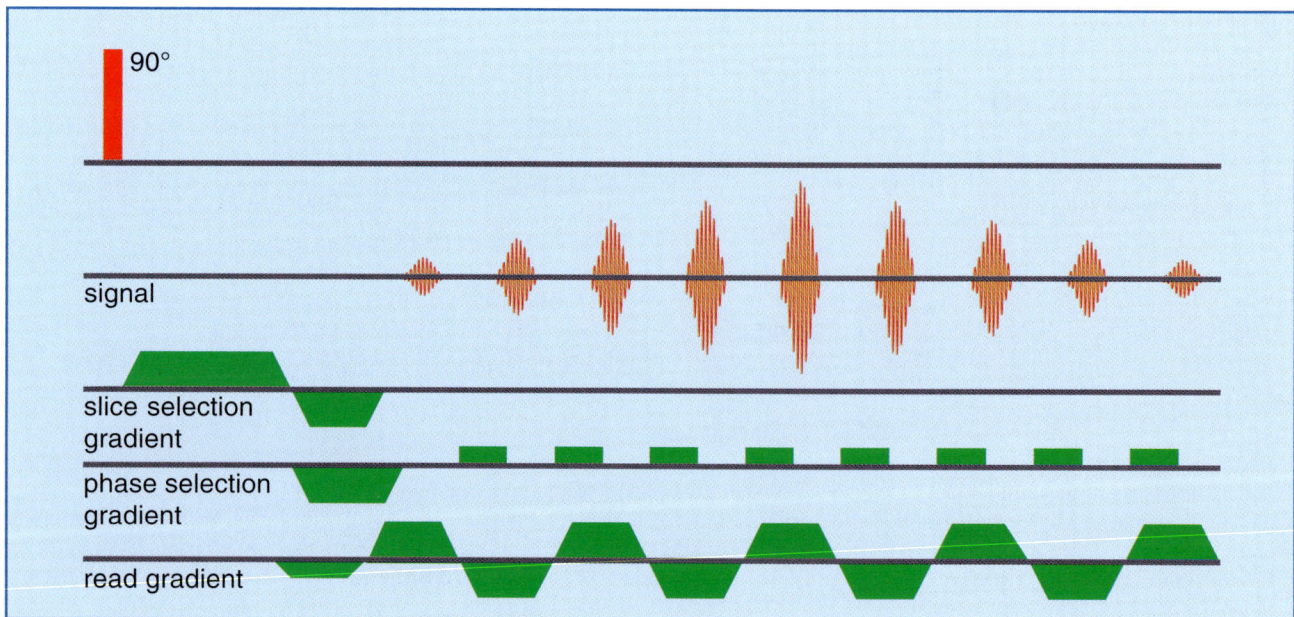

Single-shot EPI still suffers from a chemical-shift artifact since the bandwidth per pixel in the phase-encoding direction is less than the chemical shift between water and fat. The applications for EPI are the same as those cited for snapshot FLASH.

Multi-shot EPI improves image quality tremendously.

There remains some uncertainty with respect to the safety of EPI since the very rapid switching of strong gradients generates electrical currents in the body which can stimulate peripheral or cardiac nerves[1]. No definite hazard has yet been defined, but further research is currently in progress.

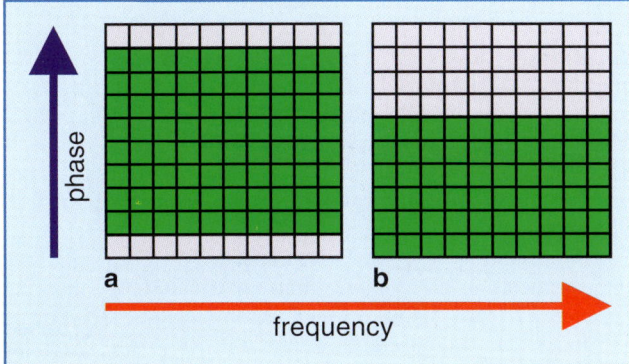

Figure 8-9:
(a) The k-space data set in reduced acquisition. (b) The k-space data set in halfscan. Slightly more than 50% of k-space is collected.

Figure 8-10:
The k-space data set in (a) full-space acquisition, and (b) rectangular field-of-view.

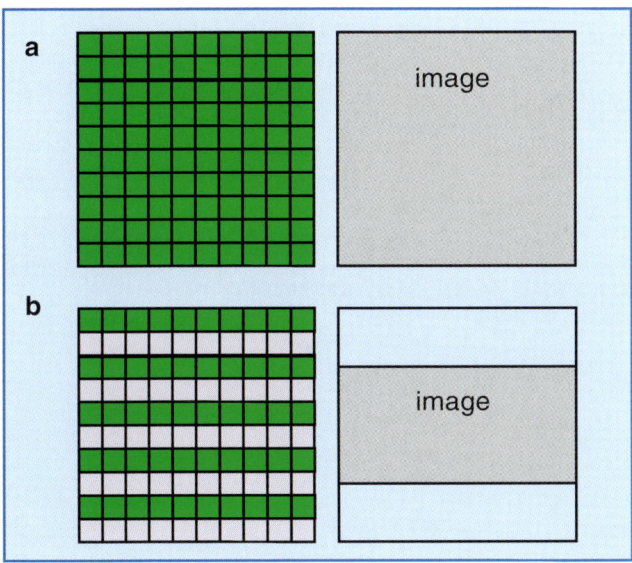

Acceleration of Image Acquisition by k-Space Manipulation

A completely different approach to accelerate image acquisition is through faster acquisition of image data rather than optimizing pulse sequences.

A number of different approaches have been proposed:

Reduced Acquisition. This approach was discussed in the Tutorial in Chapter 7 on page 86. Instead of acquiring, for instance, 256 lines, we acquire only 80% and zero-fill the remaining lines. We loose some of the spatial resolution, but for many clinical applications, the raw data are sufficient (Figure 8-9a).

Halfscan. In this case we acquire an asymmetrical fraction of the data set. The rest of the data are replaced by the symmetrical data from the other side of k-space. Spatial resolution is maintained, but there is a loss of signal-to-noise (Figure 8-9b).

Rectangular Field-of-View. The final MR image can be turned into a rectangular image by collecting only half the lines in k-space. By doing this, image time, as well as the field-of-view, will be halved, which is convenient for imaging of the extremities and the spine, or in angiography (Figure 8-10).However, the signal-to-noise ratio will also be substantially reduced.

Figure 8-11:
k-Space substitution. (a): the entire k-space of a reference image before contrast injection is collected. (b): for the dynamic part of the study, the central part of the reference k-space is removed. (c): During the uptake of the contrast agent, only the central parts of k-space are collected, and afterwards (d) are combined with the reference data.

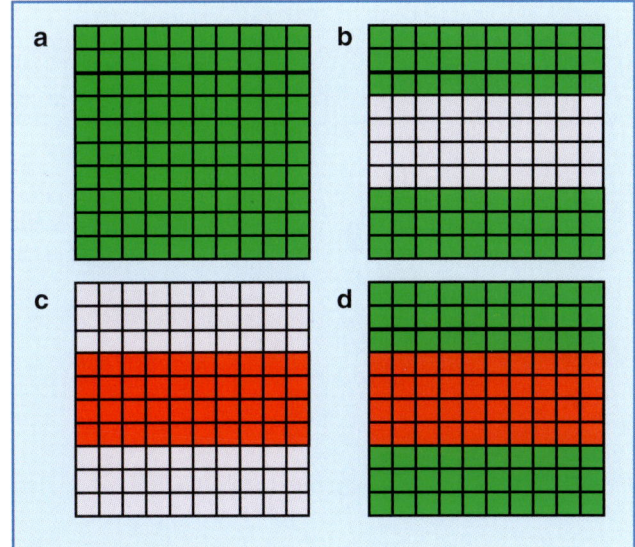

k-Space Substitution. To accelerate dynamic image data acquisition, one can apply k-space substitution, also called 'keyhole' imaging[11,18]. This technique collects the entire k-space of a reference image; for the subsequent images, however, only the central lines are recorded. These data are then combined with the outer lines of the reference data space to add information on edge definition and sharpness. In this way, the uptake of a contrast agent can be followed very rapidly (Figure 8-11).

Spiral (Helical) and Radial Scanning. Alternatives to filling k-space line-by-line are spiral (helical) or radial scanning techniques (Figure 8-12). These methods are very fast and therefore suited for dynamic imaging and, e.g., for cardiac imaging. They use projection reconstruction (backprojection) algorithms as described earlier.

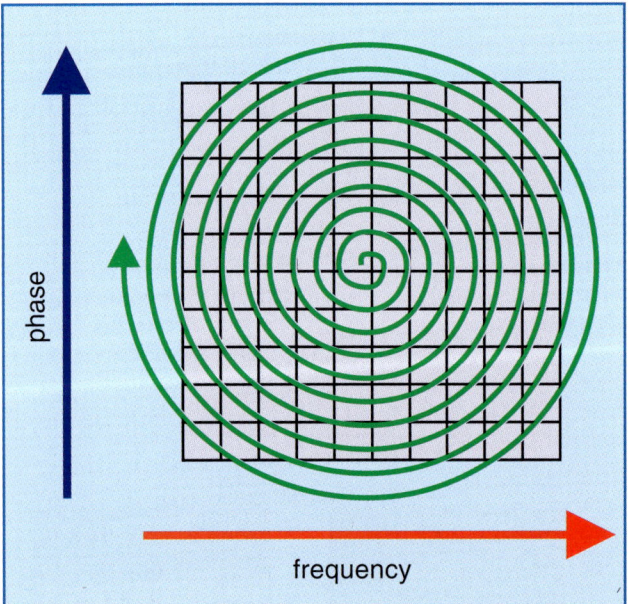

Figure 8-12:
Spiral k-space filling. Usually one starts by acquiring data for the center of k-space. Spiral filling can be performed in a single shot or interleaved as a multi-shot technique. Multi-shot techniques have a higher spatial resolution.

References

1.   Budinger TF, Fischer H, Hentschel D, Reinfelder HE, Schmitt F. Neural simulation dB/dt thresholds for frequency and number of oscillations using sinusoidal magnetic gradient fields. Book of Abstracts. Ninth Annual Meeting of the Society of Magnetic Resonance in Medicine. New York 1990; 276.

2.   Cohen MS, Weisskoff RM. Ultrafast imaging. Magn Reson Imag 1991; 9: 1-37.

3.   Ernst RR, Anderson WA. Applications of FT spectroscopy to magnetic resonance. Rev Sci Instrum 1966; 37: 93-98.

4.   Frahm J, Hänicke W, Merboldt K. Transverse coherence in rapid FLASH NMR imaging. J Magn Reson 1987; 72: 307.

5.   Freeman R, Hill HDW. Signal and intensity anomalies in FT NMR. J Magn Reson 1971; 4: 366-383.

6.   Haase A, Frahm J, Matthaei KD. FLASH imaging: rapid NMR imaging using low flip angles. J Magn Reson 1986: 67: 258-266.

7.   Haase A. Snapshot FLASH MRI - applications to T1, T2 and chemical shift imaging. Magn Reson Med 1990; 13: 77-89.

8.   Hawkes RC, Patz S. Rapid Fourier imaging using SSFP. Magn Reson Med 1987; 4: 9-23.

9.   Hennig J, Nauerth A, Friedburg H. RARE imaging - a fast imaging method for clinical MR. Magn Reson Med 1986; 3: 823-833.

10. Johnson G, Hutchison JMS, Redpath TW, Eastwood LM. Improvements in performance time for simultaneous three-dimensional NMR imaging. J Magn Reson 1983; 54: 374-384.

11. Jones RA, Haraldseth O, Müller TB, Rinck PA, Øksendal AN. K-space substitution: a novel dynamic imaging technique. Magn Reson Med 1993; 29: 830-834.

12. Mansfield P. Multi-planar image formation using NMR spin echoes. J Phys C Solid State Phys 1977; 10: 155-158

13. Melki PS, Mulkern RV, Panych LP, Jolez FA. Comparing the FAISE method with conventional dual echo sequences. J Magn Reson Imaging 1991; 1: 319-326.

14. Oppelt A, Graumann R, Barfuss H, Fischer H, Hartl W, Schajor W. FISP: a new fast MRI sequence. Electromedica 1986; 54: 15-18.

15. Pykett IL, Rzedzian RR. Instant images of the body by magnetic resonance. Magn Reson Med 1987; 5: 563-571.

16. Redpath TW, Jones RA. FADE: a new fast imaging sequence. Magn Reson Med 1988; 6: 224-234.

17. Sekihara K. Steady state magnetisation in rapid NMR imaging using small flip angles and short repetition times. IEEE Trans Med Imaging MI 1987; 6, 2: 157-164.

18. van Vaals J, Brummer M, Dixon W, et al. 'Keyhole' method for accelerating imaging of contrast agent uptake. J Magn Reson Imaging 1993; 3: 671.

19. Zur Y, Stokar S, Bendel P. An analysis of fast imaging sequences with steady state transverse magnetization refocusing. Magn Reson Med 1988; 6: 175-193.

Interlude Four

Alphabet Soup
(with comments from Hamlet)

Medical terminology cannot exist without abbreviations. For many terms there are short names, ranging from the common ones understood by everybody, like TB for tuberculosis, ECG for electrocardiogram, to SIH for somatotropin-inhibitory hormone, understood only by those who deal with it every day.

Generally speaking, abbreviations are necessary because they facilitate daily medical routine. Who wants to say 'computed tomography', 'magnetic resonance imaging' or 'endoscopic-retrograde cholangiopancreaticography' when it takes less than one second to pronounce CT, MRI, or ERCP ? Or, as William Shakespeare described it:

> *Brevity is the soul of wit.*
> (Shakespeare, Hamlet; II.ii.).

There are clear advantages in using abbreviations that are well known from everyday usage: If you sit in a bar and utter 'G & T' twelve times in half an hour you will be drunk a lot faster than the guy at the next table who says: 'Waiter, another gin and tonic, please'. He can say this only six times per half hour.

> *More matter, with less art.*
> (Shakespeare, Hamlet; II.ii.).

There has been an explosion of abbreviations and acronyms in radiology during recent years, particularly in MR imaging. Abbreviations shorten or substitute an understood or stipulated word or phrase, whereas acronyms are made up of the initial letters of a term and often sound familiar to existing words.

Unfortunately, when reading medical articles in journals or books, the abbreviations you find in the text are in many instances not explained because many authors believe that your brain works like their's. Very often, however, you have no idea what specific abbreviations or acronyms mean.

There are simple rules — not always obeyed — for the use of abbreviations in articles. No abbreviations should be used in titles and abstracts; abbreviations should be spelled out the first time they occur in the text; and if the publication is very long, a list of abbreviations should be included in an appendix to the article or book chapter.

It gets even worse if there might be a double meaning. For instance, is IQ image quality or intelligence quotient, PC phase contrast or personal computer, ADC analog-to-digital converter or apparent diffusion coefficient, ROI region-of-interest or return-on-investment, GE gradient echo or General Electric ?

The Books of Abstracts of the 1996 meeting of the International Society of Magnetic Resonance in Medicine in New York provide sufficient examples:

What on earth is ERPF ? Is it an exclamation by ducks ? We, the poor readers, have to find out ourselves.

Alas, poor Yorick !
(Shakespeare, Hamlet; V.i.)

Most abbreviations and acronyms used in radiology are rooted in the English language because, like it or not, it is today's international medical language. Different languages have different medical abbreviations and acronyms, however, and sometimes they spill over from one language to the other.

In German, the Scandinavian languages and Russian, MRT is used for magnetic resonance tomography, instead of the English MRI. When reading 'MRT' in an English text, you still can conclude that the authors mean MRI. SIDA, of course, means AIDS.

Sometimes, however, you find abbreviations you do not necessarily recognize: SEP (French: sclérose en plaque) should rather read MS in English (multiple sclerosis).

The avalanche of recent radiological acronyms was broken loose by a streak of lightning, the FLASH, which stands for 'fast low angle shot'. It was described by Haase and his collaborators as the basic gradient-echo sequence[3], and then it was taken over by Siemens. Today the company sells a different pulse sequence under the same name without having changed the acronym. Similarly, FISP has also two meanings and describes two different pulse sequences in MR imaging.

If you think that FISP describes a wasp with pronunciation problems, you are wrong. In this case, you should read the overview of such acronyms and abbreviations given by Elster in an article published in Radiology in 1993[2], the summary compiled for the MResource Guide of the Journal of Magnetic Resonance Imaging[4], or the overview by Brown and Semelka[1] (or Tables 8-1 and 10-7). Unfortunately, because of the explosive propagation of acronyms, even these thorough lexica are incomplete.

Many different acronyms describe similar procedures, which adds to the problem. Several suggestions have been made about cleaning up this disorder by creating generic names for functional groups of pulse sequences.

Spin-echo (SE) and inversion-recovery (IR) sequences would stay as they are. The gradient-echo (GRE, not GE) sequences would be grouped into S-GRE (spoiled gradient-echo), CE-GRE (contrast-enhanced gradient-echo), and R-GRE (refocused gradient-echo). All Turbo-SE, fast SE, and RARE sequences would be combined under the umbrella term RSE (rapid spin echo). However, there is no common agreement yet on this terminology.

Since the late 1980s, companies have tried to outmaneuver one another by coining new acronyms that should be easy to pronounce and have a certain marketing and sales (or sex ?) appeal — exactly how much appeal remains to be determined by the reader.

Though this be madness, yet there is method in it.
(Shakespeare, Hamlet; II.ii.).

Unfortunately, often this moderately offensive company slang is not understandable to outsiders who are not exposed to the company's products. Nevertheless, it is used for scientific publications and creeps into the scientific literature, such as into abstracts of scientific meetings.

By and large, there is no substantial difference between companies and research groups at universities or other institutions creating clever new acronyms; their purpose is the same, namely to profile and promote themselves. The result is utter confusion, followed by disregard for such presentations.

Absolute confusion can be created by not writing the acronyms in capital letters but in small letters. What do you do with a 'rare' pulse sequence ? Personally, I prefer a well-done pulse sequence. (To kill this joke completely, I believe that RARE *is* a well-done pulse sequence).

In one article, I found the acronym SELESTRA for spin echo, long echo, short TR acquisition (TR = 600 ms, TE = 50 ms). Firstly, this acronym is inconsistent, as are many other acronyms (long echo, instead of long echo time or TE). Secondly, it is completely unnecessary and irrelevant.

Just mentioning 'TE was chosen at 50 ms, TR 600 ms' gives readers all the information they need without confusing them with SELESTRA.

Some people believe that they have contributed to science just because they made up what they consider to be a 'funny' acronym. Sometimes acronyms can diminish or even completely destroy the value of a pulse sequence.

This is the very coinage of your brain.
(Shakespeare, Hamlet; III.iv.).

Guess where RODEO (rotating delivery excitation off-resonance) originates. Texas, of course.

I always thought PIPS was a chicken's disease, but it also can be applied to MR imaging. FLAT TIRE is an acronym for fluid-attenuated turbo inversion recovery, while FLAT BRAIN does not exist yet. EPISTAR describes echo-planar imaging in the stars — most likely somewhere in the Milky Way. CEPI, IEPI, and SEPI are European relatives of the yeti, the Himalayan snow man.

Let's continue:

»PRESTO, bring me my FASTCARD and be FAST, BRISK and HASTE. STIR the SPARE PASTA for DANTE with the STEAMed and SMASHed RARE SPIDER and the ROAST PEAR.

»Don't be CRITICAL with this FAIR and COSY SUPER DIET because we want to BURP and BURST. FREEZE and SHORTen your CRAZED RAGE, otherwise you might be visited by the ONG (oblique Nyquist ghost). RISE early and be SMART and not SISSI !

»What a MESS ! Let's FLASH this acronymania down the …«

There needs no ghost, my lord, come from the grave, to tell us this.
(Shakespeare, Hamlet; I.v.).

I hope that we will never see a new pulse sequence from France: magnetization-enhanced rapid double-echo, or MERDE.

To be frank and fair: some acronyms describing pulse sequences or procedures are necessary and to the point, and they give a name to a specific diagnostic tool, but Shakespeare commented on all of them:

Thou comest in such questionable shape.
(Shakespeare, Hamlet; I.iv.).

Most acronyms do not contribute anything new in terms of substance; they are only packaging. In the era of environmental protection, we do not need them. Rather, we need some orientation in the bewildering jungle of acronyms. The fundamental idea of facilitating communication between inventors and users by shortening terminology is positive, but playing around with it has no advantage for the already confused customer. Such customers or other scientists will strike back sooner or later by leaving the marketplace or scientific area.

Finally, I would like to apologize to those individuals, companies, and research groups whose acronyms were not included here, but there is not enough space for all of them. I also apologize to those whose creations have been selected.

As already mentioned, not all of them are bad or useless. My choices were made purely on the basis of emphasizing the point or just because the terms fitted nicely into the text.

I must be cruel, only to be kind.
(Shakespeare, Hamlet; III.iv.).

References

1.   Brown MA, Semelka RC. MR imaging abbreviations, definitions, and descriptions: a review. Radiology 1999; 213: 647-662.
2.   Elster AD. Gradient-echo MR imaging: techniques and acronyms. Radiology 1993; 186: 1-8.
3.   Haase A, Frahm J, Matthaei KD. FLASH imaging: rapid NMR imaging using low flip angles. J Magn Reson 1986; 67: 258-266.
4.   MResource Guide. Journal of Magnetic Resonance Imaging (J Magn Reson Imaging) 1994.

Chapter Nine — Fundamentals of Image Characteristics

The MR Image

Figure 9-1:
The characteristics of the 'mirror' image of the man with the broom will change soon. He will disturb the picture elements and drastically alter the field-of-view.

In diagnostic imaging, the contents depicted on an image should reflect the essence of the original information as objectively as possible. However, there are limitations, as we have seen earlier: the hardware and software of the MR equipment influence image content. In addition, the characteristics of the image itself affect image content (Figure 9-1).

Volume and Picture Elements

In computerized imaging, be it nuclear medicine imaging, x-ray CT or angiography, or magnetic resonance imaging, pictures are composed of elements, called *picture elements* or *pixels*, which, in turn, reflect the content of *volume elements* or *voxels*. Figure 9-2 explains this. In principle, voxels could be as small as a single cell. In reality, however, voxel size depends on a number of limiting factors, with computer capacity and the signal obtained from an individual voxel being the main obstacles.

Thus, usually 256×256×1 voxels are created of a slice of an object and turned into pixels. These 256×256 picture elements are called the *image matrix*.

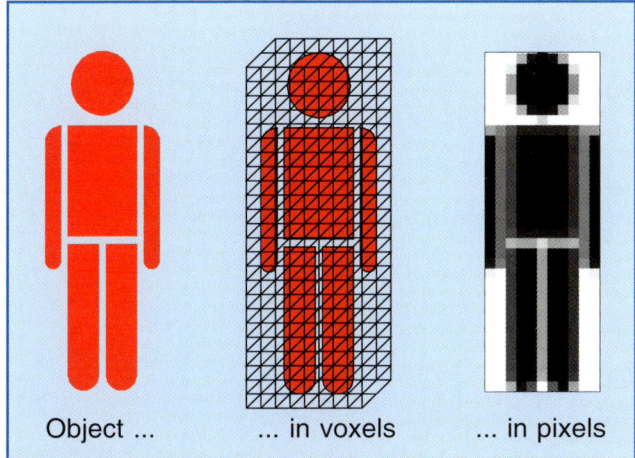

Figure 9-2:
Voxel and pixel. We want to image an entire person who for this purpose is mathematically divided into volume elements. In each volume element, the signals are averaged and turned into a number which represents a certain level on a gray scale. These numbers are used to create a picture consisting of pixels.

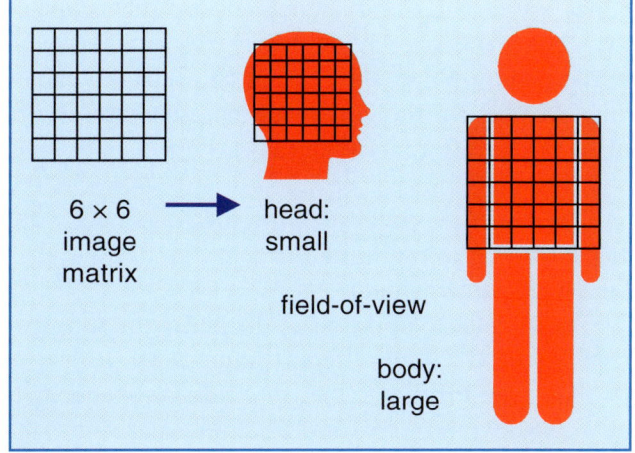

Figure 9-3:
Image matrix and field-of-view. In this case, we have an image matrix of 6×6, i.e., a grid of 6 rows and 6 columns with a total number of 36 pixels. Usually in MR imaging, the field-of-view is at least 256×256. Commonly, individual voxels and pixels are larger in body imaging than in head imaging.

Image Matrix and Field-of-View

The image matrix is characterized by the number of pixels in the *x*- and *y*-directions. It is defined by the steepness of the *x*-gradient (the frequency-encoding gradient) and the number of phase-encoding steps in the *y*-gradient. Both combined represent the field-of-view (FOV), as shown in Figure 9-3.

If the FOV is the whole head with an edge length of 25.6 cm and a matrix size of 256×256 is used, then a single pixel represents 1 mm. If the FOV is smaller (e.g., 12.8 cm) and the same matrix size is used, the spatial resolution is 0.5 mm.

Spatial Resolution and Partial Volume Effects

As in other digitized imaging methods, voxel and pixel size influence spatial resolution and thus contrast.

All anatomical structures within one voxel add to its averaged signal intensity in the final image. If the voxel has a large volume, it can contain many different structures and tissue types. In the final image pixel, they will be indistinguishable. If the voxel can be kept smaller, less structures will be represented by one single pixel, and therefore spatial resolution and contrast will be better.

Data acquisition and reconstruction methods define different voxel shapes. Isotropic reconstructions use cubes, while in anisotropic methods one side is longer than the two others. Although they may look the same in the picture plane, the content and thus the calculated number for the gray level representation in the pixel can be different (Figure 9-4).

The sometimes blurry features of these images are caused by the averaging of different structures. This is known as *partial volume effect*.

The smaller the pixel size, the better the suppression of partial volume effects (Figure 9-5).

However, the bigger the voxel size, the better will be the signal (and signal-to-noise). In general, the signal-to-noise is the determining factor for the final voxel/pixel size. Increasing the matrix size from 128×128 to 256×256, while keeping field-of-view, slice thickness and imaging constant, will reduce the signal-to-noise ratio by a factor of 4. Thus, the signal-to-noise has to be high enough to permit the increase in resolution.

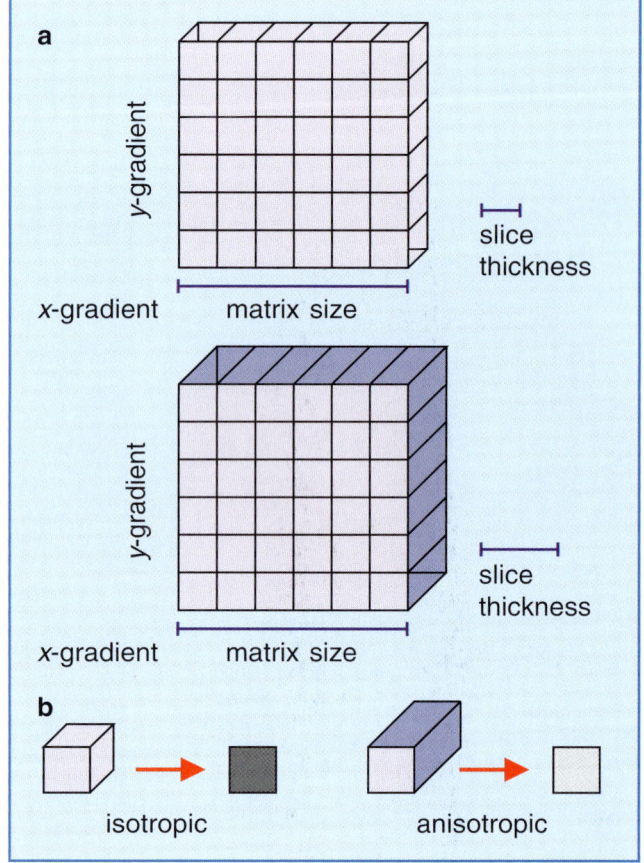

Figure 9-4:
Different slice thickness (a) can lead to isotropic or anisotropic volume elements (b) and different signal intensities.

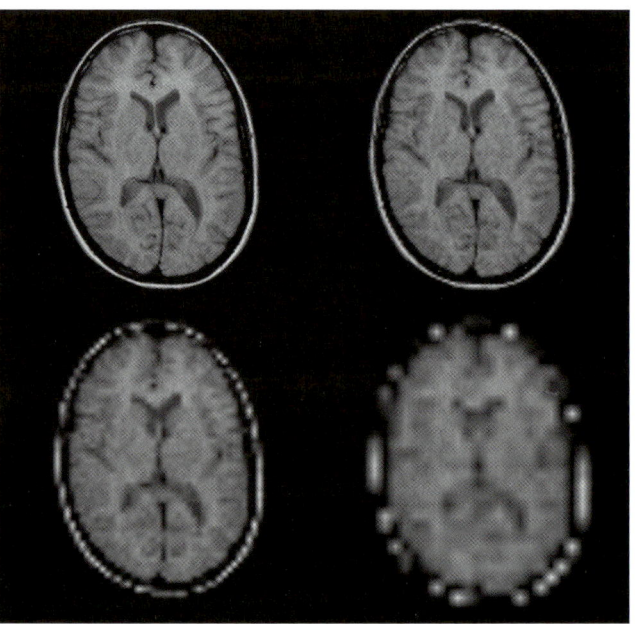

Figure 9-5:
Spatial resolution and partial volume effects: matrix size (a) 256×256, (b) 128×128, (c) 64×64, and (d) 32×32. Due to the partial volume effects, anatomic details disappear.

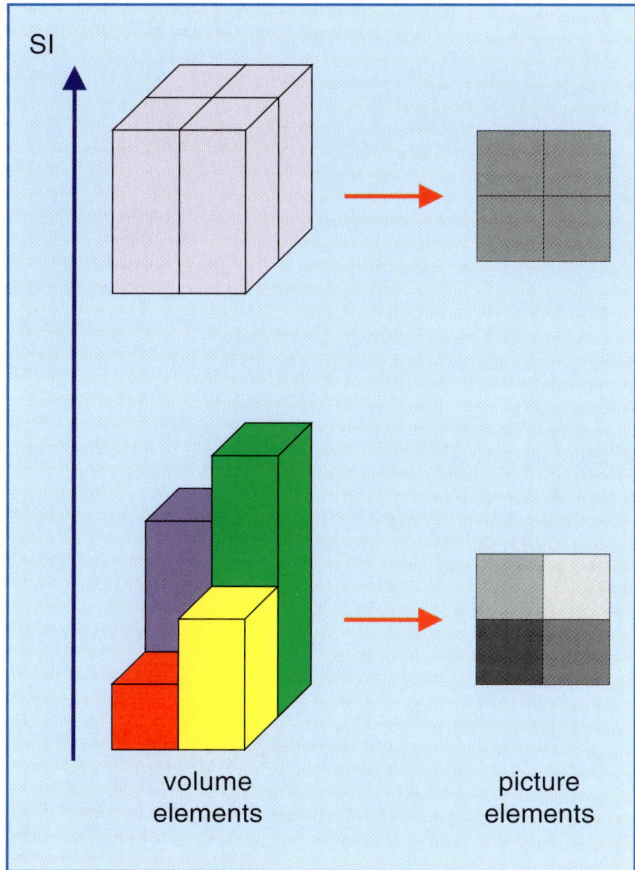

Figure 9-6:
Two examples of four neighboring volume elements. In the first case the four voxels have the same relative signal intensity (SI), and in the resulting image they cannot be distinguished from each other. In the second case, they have different relative intensities and thus they can be distinguished from each other in the final image.

Definition of Contrast

There is only one step from picture elements to image contrast.

Contrast itself is a quite controversial term in medical imaging. It describes the relative difference of intensities of two adjacent regions within an examined object on a gray or color scale. Several definitions of contrast have been proposed during recent years.

It is quite difficult to give an exact definition of contrast on a conventional x-ray image. Here, definition of contrast is merely qualitative, except when using a special measuring device.

Digitalization of images in nuclear medicine and x-ray CT opened the door to more straightforward quantitative approaches to contrast. Now, picture elements are available. Their gray-scale intensity can be expressed in numbers. The numerical difference between two intensities allows quantitative definition of contrast.

If there is no difference between two neighboring pixels, they cannot be distinguished and thus no contrast exists. The bigger the difference in the intensity of two pixels, the better will be the contrast (Figure 9-6).

A quantitative definition of contrast is given by the following equation:

$$C = (Ia - Ib) / (Ia + Ib)$$

where C is contrast and Ia and Ib are the signal intensities of two adjacent pixels or voxels.

It is important to understand that image intensity in magnetic resonance imaging is not standardized. MR imaging does not possess any correlation to Hounsfield units in x-ray CT. The signal intensity of an MR image can represent a mixture between T1-, T2-, and ρ-values, flow, diffusion, perfusion, and other factors influencing the signal emitted by structures within a volume element.

Thus, the comparison of signal intensities of two different MR images is meaningless and cannot be used for clinical diagnosis.

Only normalization of images, e.g., with a water-filled vial outside the patient's body, allows an approximation to be made and can be used to calculate relative signal intensities, which then can be compared. However, these values are only semiquantitative. They vary between different MR scanners and have no diagnostic value.

Signal-to-Noise Ratio and Data Averaging

The stronger the MR signal, the better will be the image quality. Owing to the low intensity of the magnetic resonance signal, it is often severely influenced by background noise, just like radio signals coming from remote transmitters. The quality of the signal is described as the signal-to-noise (S/N) ratio.

The aim in medical imaging is to get a combination of both the best possible signal-to-noise ratio and the best available contrast in the shortest time possible.

One method of improving S/N consists of data averaging. When a magnetic resonance experiment is repeated, and the magnetic resonance signals obtained are recorded each time, they will add up. The random or chaotic positive and negative noise signals will also add up with the number of excitations, but at a lower rate because of the statistical nature of the noise.

For many MR imaging data acquisitions, two data averages suffice for the creation of images with good S/N. For *n* data-averaging runs, the net increase of S/N will be the square root of *n*. The S/N will, for example, increase by a factor of 2 if four data averages (numbers of excitations) are performed (Figure 9-7).

Signal-to-noise increases with field strength. In analytical NMR, which uses small samples and no field gradients, signal-to-noise levels vary linearly with field strength.

In MR imaging, field gradients are used to encode spatial information. They must be large enough to compensate for the inhomogeneity of the magnet and the chemical shift between fat and water. As the field strength increases, so does the magnitude of both the field inhomogeneity and the chemical shift. Therefore, increased gradients are needed at higher fields if chemical-shift artifacts are to be satisfactorily suppressed.

If the gradient strength is doubled, then the bandwidth per pixel is also doubled. This increases image noise by a factor of $\sqrt{2}$. The net gain in S/N provided by doubling the field is therefore not twofold, but the square root of two.

This means that S/N versus field curve flattens out at higher fields, resulting in diminishing gains with increased field strength (Figure 9-8).

The intensity of noise depends on both electrical resistance in the coil and conductance losses in the human body [1,3], but is dominated by the losses in the human body at resonance frequencies above 10 MHz.

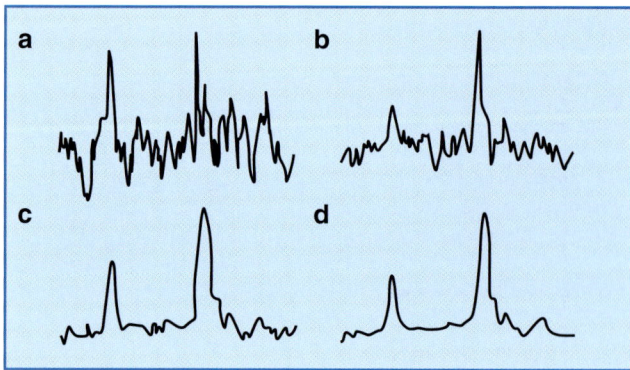

Figure 9-7:
(a) 2; (b) 8; (c) 32; and (d) 128 data averages.
The 'polluting noise' disappears slowly with higher number of averages.

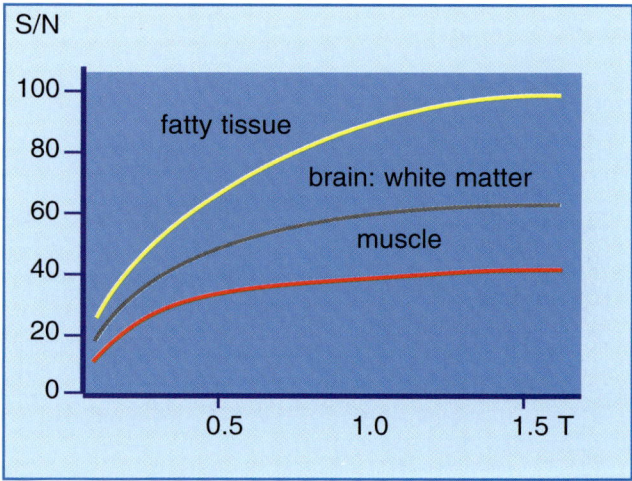

Figure 9-8:
Signal-to-noise ratio (S/N, in percent) versus field strength behavior in MRI (T1-weighted pulse sequence).

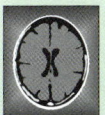

With **MR Image Expert** you can experience interactively the influence of changes in the signal-to-noise ratio.

Signal and Noise

- Select *File/Open Case* or the *Brain* icon in the **MR Image Expert** menu bar. Double-click on the file **1.5 T: Normal brain, transversal slice**.

- Select *Draw/Spoiled GRE* or click the *S GRE* icon. Set TR = 400 ms, TE = 20 ms, FA = 5° and click OK. A spoiled GRE image with these parameters will appear on the screen.

- Select *Options/Noise Level*. A dialog box will appear.

- Select *Add Noise 20%* and click OK. You will get a new image which is quite noisy. Repeat this procedure for spoiled GRE images with increasing flip angles: 20°, 35°, 50°, 65°, and 80°. The images on your screen should look like those below (Figure 1).

The highest signal-to-noise level using this kind of pulse sequence is commonly somewhere between 30° and 60°. Details of the behavior of signals intensities in spoiled gradient-echo sequences are described in Chapter 10.

Try the same with a spin-echo sequence.

Signal Averaging and Noise

- Select *File/Open Case* or the *Brain* icon in the **MR Image Expert** menu bar. Double-click on the file **1.5 T: Normal brain, transversal slice**.

- Select *Draw/Spoiled GRE* or click the *S GRE* icon. Set TR = 400 ms, TE = 20 ms, FA = 10° and click OK. A spoiled GRE image with these parameters will appear on the screen.

- Select *Options/Noise Level*. Select *Add Noise 20%* and click OK. Create a second image exactly like this one.

- Select *Options/NSA (= number of signal averages)*. Set NSA = 2. Repeat this with NSA = 3, NSA = 4, NSA = 8, and NSA = 16. You will see how the image quality improves. The images on your screen should look like those below (Figure 2).

If you use the right mouse button inside the images, you get a dialog box on which you can select *Scan Time* with the left button. Compare the scan times — the better the image quality, the longer it will take to acquire an image.

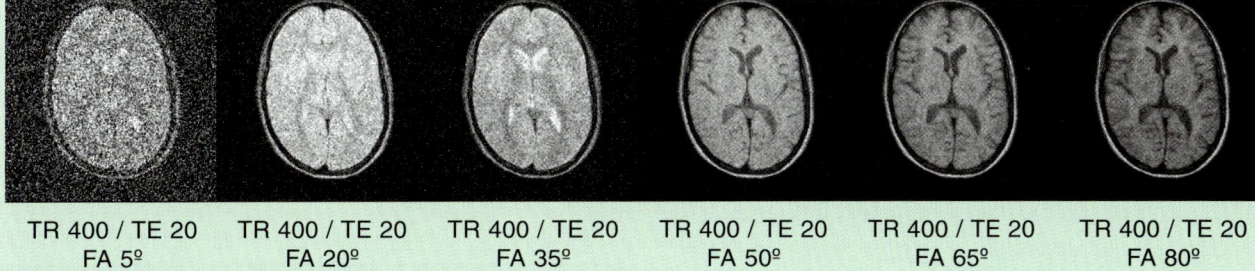

| TR 400 / TE 20 FA 5° | TR 400 / TE 20 FA 20° | TR 400 / TE 20 FA 35° | TR 400 / TE 20 FA 50° | TR 400 / TE 20 FA 65° | TR 400 / TE 20 FA 80° |

Figure 1:
Change of signal-to-noise in a spoiled GRE sequence. Note that while using the noise option, real-time change of pulse-sequence parameters is not possible. TR = repetition time, TE = echo time, FA = flip angle.

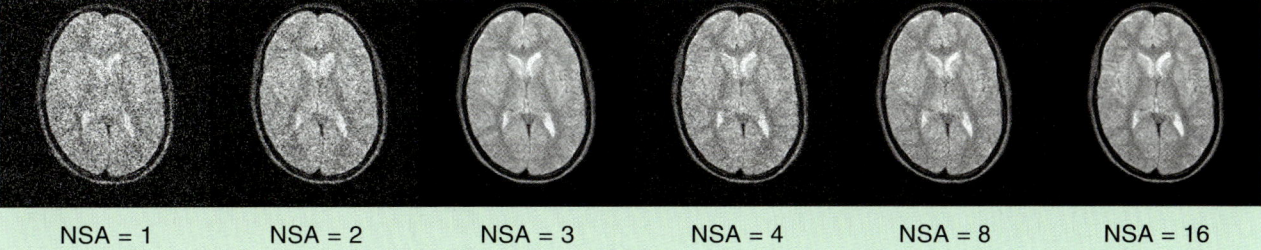

| NSA = 1 | NSA = 2 | NSA = 3 | NSA = 4 | NSA = 8 | NSA = 16 |

Figure 2:
Improvement of signal-to-noise with increasing number of signal averages (NSA; sometimes also called NEX = number of excitations, or number of acquisitions).

Contrast-to-Noise Ratio

A decisive criterion for comparing the contrast of different images and different MR machines is the contrast-to-noise ratio. Figure 9-9 explains the contrast-to-noise relationship. Even if there is sufficient contrast between two tissues, noise may obliterate this contrast and no discrimination will be possible. The figures in the **MR Image Expert** Tutorial show examples of a brain image with different noise levels.

With the increase of field strength, T1 relaxation times increase too. This also has a negative impact on signal-to-noise, because the repetition time TR has to be increased to obtain the same signal-to-noise ratio. Only when sufficient time is granted for the spin system to recover after the initial excitation pulse will signal intensity be sufficient. If we administer the next pulse after a shorter time than 5×T1, our sample will be saturated and its signal intensity will be lower (Figure 9-10). Thus, TR must be sufficiently long to receive a strong signal. However, in general the increase in signal deriving from the higher field strength compensates the T1-related signal loss.

It is worth noting, that for any given spatial resolution and contrast the signal-to-noise ratio needs only reach a level for confident detection of a lesion. Above that level, further increases in signal-to-noise make the image more pleasant but, on the other hand, 'beautiful' images do not guarantee the accuracy of the diagnosis.

When trying to optimize imaging conditions, you should never forget that there are numerous interdependencies between the different factors influencing the image and image contrast.

If we choose imaging speed as the main factor, there is a straight connection to the signal-to-noise ratio and spatial resolution. Spatial resolution is linked to contrast and artifact reduction. Contrast is also related to signal-to-noise and artifacts.

As we will discuss in the following chapter, altering one minor parameter can affect a chain of other parameters.

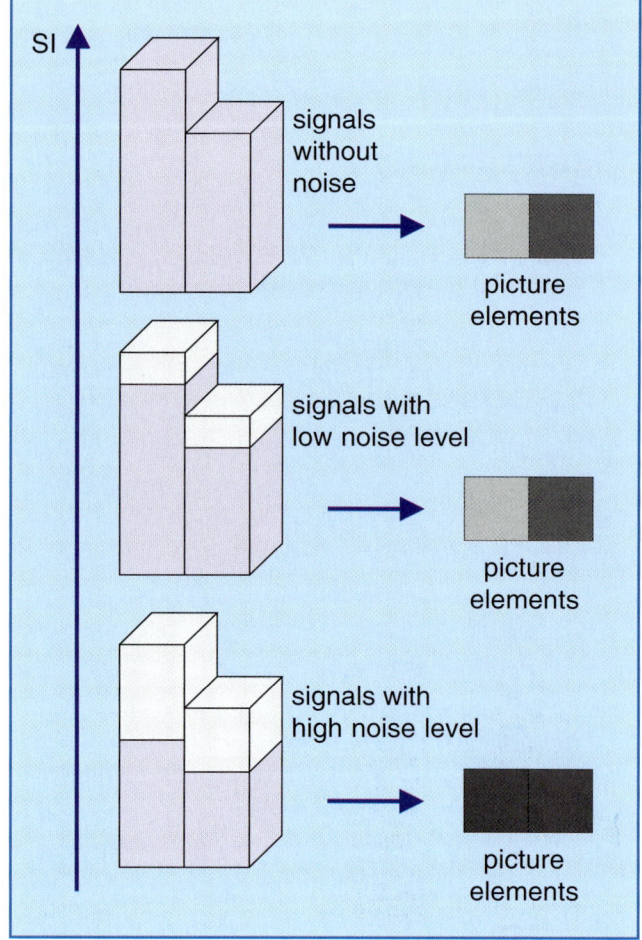

Figure 9-9:
Contrast-to-noise. Without noise, two neighboring tissues with different signal intensities can be easily distinguished from each other. If the level of noise is low (center), contrast-to-noise is sufficient and the tissues are distinguishable. In case the noise level is high and signal-to-noise is poor, it is difficult or impossible to distinguish the tissues from each other. SI = signal intensity.

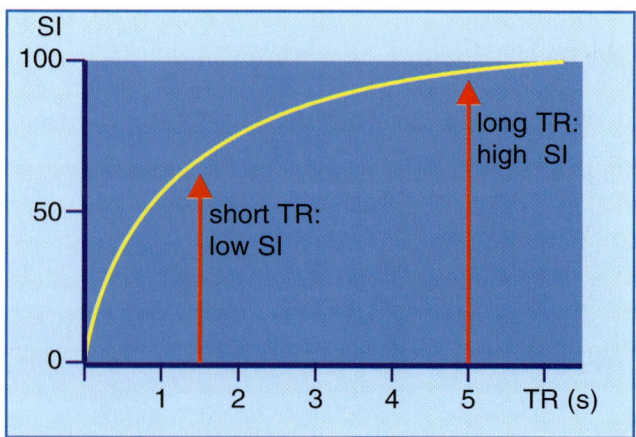

Figure 9-10:
The longer the repetition time TR, and thus the more recovered the system will be, the stronger the signal intensity.

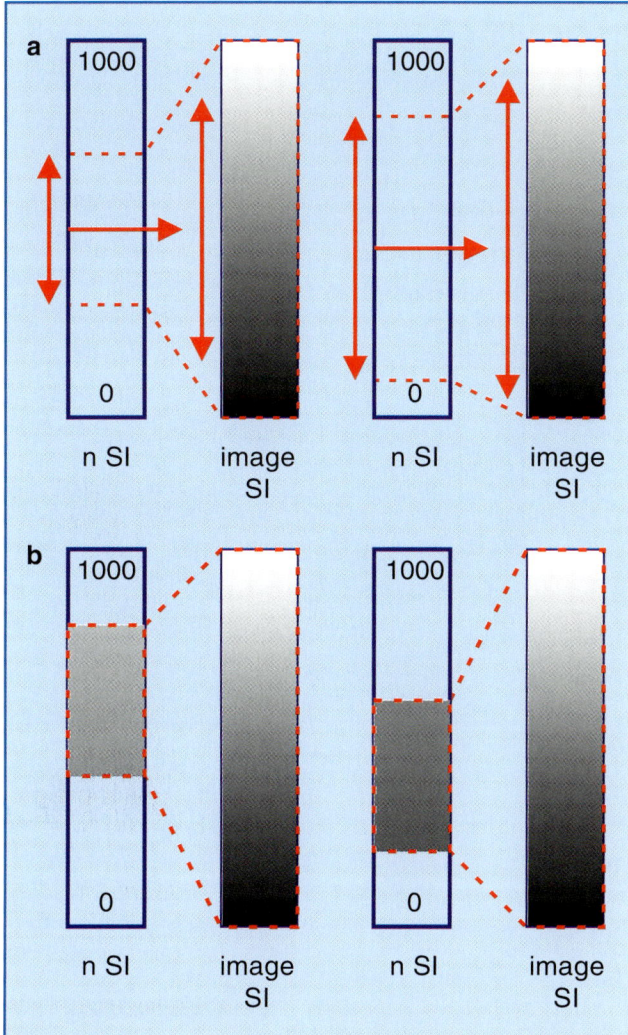

Windowing

On screen, the gray scale of the images can be adjusted. This is well known from x-ray CT and described as *windowing*. Windowing influences the image contrast by attributing certain levels on the gray scale to certain signal intensities. Windowing is completely independent of the MR image acquisition.

Figure 9-11 illustrates how the signal intensity of a pixel is determined by windowing. The image gray scale is dependent on both window center and level.

Images to be compared with each other should always have the same window level and center. If this is not the case, comparisons of structures with different signal intensities may be misleading.

Figure 9-11:
Windowing, by which the image signal intensity is adjusted so that white corresponds to the highest signal intensity and black to the lowest one. The window center can be moved up and down (a), and the window level can be narrowed or widened (b). The signal intensity scale in the image depends on both the window center and level. The original numerical signal intensity scale (n SI) does not reflect the final signal intensity gray scale (image SI).

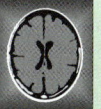

With **MR Image Expert** you can experience image windowing interactively.

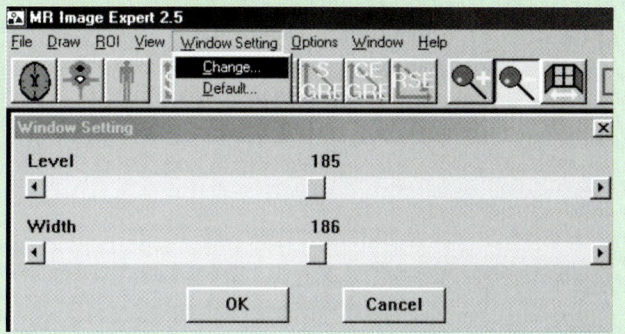

Windowing

The following functions let you change the window setting, i.e., how the intensity values are mapped on the gray scale.

Window Setting/Change...
This function allows setting both window level and window width of the currently active window by moving scroll bars. The numbers above the scroll bars indicate the current window level and width.

Window Setting/Default
This function lets you set the default window level/ width by specifying the percentage of pixels which should determine the minimum and the maximum of the de-

fault window. In addition, you can toggle the default setting on/off by clicking on the Default Window Setting. Switching off the default window setting will speed up the calculations on slow systems. Note that only the images drawn after changing the default setting will be affected.

Age

The composition of tissues in the human body changes with age. This is of particular importance in the brain, where the water content decreases and the myelin content increases dramatically during the first years of infancy. Consequently, T1 and T2 relaxation times of brain tissue decrease.

At birth, the infant brain consists of 93-95% of water and has long T1 and T2 relaxation times (Figure 9-12). There is a fast fall in water content to 82-84% during the first two years of life as myelination takes place.

Therefore, it is necessary to adjust the timing parameters of all pulse sequences accordingly. When using IR sequences at mid-field in the neonatal period, a TR of 3000 ms and TI of 1000 ms are required to produce images with useful soft-tissue contrast. The TR and TI can be halved by the time the child is two years of age. When using SE sequences, TR has to be prolonged accordingly for T2-weighted images. The use of the same pulse parameters in infants as in adults will lead to images without diagnostic value (Figure 9-13).

In children aged three to six years, the sequence parameters of adults can be used.

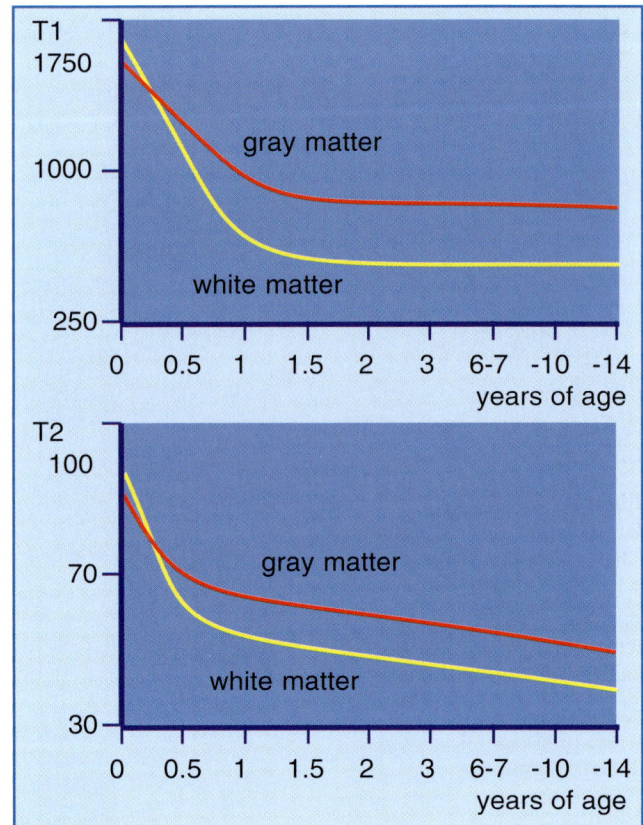

Figure 9-12:
(top) T1 relaxation times and (bottom) T2 relaxation times of gray and white matter by age in milliseconds (modified from Holland et al.[2]). Note that from birth until approximately six months of age, both T1 and T2 of gray matter are shorter than T1 and T2 of white matter. *In vivo* measurements at low field; standard deviation approximately 25%.

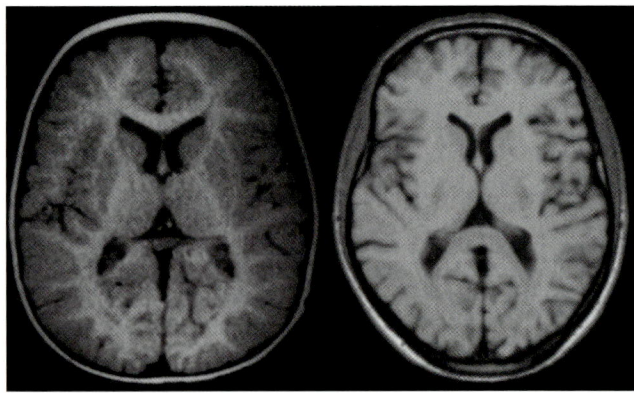

Figure 9-13:
Brain images of (a): an infant of 11 months, and (b): an adult at 0.5 T. The same pulse parameters were used (SE: TR = 500 ms, TE = 20 ms). Windowing is slightly different. Still, image contrast, in particular contrast between gray and white matter is obviously not the same because in the infant myelination has not reached the adult stage and both T1 and T2 of white matter are higher than T1 and T2 of gray matter.

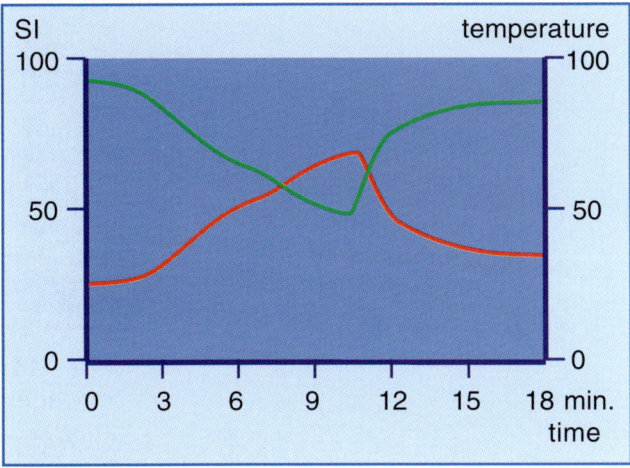

Figure 9-14:
Drastic change of temperature leads to change of signal intensity. The green curve shows the local decrease of signal intensity of a T1-weighted pulse sequence before, during and after local heating (red curve) of brain tissue in an *ex vivo* experiment. The temperature changes from 25° C to more than 60° C; relative signal intensity drops by 50%.
SI = relative signal intensity; temperature in °Celsius.

Temperature

The influence of temperature on relaxation times is well known from analytical NMR. Temperature also influences the diffusion coefficient and the chemical shift of the water peak.

Thus, the question arose if in MR imaging temperature changes in the human body may influence relaxation times of tissues and therefore contrast. This might occur, for instance in patients running high temperatures one day when undergoing MR and having normal temperatures during a follow-up examination. Relaxometric measurements proved that any differences created are within the system error and do not influence contrast in MR imaging [7].

However, it might be possible to develop MR thermometry to monitor major local changes in temperature, for instance in laser therapy of malignancies. It has been demonstrated that temperature-related effects could be mapped dynamically [4-6].

Yet to date, such kinds of measurements remain unreliable, although phantom and *in vitro* studies are promising (Figure 9-14).

References

1. Chen C-N, Sank VJ, Cohen SM, Hoult DI. The field dependence of NMR imaging. I: laboratory assessment of signal-to-noise ratio and power deposition. Magn Reson Med 1986; 3: 722-729.
2. Holland BA, Haas DK, Norman D, Brant-Zawadzki M, Newton TH. MRI of normal brain maturation. Amer J Neuroradiol 1986; 7: 201-208.
3. Hoult DI, Chen C-N, Sank VJ. The field dependence of NMR imaging. II: arguments concerning optimal field strength. Magn Reson Med 1986; 3: 730-746.
4. Hynynen K, Vykhodtseva N.I, Chung A.H., Sorrentino V, Colucci V., Jolesz FA. Thermal effects of focused ultrasound on the brain: determination with MR imaging. Radiology 1997; 204: 247-253.
5. LeBihan D, Delannoy J, Levin RL. Temperature mapping with MR imaging of molecular diffusion: application to hyperthermia. Radiology 1989; 171: 853-857.
6. Matsumoto R, Mulkern RV, Hushek SG, Jolesz FA. Tissue temperature monitoring for thermal interventional therapy: comparison of T1-weighted MR sequences. J Magn Reson Imaging 1994; 4: 65-70.
7. Rinck PA, Muller RN, Fischer H. Feld- und Temperaturabhängigkeit des Kontrastes in der Magnetresonanzbildgebung. Fortschr Röntgenstr 1987; 147: 200-206.

Chapter Ten

Image Contrast

Introduction

Figure 10-1:
Contrast is one of the major concerns in medical imaging. The ability to distinguish and characterize certain structures in the image is the goal of imaging. In conventional x-ray and in x-ray CT distinction and characterization of lesions are often based upon indirect signs. In (a) above, a glass is filled with a liquid; however, we do not know what kind of liquid this might be. Picture (b) shows the same glass with a red wine bottle next to it. Although the quality of this image is worse than (a), we can deduce that the glass contains red wine.
The aim of medical imaging, in particular of MR imaging, is going one step further. Contrast should be good enough to both highlight and characterize lesions. We do not want to rely upon indirect signs.

Everybody involved in medical imaging shares one common dream: to be able to distinguish the structures of the object examined with such sharpness and accuracy that there is no room for diagnostic speculation. Definition of normal anatomy and pathological changes should be easy and exact. This means that in addition to excellent spatial resolution, high contrast is a prerequisite for a good imaging method.

Magnetic resonance imaging has drawn the attention of many researchers, fascinated by the manifold possibilities of influencing contrast. It was believed that image contrast of such quality could be obtained that problems in lesion delineation and even lesion-typing would not occur any more.

The early enthusiasm was rapidly replaced by disillusionment and partial disappointment. It is still not clear whether the method itself is incapable of uncovering all the diseases it was intended for or whether poor understanding of the theoretical background of MR imaging led to misguided applications.

Today, many of the early mistakes and misunderstandings can be explained.

This chapter provides an overview of the main factors and parameters influencing the magnetic resonance image. We will introduce, one by one, the main pulse-sequence parameters and see how they influence image contrast.

Because looking at curves depicting signal intensity or contrast behavior leaves much to the imagination, the reader's personal computer with the software *MR Image Expert* will replace a real MR scanner in this chapter, allowing you to simulate directly the theoretical descriptions of contrast behavior.

Most simulations in this chapter are brain examples where contrast behavior can be best visualized. However, the same manner of signal and contrast changes can be observed in other parts of the body.

MR Image Expert includes additional examples which allow you to simulate imaging with different parts of the anatomy which are not described in the tutorials of this chapter.

A number of suggestions are also made on how best to utilize the method.

Main Contrast Factors in MRI

Contrast in conventional radiographs and CT images is essentially based on small density differences. It can only be changed by adding contrast agents such as barium and iodinated substances, which influence electron density within a certain organ. MR imaging possesses many more contrast-influencing factors and parameters than other imaging methods. One can compare x-ray imaging with radio broadcasting and MR imaging with color television: the former relies on one factor, sound, the latter on sound and moving color pictures.

This makes the contrast behavior of MR imaging more complex than that of any other medical imaging modality.

The numerous factors influencing contrast can be divided into two groups: intrinsic and extrinsic parameters. Table 10-1 gives an overview of the most important of these parameters.

Many of the extrinsic factors can influence the intrinsic factors. For the clinical application of MR imaging, it is necessary to be aware of all of their interactions if one is to react rapidly and efficiently in a given diagnostic question. The relative abundance of factors creates a plethora of data, which can impede rather than facilitate the diagnosis, especially if there is a lack of knowledge on how to exploit the information.

One of the main advantages of MR imaging is the possibility to change contrast by choosing special pulse sequences and pulse-sequence parameters. By emphasizing one factor or mixing several factors in a specific way, the contrast behavior of a certain morphological region or pathological lesion can be highlighted.

However, the comparison of two images of the same patient taken with two different imagers, apparently using the same parameters, often reveals different contrast patterns.

One should always bear in mind that even changing minor factors can cause severe contrast changes.

In the conventional pulse sequences, which were commonly used in clinical routine, image contrast is calculable, but there are increasingly other sequences where it is not predictable. On the following pages, we will look at the basic pulse sequences and their contrast-influencing components TR, TE, TI, FA, and others step-by-step. Many other sequences can be derived from the basic sequences; their contrast behavior follows the fundamentals described here.

Table 10-1:
Principle contrast parameters in magnetic resonance imaging.

Intrinsic	Extrinsic

• proton density
 • T1 relaxation
 • T1-ρ relaxation
 • T2 relaxation
 • cross relaxation
 • dia- and ferromagnetic perturbations
 • chemical shift
 • temperature
 • diffusion
 • perfusion
 • physiologic motion
 • bulk flow
   (e.g., blood, CSF)
 • viscosity
 • changes of tissue composition
   (e.g., age, pathological changes)
	• static and gradient magnetic field
	   strength
	 • magnetic field homogeneity
	 • hard- and software parameters
	   * type of coil
	   * number of slices, slice thickness
	     and gaps, slice location and orientation
	   * number of averages
	   * pulse shape/bandwidth
	   * pixel and matrix size, field-of-view
	   * acquisition mode (2D/3D)
	   * artifact suppression
	   * triggering/gating
	   * orientation of phase- versus
	     frequency-encoding gradients
	 • RF pulse sequences and parameters
	 • contrast-changing agents

Basic Radiofrequency Sequences

TR — the Repetition Time

The Partial Saturation Pulse Sequence. In the early days of MR imaging one plain radiofrequency pulse sequence was used to create images: the partial-saturation (PS) sequence. It consists of 90° pulses transmitted in a train. The sequence is discussed in detail in Chapter 4. Its signal intensity (SI) can be determined by the following equation:

$$SI = K \times \rho \times (1 - \exp\{-TR/T1\})$$

where K is a constant comprising bulk flow, diffusion, perfusion, and other parameters, ρ is proton density, TR the repetition time between the 90° pulses, and T1 the spin-lattice relaxation time.

The nuclei are exposed to repeated RF pulses which cause a free induction decay of a specific initial amplitude. If the repetition time between two subsequent pulses is less than 5×T1, magnetization has not completely recovered and the signal intensity will be lower than the initial amplitude. The 90° pulse saturates the spin system for a certain time. If another RF pulse is transmitted during this period, a lower signal will be received. Therefore, equilibrium signals and thus image contrast differ, depending on the length of TR.

Partial-saturation images are only slightly influenced by T2, the spin-spin relaxation, but heavily by T1, the spin-lattice relaxation, until the interpulse delay equals approximately 2×T1. At 2×T1, 90% of the magnetization has recovered. Afterwards, proton density is responsible for contrast (Figure 10-2). In general, bright regions on a PS image resemble object areas of short T1 and/or high proton density, while dark regions depict areas of long T1 and low proton density.

Hardly anybody uses PS sequences today because of their limited diagnostic value. However, partial saturation becomes a gradient-echo sequence when a field gradient is added. Most sequences dubbed *saturation recovery* today are in reality gradient-echo sequences. Their contrast behavior will be discussed below.

Figure 10-2:
Signal-intensity behavior of a partial saturation pulse sequence showing the dependence of SI of white matter, gray matter, and cerebrospinal fluid (CSF) at 0.5 Tesla. If the repetition time chosen is long enough, signal intensities differ by the factor of proton density only (WM: 72%, GM: 82%, CSF: 100%). The images are of an adult brain taken at 0.5 Tesla.

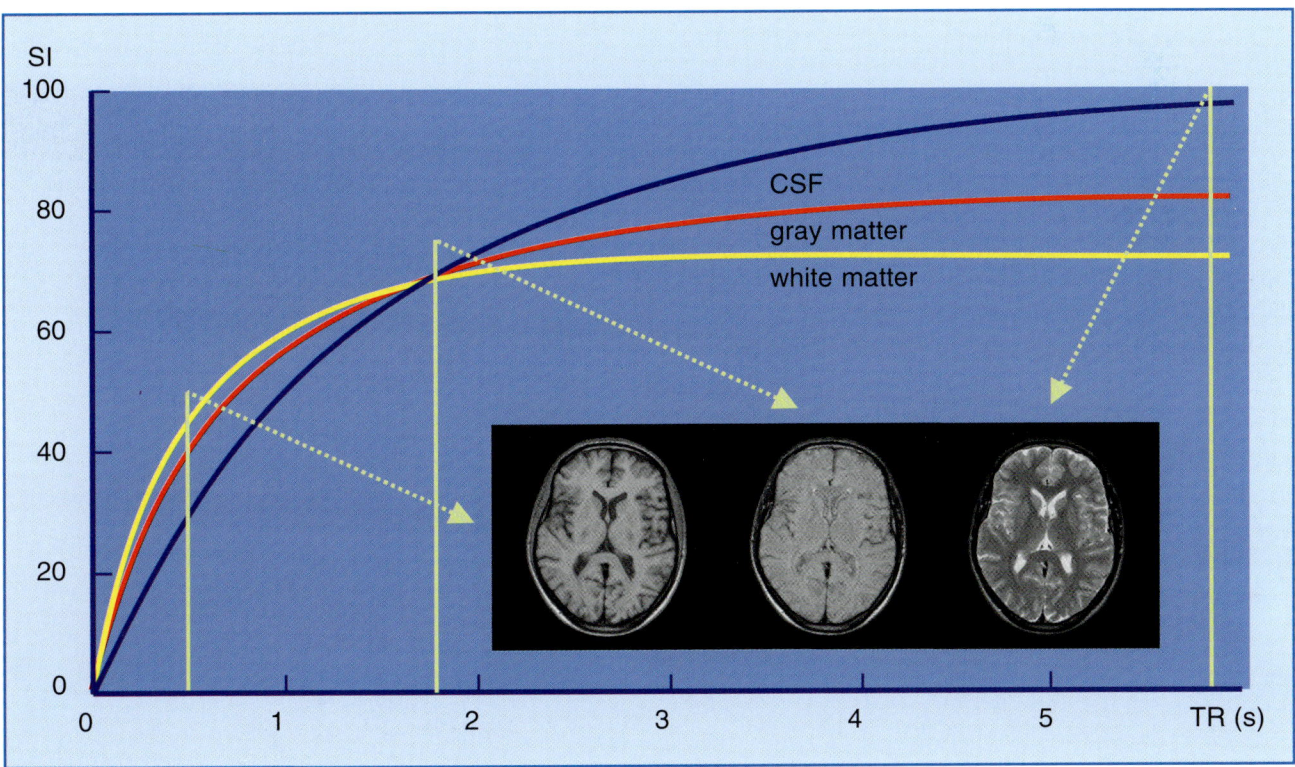

TE — the Echo Time

The Spin-Echo Pulse Sequence. The principle elements of any spin-echo sequence are a 90° pulse followed by a 180° pulse after a time interval τ which builds up a spin echo after the echo time TE (details were explained in Chapter 4).

At the beginning of magnetic resonance imaging in clinical routine, some already known brain lesions were not seen because there was a lack of contrast. In particular, SE images with a TE shorter than 60 ms sometimes did not reveal multiple sclerosis plaques, astrocytomas, meningiomas, infarctions, or other lesions. The reason for this behavior can be explained by the signal decay curves of spin-echo sequences.

You can evaluate it with **MR Image Expert**. Just follow the Tutorial overleaf.

Signal decay depends on the relaxation times and ρ of the respective tissue. The curves reach zero signal intensity faster or slower, owing to the composition of the tissue.

The spin-echo signal contains information about proton density as well as spin-lattice and spin-spin relaxation. Signal intensity of a spin-echo sequence can be calculated by the following equation:

$$SI = K \times \rho \times (1 - \exp\{-[TR-TE]/T1\} \times \exp\{-TE/T2\})$$

Again, SI stands for signal intensity, K represents the influence of flow, perfusion and diffusion, ρ is proton density, TR repetition time, TE echo time, and T1 and T2 are the relaxation times.

This equation reveals that T2-weighting of an SE image increases as the echo time advances. T1-weighting of the signal intensity depends on both TR and TE. In a single echo SE sequence, T1-weighting usually is created by both short TR and short TE. In general, the early echoes (= short TE) of an SE sequence are ρ- and T1-weighted. The later echoes are increasingly intermediately and T2-weighted [7].

Figure 10-3:
Spin-echo sequence: decay curves of gray matter, white matter, and CSF at high field strength. Relative signal intensity, SI, versus echo time, TE at a given repetition time of TR = 2000 ms. SI of the different compounds decreases with longer TE. The most interesting features of these curves, however, are the points of intersection. At these points, the respective brain tissues are isointense and there is no contrast between them: they are indistinguishable. Many pathologies possess signal intensities similar to normal brain tissue on images and therefore are invisible on images with short TE (T1- and intermediately weighted images); see also the **MR Image Expert** Tutorial on page 121.

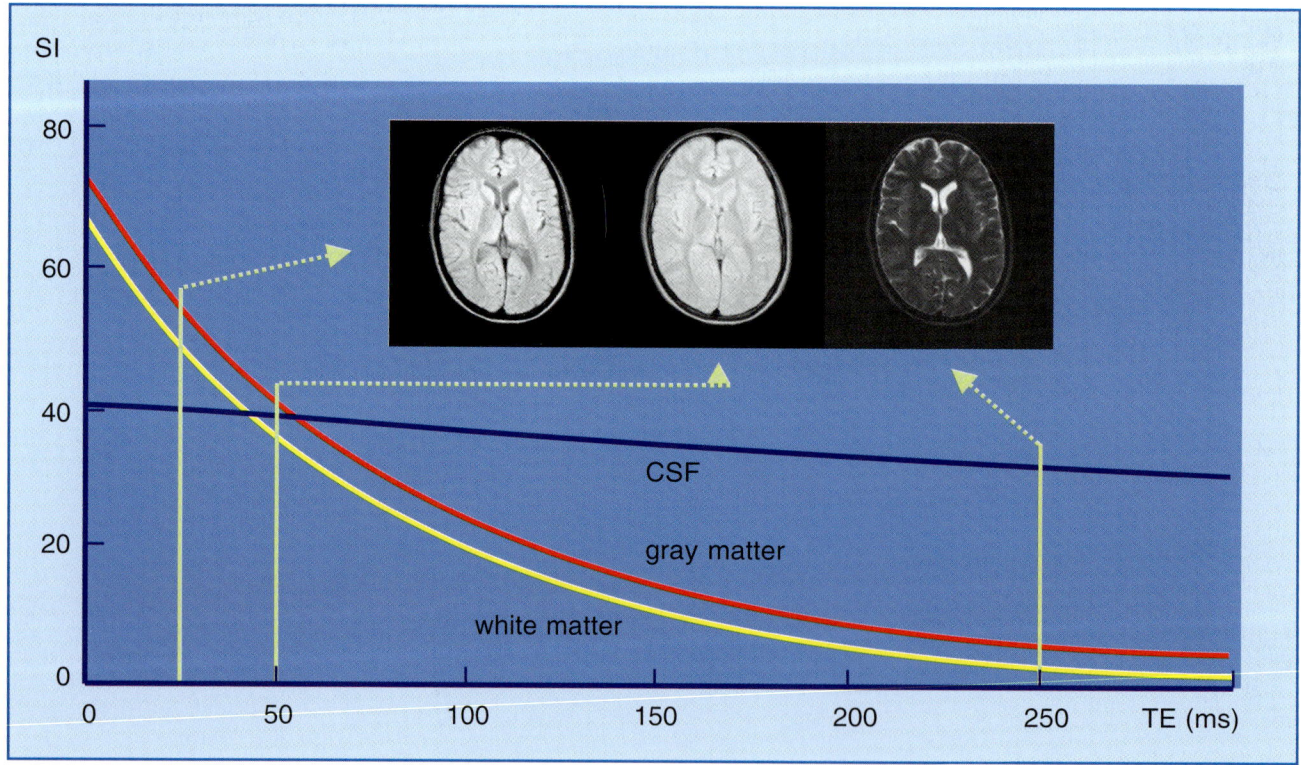

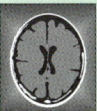

You can experience interactively the influence of TR and TE with *MR Image Expert*.

Contrast Dependence on TR

*   Start *MR Image Expert* and select *File/Open Case* or click the *Brain* icon in the *MR Image Expert* menu bar. Double-click on the file **1.5 T: Normal brain, transversal slice**. By default, a spin-echo image with TR = 500 ms and TE = 40 ms will appear on the screen.

*   Select *Draw/Spin Echo* three more times, and set TR to 1000, 3000, and 5000 ms, respectively. Do not change the echo time TE (= 40 ms).
    Each time you click OK in the *Spin Echo* menu box, a new image is displayed. You should now see on your screen the four images displayed below.

With longer TR, image contrast changes. Compare, for instance, the contrast between gray matter and white matter. With a short TR, white matter is brighter than gray matter; by prolonging TR, it becomes dark. In other words, with longer TR, contrast changes from positive to negative. Figure 10-2 explains this behavior.

Please remember that the images created can differ from those created by an MR imager.
Relaxation time and proton density values created by *MR Image Expert* are estimated values. They may differ from those of your MR machine and must not be taken as absolute values in clinical routine.

Contrast Dependence on TE

In case you have exited the program, restart as follows. Otherwise continue with step 2.

*   Select *File/Open Case* or the *Brain* icon in the *MR Image Expert* menu bar. Double-click on the file **1.5 T: Normal brain, transversal slice**.

*   Select *Draw/Spin Echo* or click the *SE* icon. Set TR = 2000 ms, TE = 10 ms and click OK. A spin-echo image with these parameters will appear on the screen. [Close the first image]

*   Select *Draw/Spin Echo* three more times, and set TE to 40, 70 and 100 ms, respectively. Each time you click OK in the *Spin Echo* menu box, a new image is displayed. With longer TE, the image gets increasingly T2-weighted. The contrast between cerebrospinal fluid (CSF) and surrounding tissue changes. CSF is dark on intermediately weighted images (these images are also called 'proton density', 'ρ-', or 'rho-weighted' images), but it turns bright on T2-weighted images.

Why does image contrast change so much ?

Figure 10-3 explains this phenomenon.

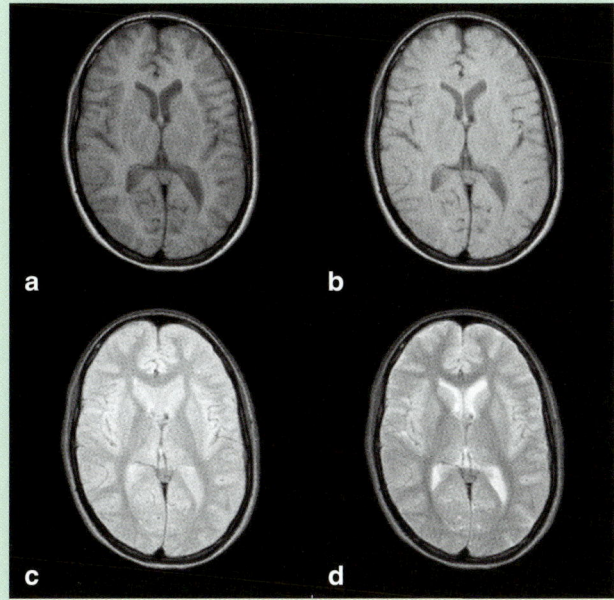

Figure 1:
Changing TR: The images created with *MR Image Expert*: (a) TR = 500 ms; (b) TR = 1000 ms; (c) TR = 3000 ms; and (d) TR = 5000 ms. TE = 40 ms in all cases.

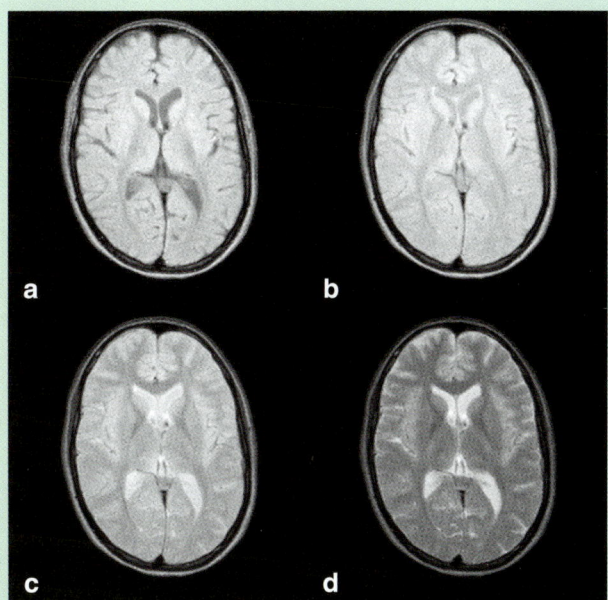

Figure 2:
Changing TE: The images created with *MR Image Expert*: (a) TE = 10 ms; (b) TE = 40 ms; (c) TE = 70 ms; and (d) TE = 100 ms. TR = 2000 ms in all cases.

Figure 10-4 explains with two examples the influence of the repetition time TR upon signal intensity and signal contrast. When a short TR is chosen, the initial signal intensity will be low and SE images with short echo times will be heavily T1-weighted. With longer repetition times, signal intensities will be higher. The relation of signal intensities to each other changes depending on TR, and thus contrast is also strongly influenced by TR.

Again, you can evaluate this behavior with **MR Image Expert**. Just follow the tutorial overleaf.

Figure 10-4:
Both TE and TR influence contrast in a spin-echo sequence. The starting signal strength (= intensity) and contrast depend on the repetition time TR. (a) Short repetition times (TR = 250 ms) emphasize T1-weighting. (b) Long repetition times (TR = 1500 ms) emphasize T2-weighting. In brain imaging, the cross-over points of no contrast move to shorter TE values when TR is increased.

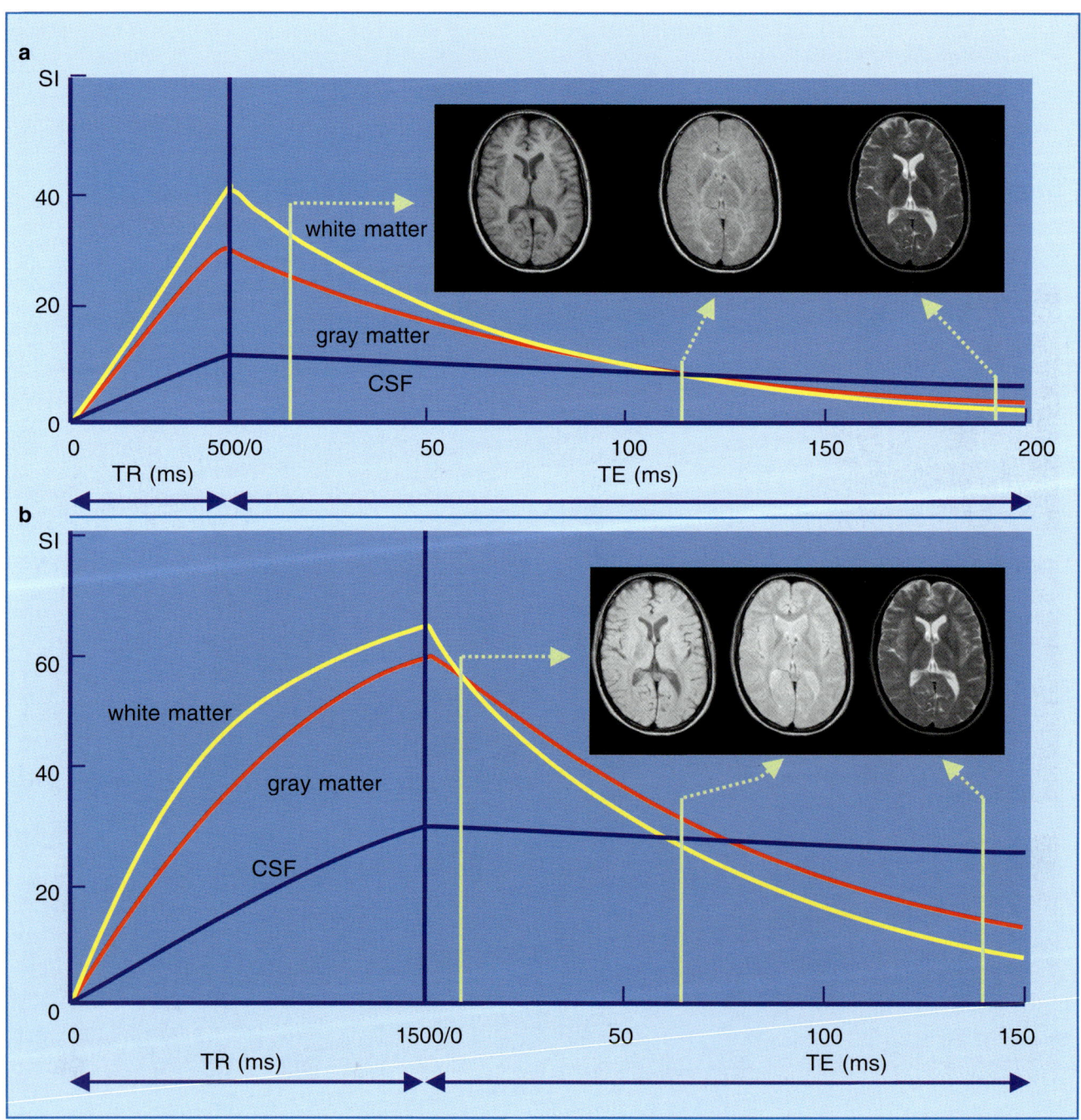

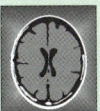 You can experience interactively the influence of TR and TE with *MR Image Expert*.

Adjusting TR and TE

- Select *File/Open Case* or the *Brain* icon in the **MR Image Expert** menu bar. Double-click on the file **1.5 T: Normal brain, transversal slice**. Now select *Draw/Spin Echo* or the *SE* icon. Set TR = 250 ms, TE = 20 ms and click OK. [Close the first image.] A spin-echo image with these parameters will appear on the screen. SE images with short TR and short TE are T1-weighted images.

- Select *Draw/Spin Echo* or the *SE* icon three more times, leave TR at 250 ms, and set TE to 30, 60 and 90 ms, respectively.

- Repeat this procedure with TR at 1500 ms (TE = 20/30/60/90 ms) and TR at 3000 ms (TE = 20/30/60/90 ms).

The four images in each set are increasingly intermediately and T2-weighted, with the first of them having short TR and short TE being more T1-weighted.

The most T1-weighted in Figure 1 is image (a) in row 1 (TR = 250 ms, TE = 20 ms); a typical intermediately weighted image is (b) in row 3 (TR = 3000, TE = 30); the most T2-weighted image in this example is (d) in row 3 (TR = 3000 ms, TE = 90 ms).

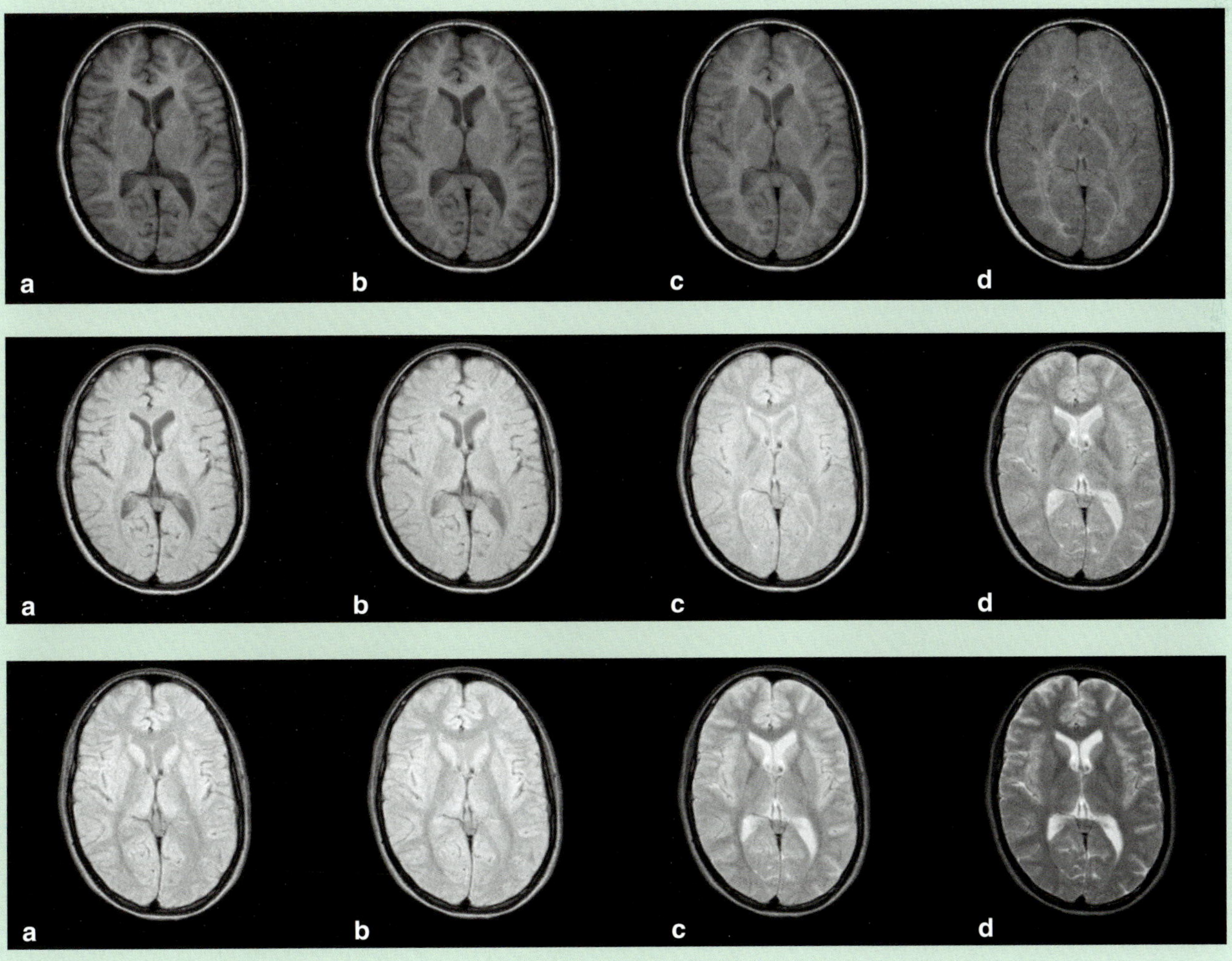

Figure 1:
Changing TR and TE. Top row: TR = 250 ms; center row: TR = 1500 ms; bottom row: TR = 3000 ms. (a) TE = 20 ms; (b) TE = 30 ms; (c) TE = 60 ms; (d) TE = 90 ms.

Contrast might be predictable with normal anatomy; however, when searching for lesions, prediction becomes impossible. The **_MR Image Expert_** Tutorial overleaf gives an example of how a pathological lesion can stay hidden or be highlighted. Sometimes the signal-decay curves of different tissues cross each other, leading to a complete extinction of contrast between them. The best contrast is seen when the relative ratio between them is highest. If the parameters (i.e., T1, T2, ρ, TR and TE) are known, the decay curves can be precalculated within a certain error range. However, if we do not know T1, T2 and ρ and use the wrong pulse-sequence parameters, we can overlook a lesion.

Spin-echo sequences are not limited to a single 180° pulse and echo. Their advantages lie in the possibility to form a multitude of echoes by transmitting a train of 180° pulses. Thus, we can receive a number of images with increasing TE values. The best-known MSE sequence is the Carr-Purcell-Meiboom-Gill sequence. The efficiency of a multiecho sequence is far higher than a single echo or an inversion-recovery sequence, both in terms of examination time and in the creation of contrast [10].

Both single and multiple SE sequences can be acquired in single-slice, multiple-slice and 3D modes. Multislice imaging is limited by TE and TR. Its contrast is also influenced by the gaps between the slices and flow in vessels. In 3D imaging, the entire sample volume is excited simultaneously and slices are obtained by the use of an additional phase-encoding gradient. However, 3D imaging with an SE sequence is time-consuming and prohibitively long in clinical settings.

As we have seen in Chapter 4, one must clearly distinguish between T1- and T2-images ('pure T1- and T2-images') on the one hand, and T1-, T2- and intermediately (proton-density weighted) images on the other. Whereas in the former images calculated relaxation times are depicted, the latter images show signal intensities with, e.g., in the case of a T1-weighted image, a high influence of T1, but at the same time also with T2 and proton-density contributions.

Table 10-2 gives an approximation to image weighting of SE sequences. Short TE and TR emphasize T1 influence, long echo and repetition times emphasize T2 influence. The combination of long repetition times and short echo times eliminate part of the T1 and T2 effects, thus emphasizing ρ. Table 10-3 depicts the signal intensity behavior in 'weighted' images.

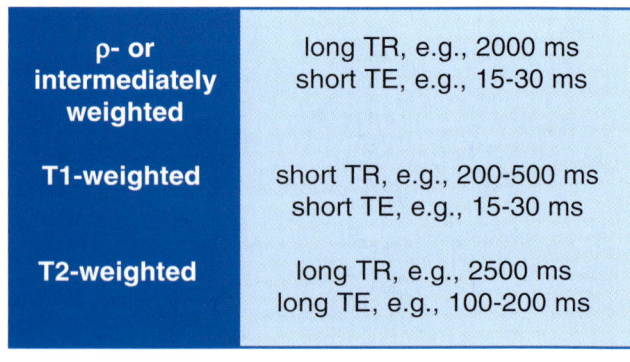

ρ- or intermediately weighted	long TR, e.g., 2000 ms short TE, e.g., 15-30 ms
T1-weighted	short TR, e.g., 200-500 ms short TE, e.g., 15-30 ms
T2-weighted	long TR, e.g., 2500 ms long TE, e.g., 100-200 ms

Table 10-2:
Image weighting. Proton-density (ρ-) weighted images should preferably be called _intermediately_ weighted images, because the highest signal intensity in these images might not present the highest water content. TR and TE depend on field strength. At high fields, TR and TE for both T1- and T2-weighting are shorter than at medium or low fields.

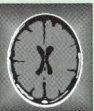

Highlight a lesion by choosing the correct TR and TE with *MR Image Expert*.

The Necessity of Different TR and TE

- Select *File/Open Case* or the *Brain* icon in the *MR Image Expert* menu bar. Double-click on the file **1.5 T: Old MCA infarction, plain, transversal slice**. An image with the following SE pulse parameters will appear on screen: TR = 500 ms, TE = 40 ms. Now select *Draw/Spin Echo* or the *SE* icon. Set TR = 2000 ms, TE = 20 ms and click OK. Repeat this procedures with TR = 2000 ms / TE = 60 ms, TE = 100 ms, and TE = 180 ms.

You now see five SE images with different contrast weighing on screen (Figure 1).

On the more T2-weighted images, there is a clearly visible lesion. However, if you just look at the T1-weighted image on the left, only very suspicious radiologists will describe possible pathological changes.

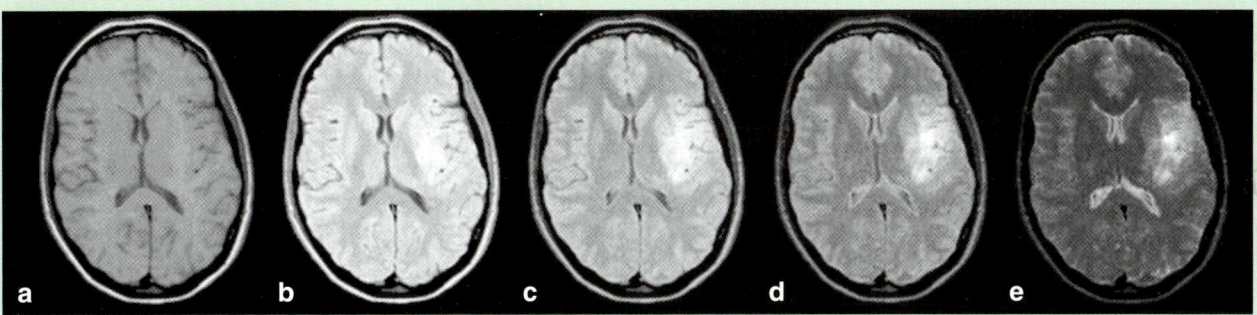

Figure 1:
(a) TR = 250 ms, TE = 40 ms; for all other images, TR = 2000; (b) TE = 20; (c) TE = 60 ms; (d) TE = 100 ms; (e) TE = 180 ms.

Gray Scale	Intermediately weighted	T1-weighted	T2-weighted
	Gray and white matter	Fat Bone marrow	CSF
			Gray and white matter
	Fat	Gray and white matter	
			Fat
	CSF		
		CSF	
	Cortical bone Flowing blood	Cortical bone Flowing blood	Cortical bone Flowing blood

Table 10-3:
Signal-intensity behavior in weighted images. Only weighted spin-echo images are used in routine clinical examinations. Note that the respective signal intensities on weighted images vary according to the TE and TR chosen and with the strength of the magnetic field. The easiest way to distinguish T1- and T2-weighted images is by looking at water-like liquids: on T1-weighted images, they are dark; on T2-weighted images, they are bright.

From TE to Effective TE and Echo Train Length

RSE (Rapid Spin-Echo); RARE (Rapid Acquisition with Relaxation Enhancement); FSE (Fast Spin-Echo); TSE (Turbo Spin-Echo). This pulse sequence is based upon the multiple-echo spin-echo sequence. However, rather than using the same amount of phase-encoding for each echo and each echo as one line for an image associated with a particular TE, different amounts of phase-encoding can be applied to each echo. Thus, these echoes can be implemented as different lines in k-space in a single image.

The numbers of echoes per excitation which are incorporated into one image determine the time-saving factor of the sequence.

If you have a spin-echo sequence with a 256×256 image matrix and a data acquisition time of 256 seconds (4 minutes and 16 seconds), in an RSE sequence with 8 echoes data acquisition will take 256/8 = 32 seconds; with 16 echoes, data acquisition will only take 256/16 = 16 seconds. Multiple-slice sequences will take longer, according to the number of slices.

It is not as obvious as with conventional imaging techniques what sort of contrast we can obtain in RSE sequences. A straightforward signal-intensity calculation, similar to those in conventional pulse sequences, is not possible. Basically, the contrast depends on the order in which we apply the phase-encoding. Contrast manipulation is achieved by different ordering of the contributions in k-space; neither TR nor TE are changed, but different echoes are assigned to the reconstruction in k-space. Images can be proton density- or T2-weighted.

There is less speed advantage in T1-weighted RSE compared to conventional SE images; T1-weighted contrast can only be obtained through inversion recovery or partial saturation with the application of shorter TR. However, for a large number of clinical questions ρ-weighted images can substitute T1-weighted images.

The trade-off of RSE is several, although subtle, differences in contrast, most importantly due to the different signal intensity of fatty tissues. In RSE images, the lipid signal is usually higher than on similar SE images. This is claimed to be caused by several factors, including spin coupling among glyceride protons. There are also magnetization-transfer phenomena that cause protein-containing tissues to appear darker than on similar SE images, whereas hemorrhage with hemosiderin will appear less dark and CSF will appear relatively brighter, obliterating contrast between the ventricles and, e.g., periventricular multiple sclerosis plaques[8]. One also might exchange image sharpness for either enhanced or blurred edges, depending on whether k-space is acquired first or last.

RSE sequences add two more parameters to TR and TE: the number of echoes per excitation (also called *echo train length*, *TSE factor*, or *turbo factor*) and the echo spacing. Since the echo time is close to a time average in RSE sequences, TE is called *effective TE* (Figure 10-5).

A recent addition to the RSE armory can use longer echo trains (HASTE, GRASE and similar sequences).

Rapid Spin-Echo Pulse Sequence Parameters	
Echo Train Length (ETL)	number of echoes per excitation
Effective Echo Time (eff. TE)	echo time of the central line in k-space
Echo Spacing (ES)	time between the 180º refocusing pulses
Repetition Time (TR)	repetition between the 90º pulses

Table 10-4 (left):
The pulse-sequence parameters of an RSE sequence are different from those of a conventional SE sequence (with the exception of TR). In the case shown in Figure 10-5 and the example simulated with **MR Image Expert** the parameters are as follows: ETL = 8; eff. TE = 64 ms; ES = 16 ms; TR = 3000 ms.

Figure 10-5 (right):
A rapid spin-echo sequence utilizes an initial 90º pulse followed by multiple 180º refocusing pulses, producing an echo train. In our example, all echoes are used for one k-space. The echo with the 'effective TE' is assigned to the center slab of k-space and determines overall image contrast; the data of other echoes are placed in the slabs further away.

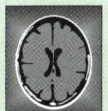

Simulate RSE, RARE, FSE, TSE with *MR Image Expert*.

Rapid Spin-Echo Imaging

- Select *File/Open Case* or the *Brain* icon. Double-click on the file **1.5 T: Normal brain, transversal slice**. Then select *Draw/RSE* or the *RSE* icon. Set TR = 3000 ms, (effective) TE = 64 ms, and Echo spacing = 16. Click OK.

The image you should get is shown below (Figure 1a). Compare it with a spin-echo image with the same repetition and echo time values (Figure 1b).

RSE is not a steady-state technique which can be simulated with equations. Therefore, an exact simulation of RSE images is far more difficult than of SE, IR, or GRE images. Physiological contributions caused by blood flow, susceptibility effects, and interactions between multiple slices add to unpredictable contrast changes. There are also differences in reconstruction between different MR equipment manufacturers. These are the reasons why RSE images in *MR Image Expert* may look different from those created on your MR machine.

Since *MR Image Expert* produces only one slice per case, you could be lured into the idea that it is easy to create T1-weighted images with RSE — one just chooses a short TR-, TE-, and echo spacing; with only two echoes, you will get a T1-weighted image. But in reality you want to acquire several parallel slices; therefore, you need a longer TR. Thus, the result is either several T1-weighted slices with the same imaging time you need for a conventional SE sequence or one T1-weighted slice — but faster.

For the sake of simplicity, you only have to change echo spacing as additional pulse sequence parameter for RSE in *MR Image Expert*. The according number of echoes is calculated by the program automatically.

If you try other RSE parameters, always keep in mind that TE must be a multiple of echo spacing.

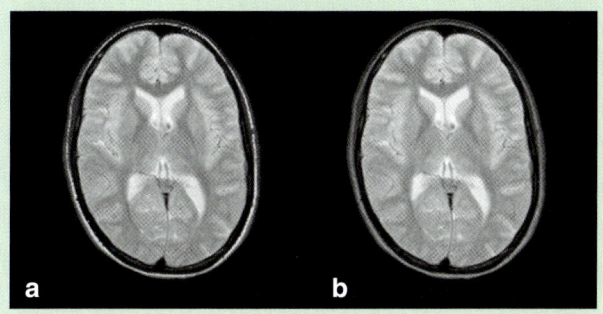

Figure 1:
(a) RSE: TR = 3000 ms, (effective) TE = 64 ms; echo spacing: 16 ms; (b) SE: TE = 64, TR = 3000 ms.
With the exception of the signal from subcutaneous fat, contrast is very similar.

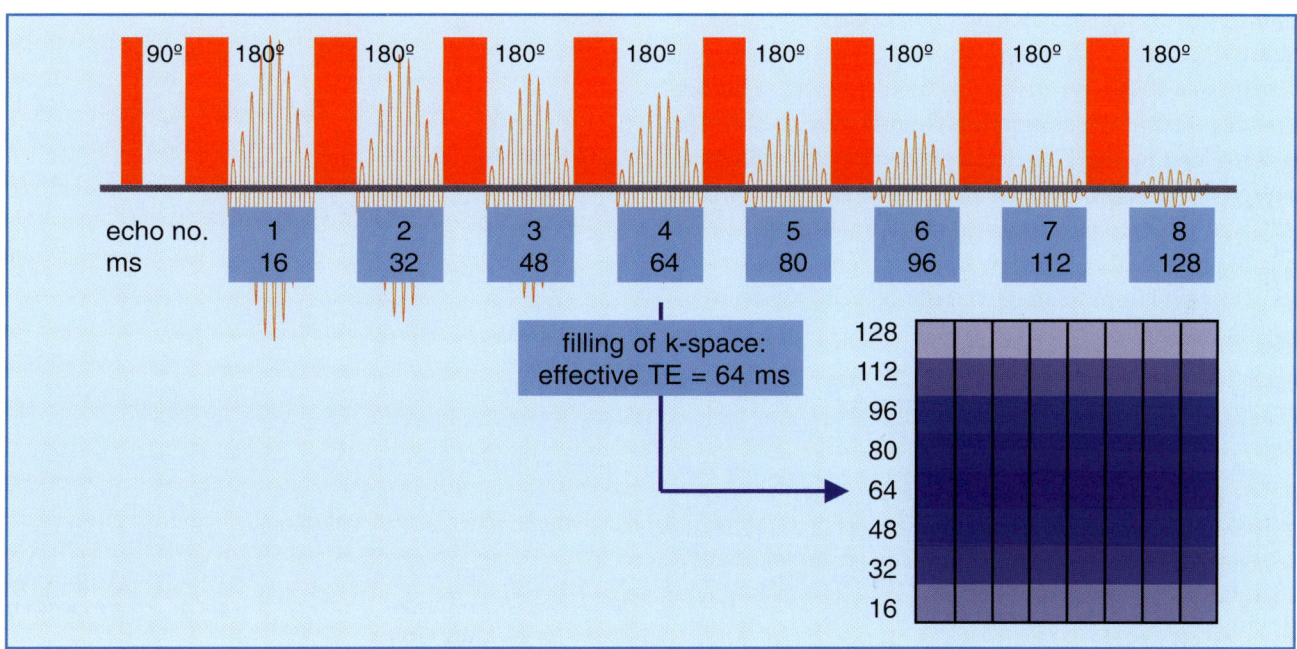

TI — the Inversion Time

The Inversion-Recovery Pulse Sequence. In the early 1980s, the research group at Hammersmith Hospital in London succeeded in imaging the brain with such an excellent spatial and contrast resolution that they could compare their images with slices seen before only in necropsy [1]. They used inversion-recovery (IR) pulse sequences.

In IR sequences, the initial magnetization is turned around by a 180° pulse. Thus, the recovery starts with a negative value, reaches 0 after 0.69 times the respective T1, then becomes positive and returns to its equilibrium after approximately five times T1 (Figure 10-6a).

At some stage during this period, a 90° pulse is transmitted, which changes the partly recovered longitudinal magnetization into observable transverse magnetization. Commonly this is followed by another 180° pulse which creates a spin echo. A detailed description of this pulse sequence can be found in Chapter 4.

In reality, the initial negative magnetization is lost due to the calculation of the magnitude or modulus image (solid lines in Figure 10-6a) from the real and imaginary signals, which is the standard procedure on most systems. Therefore, negative signals are recorded as positive signals of the same strength, and the signal is positive on either side of the null point.

Image contrast in IR sequences reflects this behavior (Figure 10-6b).

To retain the signal information, a reference image is needed so that the phase in each pixel can be compared. An interlaced IR/PS sequence is one way of achieving this.

Images with long inversion times (TI) have hardly any contrast between neighboring tissues, except the one created by differences in proton density, whereas images with short TI show high contrast.

Like partial saturation sequences, IR emphasizes the longitudinal relaxation time T1. The intensity of an averaged signal can be calculated with the following equation:

$$SI = K \times \rho \times M_0 (1 - 2\exp\{-TI/T1\} + \exp\{-TR/T1\})$$

where SI is signal intensity, K is a constant comprising bulk flow, diffusion, perfusion and other parameters, ρ is proton density, M_0 is magnetization at time 0, TI is the inversion time, TR is the repetition time, and T1 is the longitudinal relaxation time.

The signal intensity for a given T1 is strongly dependent on inversion and repetition times.

The IR sequences implemented in clinical scanners add a second 180° pulse after the 90° pulse to rephase T2 influences. Thus, clinical IR images are also affected by T2; however, T2 is not a primary source of contrast in IR imaging and will be neglected in this context.

If a shorter TR is chosen than the T1 of a particular component of the sample, it can happen that — at a certain TI, which is shorter than T1 — the relative signal intensity of this tissue component will be higher than that of a neighboring component with a shorter T1. This complicated and ambiguous contrast behavior is best explained in the *MR Image Expert* Tutorial on page 126.

Figure 10-6:

(a) Signal-intensity (SI) behavior of an inversion-recovery sequence with a repetition time TR = 2000 ms, B_0 = 1.5 T. Note that until 0.69×T1 has been reached, the signal is negative (dotted lines). The solid lines depict the IR signal intensity behavior in reality (magnitude image): the (positive) signal decreases to zero, then increases again until it reaches equilibrium.

(b) Relative contrast (C) between gray and white matter and CSF in the same inversion-recovery experiment.
There is poor contrast with short inversion times. It then increases and reaches a peak at approximately 400 ms. Immediately afterwards, there is a sharp drop in gray/white-matter contrast; it disappears completely, and then turns negative, reaching another peak. With long inversion times, it disappears again. CSF/white-matter contrast behaves similarly.
Note that peaks of optimal contrast can be very close to zero contrast. When looking for pathology, this means that lesions may easily be overlooked if wrong inversion times are chosen.

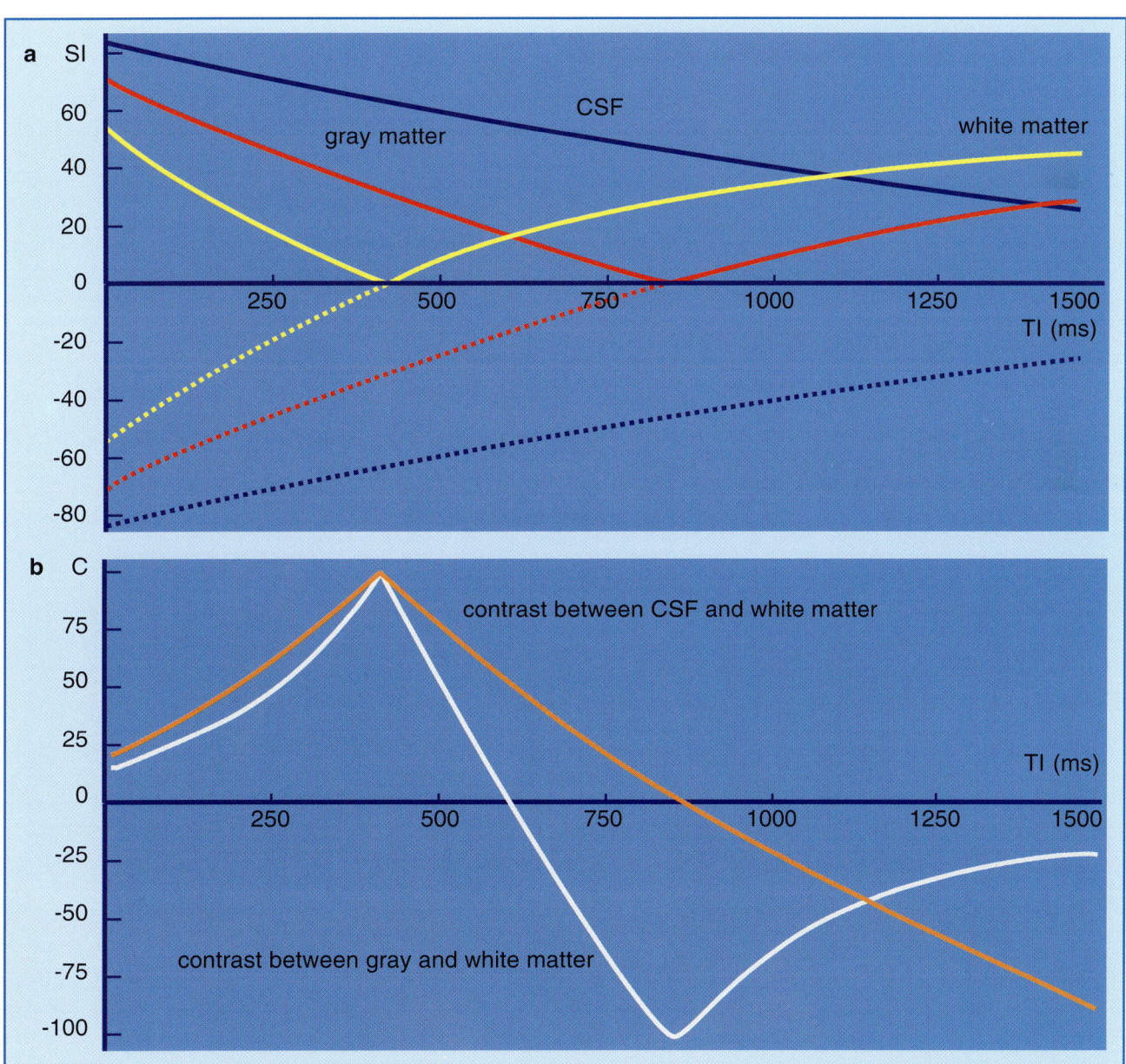

The images in this tutorial show that the gaps between no contrast and high contrast are very small. If there are no unpredictable pathological changes present in the sample, contrast can be foretold. Problems arise when unknown lesions have to be found.

To avoid false interpretations of MR images, several pictures at different TI are desirable. This, however, prolongs imaging and examination times. It is possible to produce several images with different TI values in a single interval using a special IR sequence, but the signal-to-noise ratio of this sequence is reduced because flip angles of < 90° are applied for each excitation [5,12].

Usually, IR images are acquired as multislice pictures. This means that several parallel slices, rather than one single slice, are imaged at a time. This is an elegant way to save time and shorten an examination, and greatly increases the efficiency of IR examinations, which are very time-consuming.

The pulse sequences used for multislice IR are the BIR (balanced IR) sequence, which is also known as MDEFT (modified driven-equilibrium Fourier transformation). These sequences provide superior T1-weighted contrast to T1-weighted SE sequences.

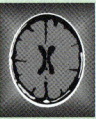 You can experience interactively the influence of TR and TI with *MR Image Expert*.

Contrast Dependence on TI

One of the main problems with the IR sequence is that its contrast behavior can change dramatically with only minimal changes of the inversion time, TI.

- Select *File/Open Case* or click the *Brain* icon in the *MR Image Expert* menu bar. Double-click on the file **1.5 T: Normal brain, transversal slice**. Then select *Draw/Inversion Recovery* or click the *IR* pulse sequence icon. Set TR = 4000 ms, TI = 200 ms. TE = 10 ms. Click OK.

- Select *Draw/Inversion Recovery* three more times. Keep the TR and TE values, while setting TI to 400, 600 and 800 ms, respectively. Note that the contrast between gray and white matter changes when TI is increased and that contrast between gray and white matter becomes negative after a TI of approximately 500 ms. By now, you should have the four IR images below on screen.

For easier and better understanding of the contrast behavior of inversion recovery, we will not alter the echo time (TE) in this tutorial.

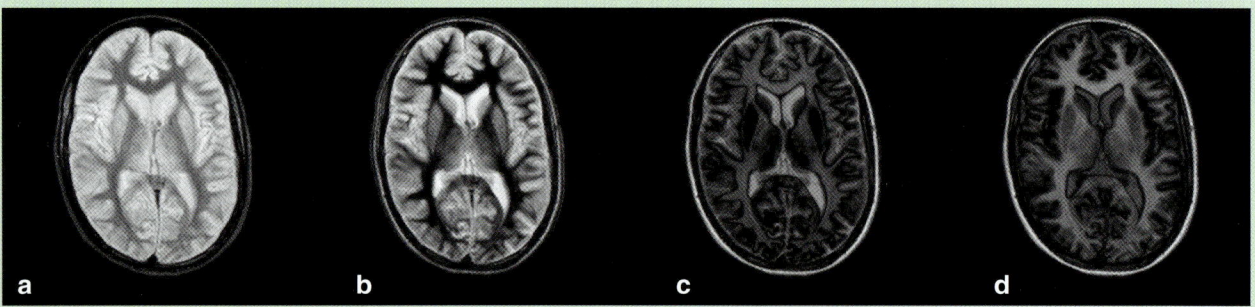

a   b   c   d

Figure 1:
Changing TI: in all images, TR = 4000 ms, TE = 10 ms; (a) TI = 200 ms, (b) TI = 400 ms, (c) TI = 600 ms, (d) TI = 800 ms.
Contrast between gray and white matter and CSF changes markedly. Try the *Real-time Simulation* feature of *MR Image Expert* to follow contrast changes by subtly altering TI (press the right mouse button with the arrow inside the image frame; then select *Sequence* with the left button).

Fat and Water Suppression (STIR and FLAIR)

Sometimes it is helpful to dispose of the high-intensity signal of fat or fluids. There are numerous suppression techniques. Two of them are inversion-recovery methods; additional techniques are discussed in the following chapter.

Images with long inversion times (TI) have little contrast between neighboring tissues, except the one created by differences in proton density, whereas images with short TI can show high contrast. This feature is exploited in a special inversion-recovery pulse sequence, the STIR sequence (Short Time Inversion-Recovery).

STIR sequences are often used when looking for high signal intensity lesions such as contrast-enhancing tumors close to or within fatty tissue because this inversion-recovery sequence facilitates the suppreession of the signal stemming from fat. The fat signal reaches zero signal intensity at TR > 3000 ms, TI ~ 200 ms, and TE 20 ms (at 1.5 T); TI is lower at lower field strength.

FLAIR (Fluid Attenuated Inversion-Recovery) eliminates the signal from cerebrospinal fluid by using very long inversion times (2000 - 2500 ms). It is especially useful in brain lesions with low contrast. As seen in Figure 10-6, CSF reaches the null point of no signal at an inversion time of ~ 2000 ms (TR > 8000 ms; TE > 100 ms)[3].

Both techniques can be combined with RSE sequences; they then become Fast STIR and Fast FLAIR.

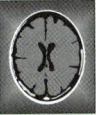 You can experience interactively signal elimination with **MR Image Expert**.

Fat Suppression with STIR

- Select *File/Open Case* or click the *Brain* icon in the **MR Image Expert** menu bar. Double-click on the file **1.5 T: Orbit, sagittal slice**. Then select *Draw/Inversion Recovery* or click the *IR* pulse sequence icon. Set TR = 4000 ms, TI = 50 ms, TE = 10 ms. Click OK.

Use the real-time display feature to see how the retrobulbar fat signal evolves (three possibilities are shown below). A tumor next to the optic nerve (plain or Gd-enhanced) would be best seen with a STIR sequence at TI ~ 200-250 ms.

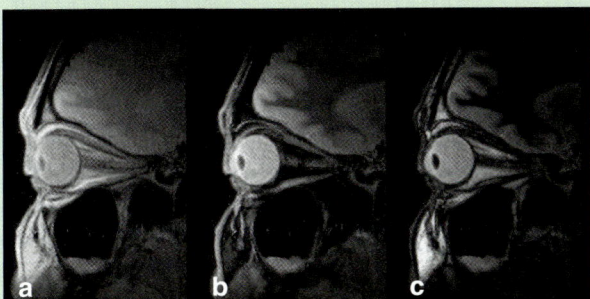

Figure 1:
STIR. In all images, TR = 4000 ms, TE = 10 ms; (a) TI = 50 ms, (b) TI = 240 ms, (c) TI = 450 ms.
The fatty tissue close to the optic nerve disappears with a TI of approximately 240 ms.

Fluid Suppression with FLAIR

- Select *File/Open Case* or click the *Brain* icon in the **MR Image Expert** menu bar. Double-click on the file **1.5 T: Meningioma, plain, transversal slice**. Then select *Draw/Inversion Recovery* or click the *IR* pulse sequence icon. Set TR = 8000 ms, TI = 2500 ms. Click OK.

Create a T1-weighted and a T2-weighted SE image to compare with the FLAIR image. The brain lesions caused by the pressure of huge meningioma periventricularly and in other parts of the brain are best seen on the FLAIR image.

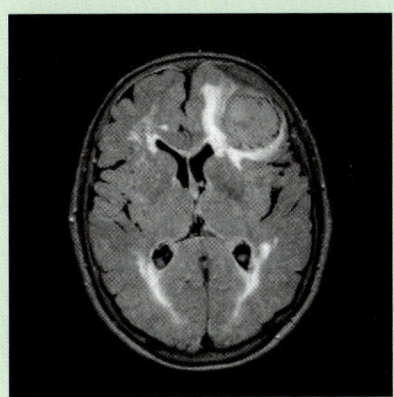

Figure 2:
FLAIR. TR = 8000 ms, TE = 120 ms, TI = 2500 ms.

FA — the Flip Angle

Gradient-Echo Sequences. For many clinical indications, rapid imaging sequences are essential to avoid long imaging times, which can cause motion artifacts and reduce patient throughput. Imaging time can drop from several minutes per standard SE image to seconds or even milliseconds. The number of specific indications of particular rapid pulse sequences has steadily increased over the last few years.

For faster imaging, the number of averages, the number of matrix points or image lines, or the repetition have to be shortened. In general, signal-to-noise ratio and spatial resolution will worsen with faster imaging methods, but as stated, a good clinical diagnosis does not necessarily require beautiful image quality, but sufficient image quality.

The main contrast parameters of conventional and rapid sequences are summarized in Table 10-5. Actual weighting of the sequences depends on a number of factors and might not be available on all MR imaging equipment.

PS	IR	SE	RSE	GRE
ρ	ρ	ρ	ρ	ρ
T1	T1	T1	[T1]	T1
(T2*)	[T2]	T2	T2*	T2*
[flow]	[flow]	[flow]	[flow]	flow

Table 10-5:
Pulse sequences and their contrast dependence. All pulse sequences are influenced by ρ and by bulk flow. The T2*-dependent PS sequence equals a GRE sequence.

Pulse para-meters	T1 (weighted)	T2* (slightly weighted)	T2* (heavily weighted)	ρ (intermed. weighted)
TR (ms)	200-400	20-50	200-400	200-400
TE (ms)	12-15	12-15	30-60	12-15
α (º)	45-90	30-60	5-20	5-20

Table 10-6:
Contrast characteristics in a standard GRE sequence at high field (approximation).

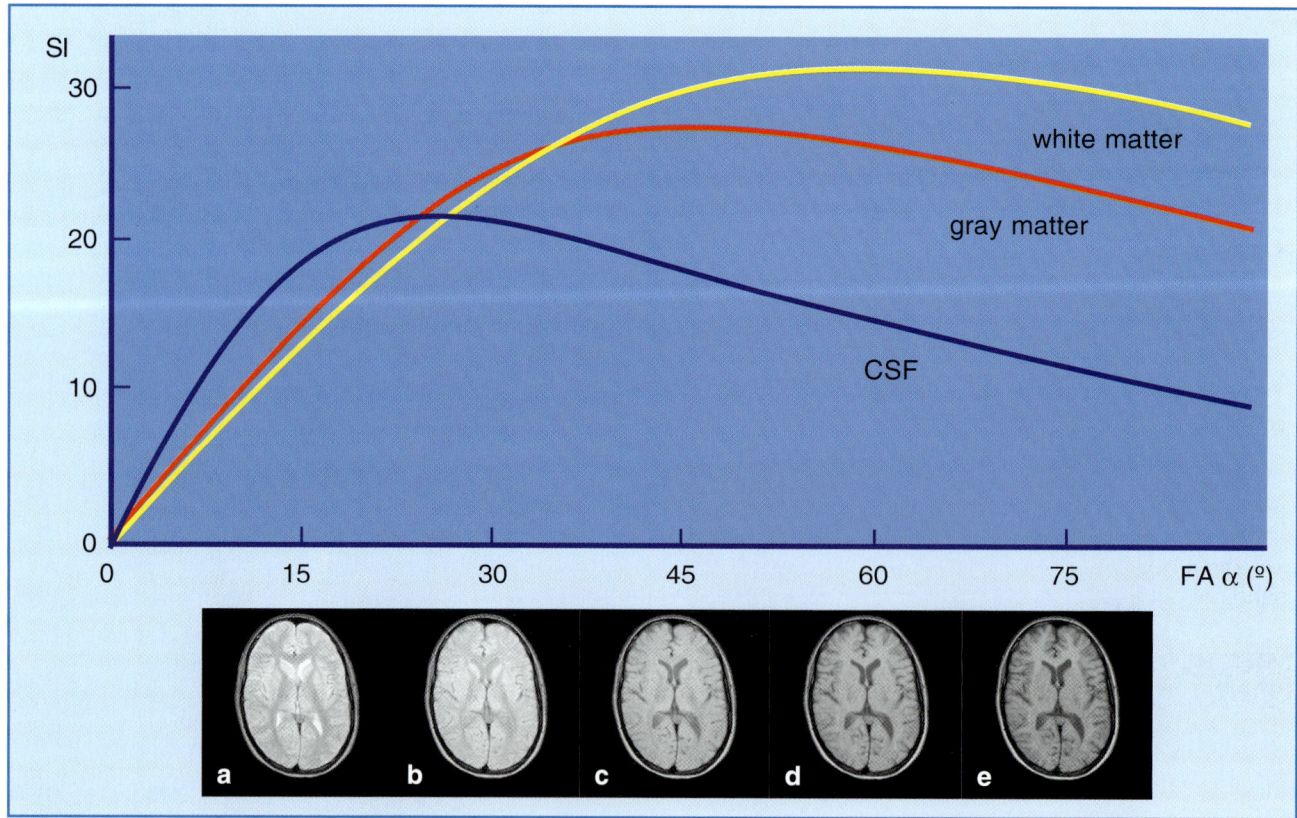

Figure 10-7:
Gradient-echo sequence (spoiled GRE) through the brain of a normal volunteer. TR = 400 ms; TE = 20 ms; α = 15°in (a), 30° in (b), 45° in (c), 60° in (d), and 75° in (e). At high flip angles, T1 dominates; at low flip angles, proton density dominates. Compare the contrast curves of white matter versus CSF and white matter versus gray matter.

GRE sequences were the favorite rapid imaging sequences; however, the popularity of sequences in the RSE family is constantly increasing. GRE sequences take advantage of the saturation of the spin system when TR is shortened. The signal intensity after a series of 90° pulses becomes weaker, until an equilibrium (i.e., the saturation) is reached. Under these conditions, pulse angles smaller than 90° are more effective.

It was unexpected that shortening TR below 100 ms, even below 10 ms, still provided images with a signal-to-noise ratio which was sufficient and allowed diagnostic assessment. Gradient-echo sequences of this kind, with such short TR, have been dubbed FLASH sequences[6]. They are commercially available under several different trade names (see Tables 8-1 and 10-7).

Signal intensity in rapid imaging sequences can be calculated with the following equation, if TR is shorter than T1 but longer than T2* or when gradient/RF spoiling is applied to remove transverse coherences:

$$SI = \sin a \times [1-\exp(-TR/T1)] \times \exp(-TE/T2)/ \ 1-\cos a \times \exp(-TR/T1)$$

where a is the flip angle, TR the repetition time, T1 the longitudinal and T2 the transversal relaxation time.

FLASH sequences add a fourth parameter to TR, TE, and TI: the pulse or flip angle (FA). Similar to SE sequences, GRE sequences can be weighted depending on repetition and echo times, the exact pulse sequence and the pulse angle.

However, there is one big difference: whereas SE and RSE sequences reflect true T2 in their T2-weighted images, GRE sequences show only T2* contrast. Figure 10-7 depicts the typical signal intensity behavior

---

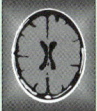 You can experience interactively the influence of the flip angle upon contrast with *MR Image Expert*.

Gradient-Echo Imaging

- Select *File/Open Case* or the *Brain* icon. Double-click on the file **1.5 T: Normal brain, sagittal slice**. Then select *Draw/Spoiled GRE* or the *S GRE* icon. Set TR = 200 ms, TE = 60 ms, and *Flip Angle* = 5º. Click OK.

- Select *Draw/Spoiled GRE* five more times. Keep the TR and TE values while setting *Flip angle* to 20º, 35º, 50º, 65º and 80º, respectively. Note how the contrast between brain parenchyma and CSF changes when the flip angle is increased.

The six spoiled GRE images you should see on your screen now are depicted below.

Because of the three variables TR, TE, and FA, there are nearly unlimited possibilities for changing contrast.

With *MR Image Expert*, you can experiment without any restrictions. Figure 10-7 shows only one of the thousands of different possibilities of signal behavior. Generally, at high flip angles T1 dominates the contrast, while at low flip angles proton density becomes more important.

This version of *MR Image Expert* offers you three different GRE sequences: refocused GRE (R-GRE), spoiled GRE (S-GRE), and contrast-enhanced GRE (CE-GRE). These gradient-echo sequences differ by a number of additional components built into the sequences. Spoiled GRE is the most popular gradient-echo pulse sequence. For details refer to Table 10-7. We suggest you play around with the sequences and their parameters to get a feeling for their contrast behavior.

Unfortunately we cannot simulate one of the major contributors to contrast in gradient-echo sequences, namely blood and CSF flow.

On the other hand, your images at low flip angles will be far better than those produced by your MR machine. Usually such images are of poor quality because of a very low signal-to-noise ratio; there is hardly any signal, as you can see in the graph on the preceding page.

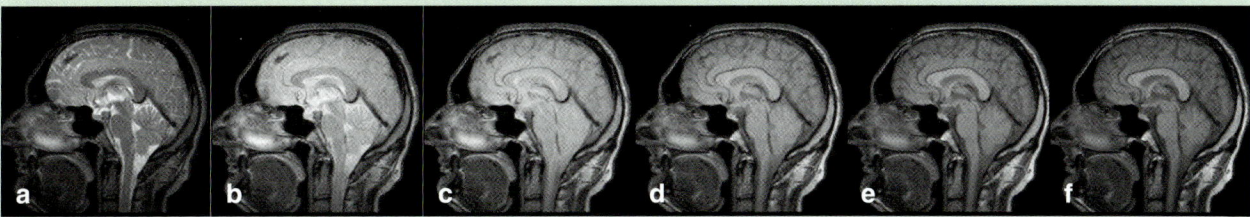

Figure 1:
S-GRE. In all images, TR = 200 ms, TE = 60 ms; α = 5º in (a), 20º in (b), 35º in (c), 50º in (d), 65º in (e), and 80º in (f).

of a GRE sequence, in this case a spoiled FLASH sequence. Commonly, the signal intensities reach a maximum at between 30° and 60°. As we have seen with the signal intensity and contrast behavior of SE sequences, best contrast is not necessarily obtained at the point of highest signal intensity. This is also the case in GRE sequences, as the contrast behavior of the brain images at the bottom of Figure 10-7 shows. At the greatest signal intensity, there is poor or no contrast. It turns out that images acquired using the Ernst angle tend to have rather poor contrast. Higher flip angles have to be used to improve the contrast. The effect of this is a reduction of the signal left along the z-axis after the RF pulse.

Thus, the signal level depends on the rate at which the signal recovered during TR, and is hence strongly T1-dependent. The image series in Figure 10-7 and the **MR Image Expert** Tutorial on page 129 give an overview of how contrast changes with increasing flip angle. GRE sequences can provide sharp contrast between the CSF compartment of the spine, the spinal cord, and the peripheral spinal column. The myelogram effect of the T2*-weighted images allows a fast screening for disk protrusions and is one example of clinical applications of GRE. However, SE and, in some instances, RSE sequences commonly yield sharper spatial detail, and contrast of GRE sequences generally is inferior to that of SE sequences. Contrast can be enhanced by contrast agents; by applying a high flip angle, the T1 effect of paramagnetic contrast agents can be emphasized.

The creation of T2*-weighted contrast is hampered by field inhomogeneities, which are not refocused by the gradient echo. The inhomogeneities solicit short echo times and limit the use of long echo times necessary for T2*-weighting.

Reduction of TR shorter than T2 leads to the generation of transverse coherences which can either be spoiled or refocused, as described in Chapter 8 (see also Table 10-7). The spoiled FLASH sequence removes the effect of the transverse coherences, usually by the application of spoiler gradients, to give genuine partial saturation contrast.

Refocusing sequences incorporate the transverse coherences into the observed signal, and thus have a better signal-to-noise ratio. However, the basic refo-

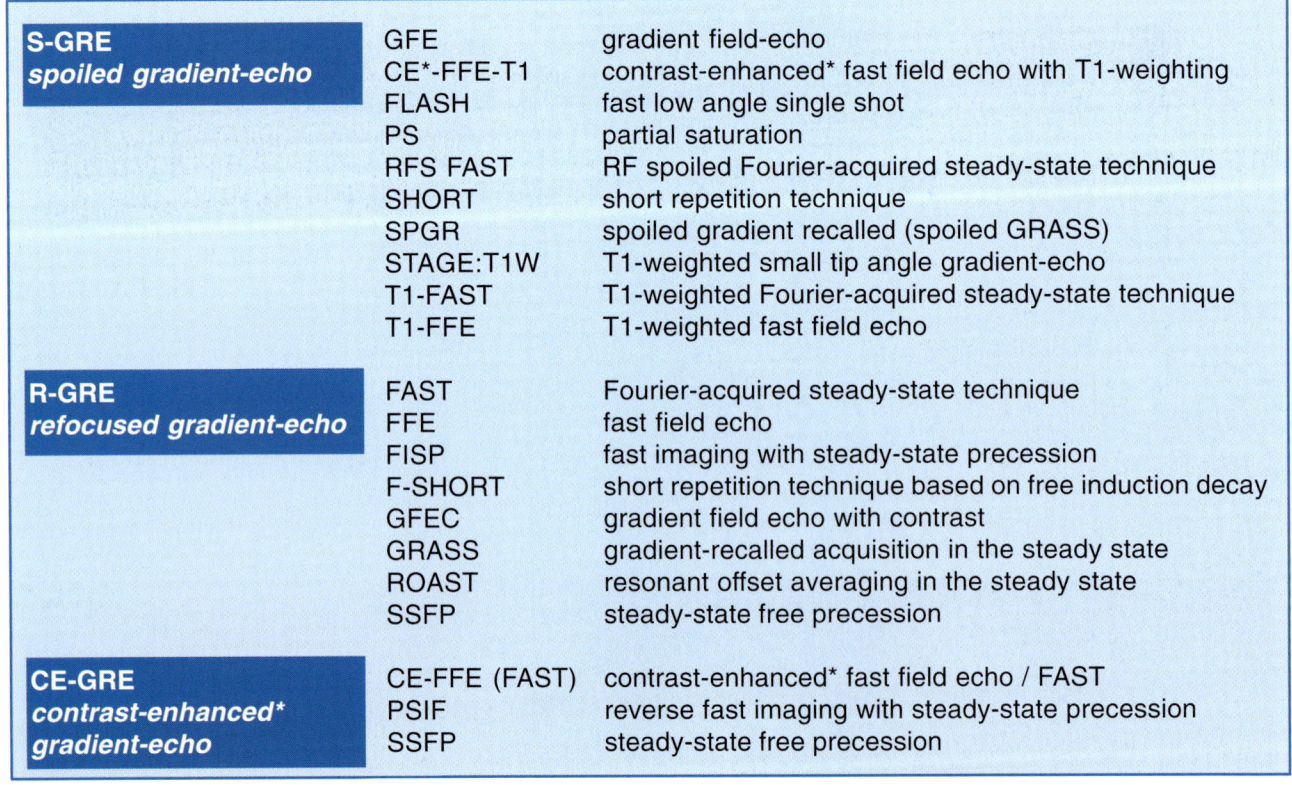

S-GRE spoiled gradient-echo	GFE	gradient field-echo
	CE*-FFE-T1	contrast-enhanced* fast field echo with T1-weighting
	FLASH	fast low angle single shot
	PS	partial saturation
	RFS FAST	RF spoiled Fourier-acquired steady-state technique
	SHORT	short repetition technique
	SPGR	spoiled gradient recalled (spoiled GRASS)
	STAGE:T1W	T1-weighted small tip angle gradient-echo
	T1-FAST	T1-weighted Fourier-acquired steady-state technique
	T1-FFE	T1-weighted fast field echo
R-GRE refocused gradient-echo	FAST	Fourier-acquired steady-state technique
	FFE	fast field echo
	FISP	fast imaging with steady-state precession
	F-SHORT	short repetition technique based on free induction decay
	GFEC	gradient field echo with contrast
	GRASS	gradient-recalled acquisition in the steady state
	ROAST	resonant offset averaging in the steady state
	SSFP	steady-state free precession
CE-GRE contrast-enhanced* gradient-echo	CE-FFE (FAST)	contrast-enhanced* fast field echo / FAST
	PSIF	reverse fast imaging with steady-state precession
	SSFP	steady-state free precession

Table 10-7:
Different terms for the gradient-echo pulse sequences. * In this context, 'contrast-enhanced' refers to the radiofrequency pulse sequence; it does not mean enhancement with a contrast agent.

cused FLASH sequence generally has rather poor contrast (which depends on T1/T2). The contrast-enhanced version, CE-FLASH, offers extra T2 contrast, the amount of T2-weighting being determined by TR and T2. The T2-weighting is greatest at longer T2 values (e.g., 30-60 ms), but the signal-to-noise ratio is poorer than at short TR values.

GRE sequences are exquisitely sensitive to magnetic susceptibility (e.g., depicting hemorrhage and blood degradation products) and to flow phenomena (angiography). Table 10-6 summarizes the features of a standard FLASH sequence at high field.

In *snapshot gradient-echo* scans, the signal evolves to different levels during the scan. Therefore, by manipulating the starting value one can alter the form of the evolution and thus the image contrast. The most commonly used preparation pulse is a 180° inversion pulse, followed by a recovery delay and then the scan (MP-RAGE).

As we have seen in the SE sequences, one can hide pathological changes by choosing the wrong pulse sequence. This also holds for rapid sequences. If we select a T1-weighted sequence, we cannot distinguish a lesion which possesses a similar T1 to its neighboring tissues. If we apply a T2*-weighted sequence, we cannot delineate a lesion with a T2 close to the T2 of its surroundings. The signal intensity of the vascular malformation in Figures 10-8 and 10-9 is a good example of this problem.

Other rapid imaging sequences. Chapter 8 has described a number of different newly developed fast imaging sequences, such as EPI.

Contrast in EPI depends on the preparation module used before the EPI module. This can be an SE module, a GRE module, or an IR module. The contrast of the EPI sequence will behave accordingly.

Table 10-1 lists a large number of factors influencing contrast. As seen in the discussion of the main pulse sequences, among the main contrast parameters are the relaxation times T1 and T2. If they change, contrast also changes.

In the following paragraphs, we will discuss how relaxation time changes also occur, for instance, with alterations of the magnetic field strength.

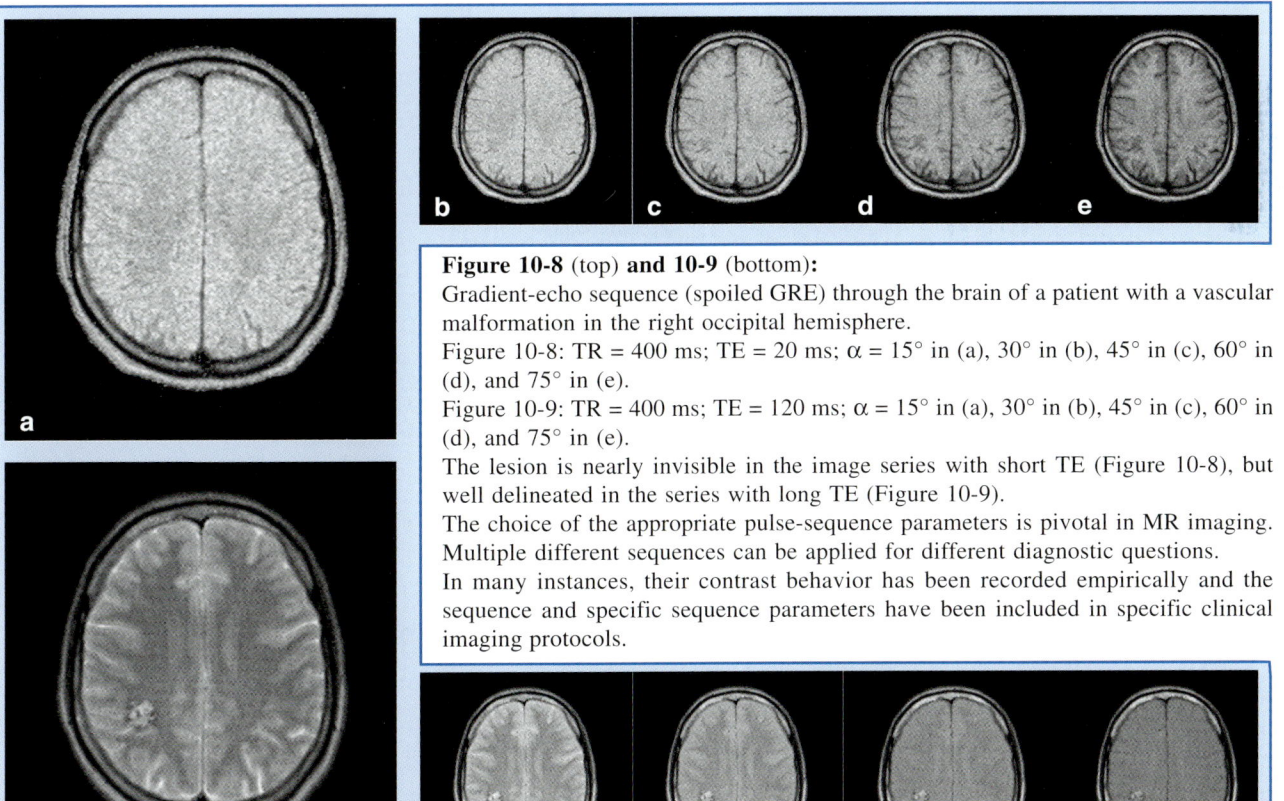

Figure 10-8 (top) **and 10-9** (bottom):
Gradient-echo sequence (spoiled GRE) through the brain of a patient with a vascular malformation in the right occipital hemisphere.
Figure 10-8: TR = 400 ms; TE = 20 ms; α = 15° in (a), 30° in (b), 45° in (c), 60° in (d), and 75° in (e).
Figure 10-9: TR = 400 ms; TE = 120 ms; α = 15° in (a), 30° in (b), 45° in (c), 60° in (d), and 75° in (e).
The lesion is nearly invisible in the image series with short TE (Figure 10-8), but well delineated in the series with long TE (Figure 10-9).
The choice of the appropriate pulse-sequence parameters is pivotal in MR imaging. Multiple different sequences can be applied for different diagnostic questions.
In many instances, their contrast behavior has been recorded empirically and the sequence and specific sequence parameters have been included in specific clinical imaging protocols.

Static Field Strength

Unfortunately, many contrast features we have seen so far change if we move from one magnetic field strength to another. T1 is strongly field-dependent, while within today's imaging range, T2 is only minimally influenced by field alterations.

It is extremely difficult to extrapolate the T1 values of a particular tissue acquired at one field strength to another field strength, because T1 is determined by several factors which vary at different fields.

The best method to examine these features of relaxation is relaxometry. Relaxometry is the study of the behavior of longitudinal and transversal nuclear relaxations and of their dependence on internal and external parameters.

Among them are molecular and supramolecular structures, temperature, viscosity, pH, magnetic field strength, and paramagnetic and ferromagnetic agents. Field-cycling relaxometry deals with the relaxation behavior and its changes with the strength of the magnetic field[11].

Figure 10-10 depicts the T1 relaxation times of white and gray matter of an adult versus field strength. They both increase with field, but with a different ratio to each other. This increase is the reason why images taken with the same pulse parameters, but at different fields, change their contrast appearance. Thus, they cannot be directly compared with each other, as the **MR Image Expert** Tutorial on the next page shows.

When comparing the contrast behavior of different tissues within the imaging range of MR fields, one finds that pure T1-contrast at low and ultra-low fields in many cases is poor. It increases and reaches a peak in medium fields, and worsens again in high fields (Figure 10-11).

The relevance of this finding for clinical imaging is relatively unimportant because pure T1 images are not used. Usually, clinical images (i.e., T1-, T2- or intermediately weighted) possess sufficient contrast even at very high fields if parameters and pulse sequences are chosen properly [2,4,9,11].

However, this feature becomes important for heavily T1-weighted images of any pulse sequence because it means that diseases cannot be detected at certain fields with T1-weighted sequences (Figure 10-12).

This has been known from imaging experience for a long time, but the implications of this phenomenon only become visible in field-cycling magnetic resonance relaxometry.

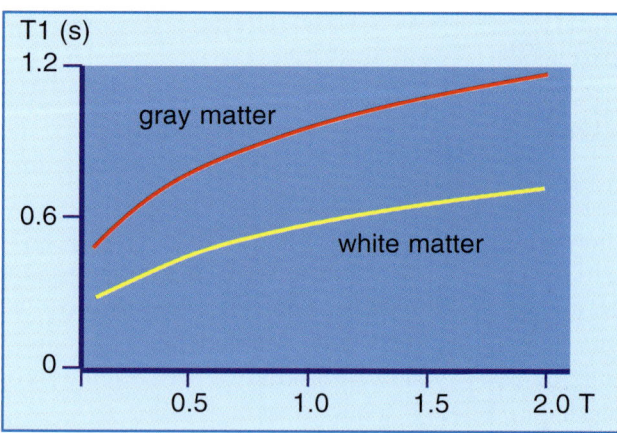

Figure 10-10:
T1 relaxation-time values of gray and white matter versus field strength. Note the increase of T1 with field strength.

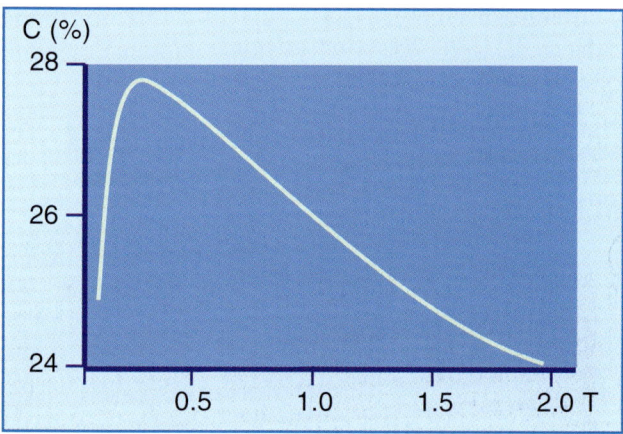

Figure 10-11:
Pure T1 contrast changing with field. The contrast is relatively poor at low and high fields but, depending on the brain tissues, reaches a peak at fields between 0.2 and 0.8 T.

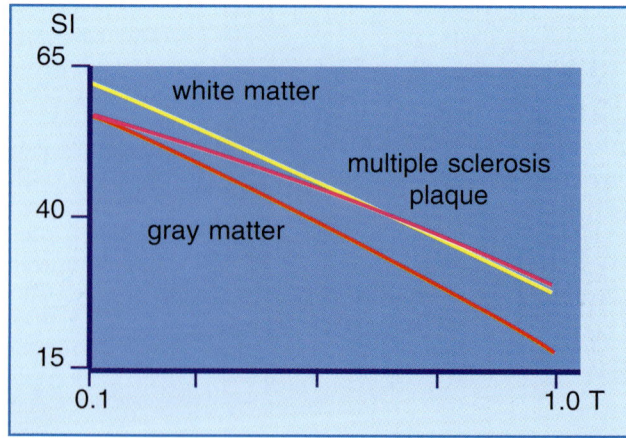

Figure 10-12:
Multiple sclerosis plaques and other brain lesions have a similar signal intensity as their surrounding tissue at medium and high fields when imaged with a T1-weighted conventional SE sequence. Contrast exists only at ultra-low fields (log scale).

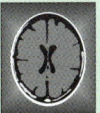 You can experience interactively the influence
of changes of the field strength upon contrast with **MR Image Expert**.

Field Strength and Contrast

- Select *File/Open Case* or the *Brain* icon. Double-click on the file **1.5 T: Normal brain, transversal slice**.

- Select *Draw/Spin Echo* or the *SE* icon. Set TR = 2000 ms, TE = 25 ms. Click OK. Repeat this three times setting TE to 50, 100, and 200 ms respectively. Look at the images and note that the contrast between CSF and surrounding white matter changes from CSF being black on the intermediately weighted image with short TE to becoming increasingly brighter on the T2-weighted images with long TE.

At an echo time somewhere between 50 and 100 ms, the signal intensities of both white matter and CSF are the same. This crossing-over point is at TE ~ 65 ms. At this point, the contrast between these two brain components has disappeared and they cannot be distinguished from each other.

Now, let's continue with this Tutorial:

- Move all four images in a row to the lower part of the screen.
- Select *File/Open* (or the *Brain* icon) in the **MR Image Expert** menu bar. Double-click on the file **0.5 T: Normal brain, transversal slice**. Now you have chosen a transverse slice through a brain similar to the one before, but acquired at 0.5 Tesla.
- Select *Draw/Spin Echo*. Set TR = 2000 ms, TE = 25, 50, 100, and 200 ms respectively.

You now have created four images with the same pulse-sequence parameters as the 1.5 Tesla series before. But the contrast behavior is different. The crossing over point between white matter and CSF is reached at a longer TE.

The reason for this different behavior is the increase of the T1 relaxation time with field strength. Therefore, contrast at different fields varies, in spite of the use of the same pulse-sequence parameters.

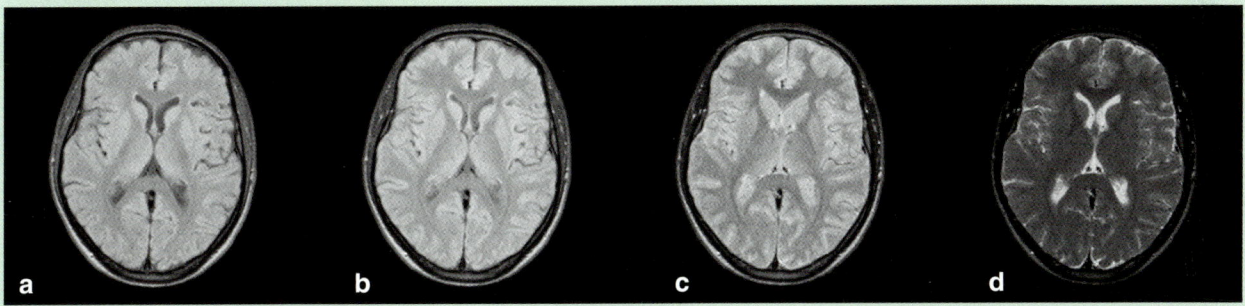

0.5 Tesla

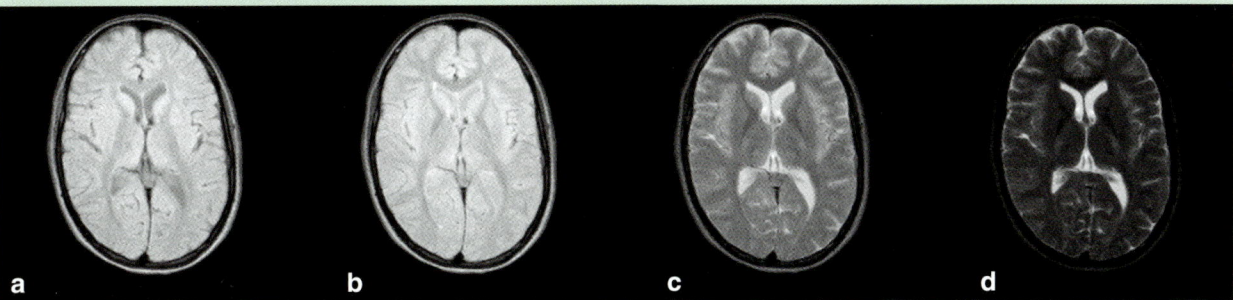

1.5 Tesla

Figure 1:
In all images TR = 2000 ms; (a) TE = 25 ms, (b) TE = 50 ms, (c) TE = 100 ms, (d) TE = 200 ms.
Top four images are at 0.5 Tesla, bottom four images at 1.5 Tesla; compare how contrast changes and differs between the two fields strengths.

References

1. Bydder GM, Steiner RE, Young IR, et al. Clinical NMR imaging of the brain: 140 cases. Amer J Roentgenol 1982; 139: 215-236.

2. Chen C-N, Sank VJ, Cohen SM, Hoult DI. The field dependence of NMR imaging. I: laboratory assessment of signal-to-noise ratio and power deposition. Magn Reson Med 1986; 3: 722-729.

3. De Coene B, Hajnal J, Gatehouse P, Longmore D, White S, Oatridge A, Pennock J, Young I, Bydder G. MR of the brain using fluid-attenuated inversion recovery (FLAIR) pulse sequences. Amer J Neuroradiol 1992; 13: 1555-1564.

4. Fischer HW, Rinck PA, van Haverbeke Y, and Muller RN. Nuclear relaxation of human brain gray and white matter: analysis of field dependence and implications for MRI. Magn Res Med 1990; 16: 317-334.

5. Graumann R, Barfuß H, Fischer H, Hentschel D, Oppelt A. TOMROP: a sequence for determining the longitudinal relaxation time T1 in magnetic resonance tomography. Electromedica 1987; 55: 67-72.

6. Haase A, Frahm J, Matthaei D, Hänicke W, Merboldt K-D. FLASH imaging. Rapid NMR imaging using low flip-angle pulses. J Mag Reson 1986; 67: 258-266.

7. Harms SE, Morgan TS, Yamanashi WS, Harle TS, Dodd GD. Principles of nuclear magnetic resonance imaging. Radio Graphics 1984; 4: 27-47.

8. Henkelman RM, Hardy PA, Bishop JE, Poon CS, Plewes DB. Why fat is bright in RARE and fast spin-echo imaging. J Magn Reson Imaging 1992; 2: 533-540.

9. Hoult DI, Chen C-N, Sank VJ. The field dependence of NMR imaging. II: arguments concerning optimal field strength. Magn Reson Med 1986; 3: 730-746.

10. Rinck PA, Bielke G, Meves M. Modified spin-echo sequence in tumor diagnosis. in: Society of Magnetic Resonance in Medicine. Proceedings of the Second Annual Meeting. San Francisco 1983; 302-303. Also in: Magn Reson Med 1984; 1: 237.

11. Rinck PA, Fischer HW, Vander Elst L, Van Haverbeke Y, Muller RN. Field-cycling relaxometry: medical applications. Radiology 1988; 168: 843-849.

12. Young IR, Hall AS, Bydder GM. Design of a multiple inversion-recovery sequence for T1 measurement. Magn Reson Med 1987; 5: 99-108.

Chapter Eleven

Advanced Imaging and Contrast Concepts

Figure 11-1:
What are the bright spots in the sky on these images: sun or moon ? Answers to this question are on the next page.
Sometimes one cannot determine exactly what is seen on a picture — even when the details are clearly visible. Then, additional information or specific approaches are helpful.

Over the last few years, several new ideas and concepts have been developed on how to influence and enhance contrast by either suppressing or highlighting certain tissue structures. These concepts have added to the diagnostic options of MR imaging and are commonly used to solve specific questions or particular research tasks (not similar to those in Figure 11-1).

We introduce some of these techniques in the following paragraphs.

Suppression Techniques

Fat and, in a similar way, water can create contrast problems for a number of clinical issues. It possesses high signal on T1-weighted SE images, which can obscure other tissues or pathologies with high signal adjacent to the fatty tissue.

In certain cases, it would be of great advantage to eliminate its signal. This includes lesions in fatty tissues such as the orbit or in examinations of fatty livers, in heart examinations, and in the differentiation of bone and marrow diseases.

We have already described two of the suppression techniques in Chapter 10: fat and fluid suppression with STIR and FLAIR. We will discuss three different approaches below.

Phase-Sensitive Methods

In Chapter 5, we have introduced chemical shift: the molecular difference between fat and water makes them precess at slightly different frequencies. If MR imaging is performed at high fields, chemical shift can lead to two different images of the same anatomical structure, which is known as *chemical-shift artifact*. Figure 11-2 explains the origin of this artifact.

There is a positive side to this feature: it can be used to eliminate the unwanted fat signal.

In gradient-echo sequences chemical-shift effects are not refocused and will depend on the echo time, as the following description exemplifies.

Water and fat have a chemical shift of 145 Hz at 1.0 T or of 225 Hz at 1.5 T. At the latter frequency, the off-resonance fat signal rotates through 360° every 4.4

ms. Thus, at echo times which are even multiples of 4.4 ms, the fat and water signals are in phase, while for echo times which are odd multiples of 2.2 ms, the signals are out of phase (Figure 11-3). ΔB_0 effects cause local variations in the exact phase of each component, but their phase difference is preserved. By choosing an appropriate echo time, we can emphasize or minimize the contribution of the fat signal and by adding two averages which use in-phase and out-of-phase echo times respectively, the fat signal can be removed.

This kind of *fat suppression sequence* is also known as the *Dixon* method. It is similar to *chemical-shift imaging* or *phase contrast* [3,4,18].

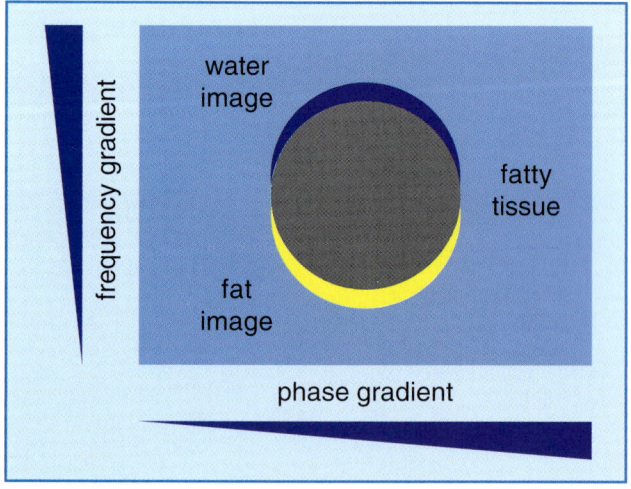

Figure 11-2:
Because of the chemical shift between water and fat signals, the image representation of fat (yellow) is shifted in the frequency-encoding direction with respect to the neighboring water image (blue); in other words, there are two images from the same tissue. In high-field MR imaging, this is an unwanted feature known as *chemical-shift artifact*. For fat-suppression imaging, this feature can be exploited to eliminate the fat signal.

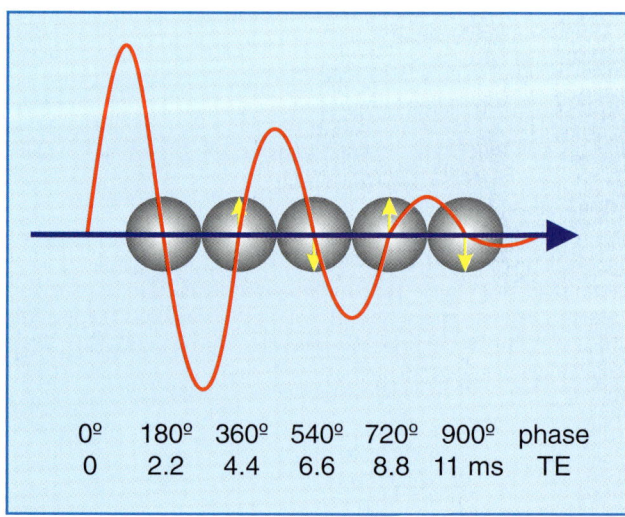

Figure 11-3:
Dephasing of fat. By choosing an appropriate echo time in a GRE sequence, the fat signal is either in phase with the phase of water or out of phase. This example displays the phase-contrast behavior at 1.5 T where the frequency difference between water and fat is 225 Hz. The fat signal rotates through 360° every 4.4 ms (1/225 s). This means that the water and fat signal are in phase at 0.0, 4.4, 8.8, etc. ms (arrow up) and 180° out of phase at 2.2, 6.6, 11.0, ... ms (arrow down).

Figure 11-1: Solution: (a) sunset in Manhattan; (b) moonrise in Switzerland.

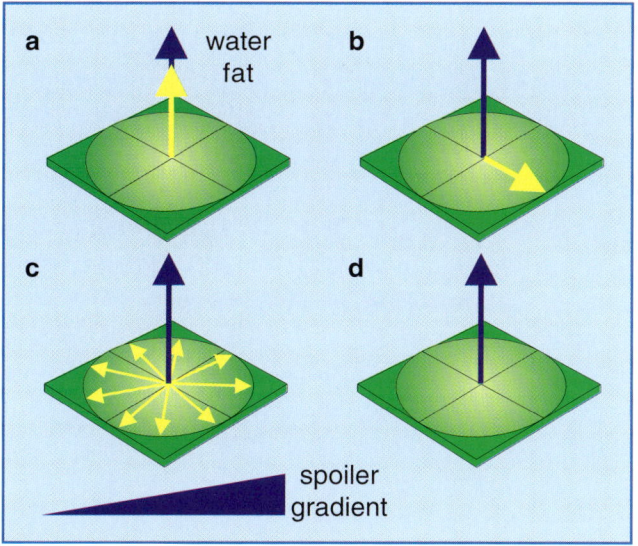

Figure 11-4:
Selectively saturating the fat component: (a) A fat-saturating RF pulse is transmitted and (b) rotates the yellow fat magnetization into the transverse plane. (c) The fat spins start dephasing in the *x-y* plane, accelerated by a dephasing gradient. (d) Only the blue water magnetization remains.

Presaturation

By applying an RF pulse of the appropriate frequency before the regular imaging pulse sequence, one can eliminate the signal of a specific tissue. Again, this method is field strength-dependent and best used at high fields where water/fat shifts are high.

A presaturation pulse is applied at the precession frequency of fat (or the compound to be saturated); this pulse does not influence the water component of the tissue.

Usually a chemical-shift selective pulse sequence (CHESS) or a variation of this sequence is used. With a frequency-selective 90° pulse, the magnetization of fat is rotated into the transverse plane where its dephasing is accelerated by a spoiler (or 'crusher') gradient. Then the regular pulse sequence follows, but it only excites the water in the sample.

Figure 11-5 shows an example of the application of fat suppression.

A different kind of presaturation is used in artifact suppression in flow imaging (see Chapter 17).

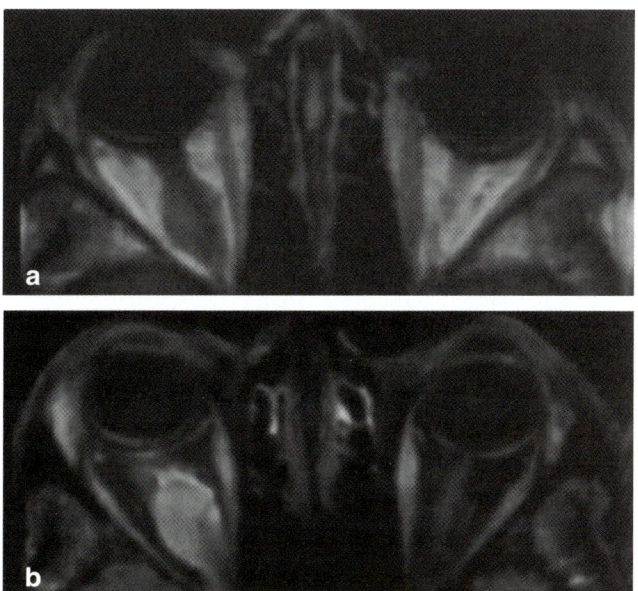

Figure 11-5:
Example of fat suppression — tumor in the right orbit. T1-weighted SE images. (a) Plain image. (b) Contrast enhancement of the tumor after Gd-DTPA. The tumor has become bright. The fat signal has been eliminated; both orbits now are dark and the enhancing parts of the tumor are easily delineated.

Suppression of Protein-Bound Water: Magnetization Transfer

The idea of altering contrast by off-resonance irradiation of the sample was first described by Muller and collaborators in 1983 [15].

Wolff and Balaban coined the term 'magnetization transfer' (magnetization transfer contrast = MTC) for this kind of alteration of image contrast [20]. Lipton, Sepponen and collaborators improved contrast enhancement of the method [13]. MTC is related to spin-lock imaging (page 33).

MTC is based on the fact that in most biological tissues there is a cross relaxation between the free proton pool (H_f) representing mobile water protons and the restricted proton pool (H_r) representing the protons associated with macromolecules or immobile water [5,13]. The restricted H_r pool has a much shorter T2 than the mobile H_f pool, and consequently is not directly observed with standard MR techniques.

Thus, its influence upon image contrast cannot be exploited with standard pulse sequences. The cross relaxation and/or chemical exchange between these two pools means that saturating the resonance corresponding to one of them also affects the second pool (Figure 11-6).

Saturating the H_r pool leads to a loss of signal from the H_f pool. The cross relaxation is a short range process and, therefore, the direct effect is limited to interfaces between the two pools, although diffusion relays the effect to the bulk of free water. The H_r pool is known to have a very short T2 value; thus, the behavior of the magnetization during the RF pulse is dominated by relaxation.

The majority of sequences developed to date for MTC imaging [9,20] use a relatively long, low-power, off-resonance saturation pulse to selectively saturate H_r. New pulse sequences have been proposed to optimize MTC [10].

To date, the clinical applications of MTC have been limited; it can be used in time-of-flight MR angiography to suppress background tissue. In T2-weighted images, MTC may help to detect early demyelination.

A combination of MTC and contrast agent application enhances contrast in cases where one of the techniques alone does not create sufficient enhancement, for instance in multiple sclerosis and other brain lesions, in brain infarction, and in the detection of recent myocardial infarction (Figure 11-7) [8,19].

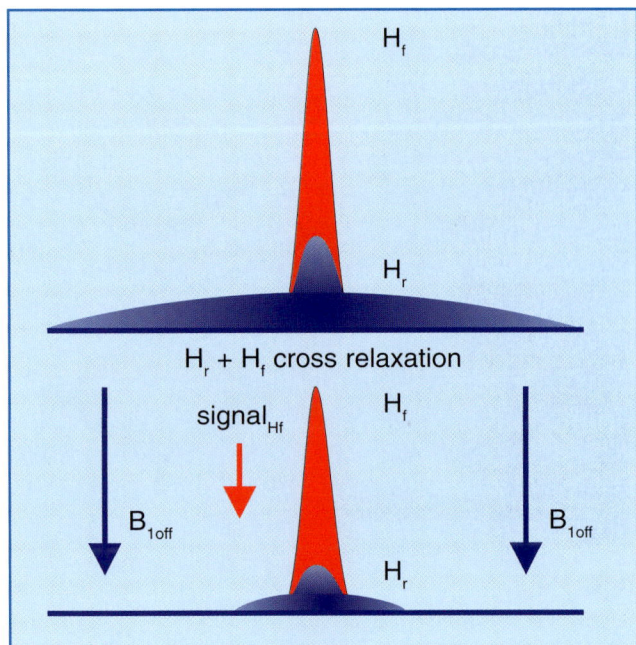

Figure 11-6:
The signal in a conventional MR examination consists of the part created by the narrow peak of the mobile protons (free protons: H_f) and the broad peak of the immobile protons (restricted protons: H_r). Both pools interact and exchange information. The restricted proton pool can be saturated by off-resonance irradiation, which reduces its magnetization to zero (in the best case). The exchange between the two pools then leads to a reduction in the free water signal.

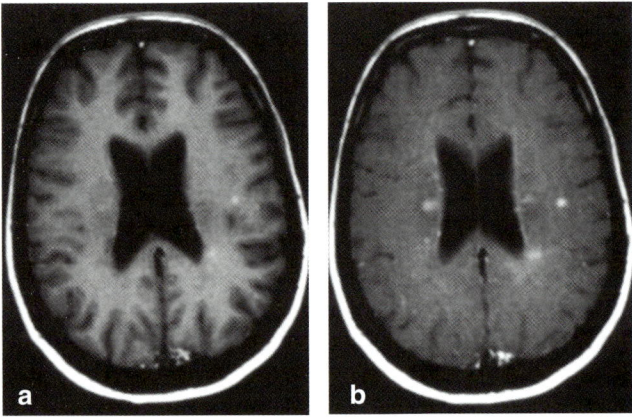

Figure 11-7:
Example of magnetization transfer contrast. Patient with multiple sclerosis. (a) T1-weighted brain images after enhancement with a gadolinium-based contrast agent; (b) image with additional magnetization transfer contrast.
The combination of contrast agent and MTC clearly enhances contrast and shows more lesions, although it remains unclear whether all of these lesions are active.

Flow	usually bulk flow of blood or CSF, is defined as (blood) volume per time unit, i.e., the macroscopic physiological motion of blood. See *flow and angiography* in Chapter 14.
Perfusion	relates to blood delivery to tissues, which usually is motion at the capillary level. See *functional MRI* (next page) and *dynamic MRI* in Chapter 16.
Diffusion	is the random (Brownian) motion of tissue water molecules within cells. They collide with each other and also pass through cell membranes.

Table 11-1:
Forms of fluid motion in the human body.

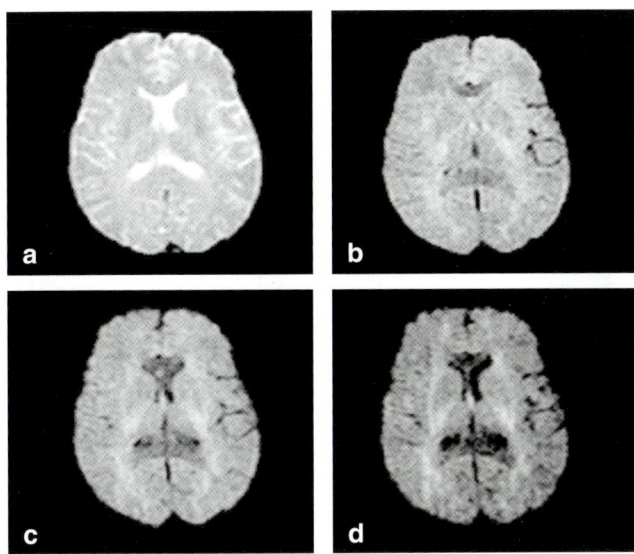

Figure 11-8:
Transverse, increasingly diffusion-weighted images. (a) no dissusion weighting, b = 0 s/mm²; (b) b = 600 s/mm²; (c) b = 900 s/mm²; (d) b = 1200 s/mm². The *b value* is a term describing the diffusion sensitivity.

Diffusion Imaging

Fluids in the human body move in different ways (Table 11-1). Tissue water can diffuse randomly, but barriers such as cell membranes can influence its diffusion and alter its random motion to a partly directed motion. For instance, diffusion in white matter shows a clear directional dependence because the myelin envelope covering the nerve fibers is virtually impenetrable for diffusing water molecules. This leads to an anisotropically restricted motion.

The feasibility to visualize diffusion has been discussed for a long time because it would allow differentiation between tissues according to their cellular structure.

Diffusion can be characterized by the diffusion coefficient, D. This coefficient depends on several factors, the most important being viscosity. Changes of intra- or extracellular viscosity induce alterations of diffusion, and thus can change image contrast in diffusion-weighted MR imaging (DWI). Diffusion is independent of the relaxation times and thus adds another factor to contrast [2,12].

The basic principle of diffusion imaging is that the small random motion of the molecules results in a Gaussian distribution of phases. The effect of these variations is enhanced by using a T2-weighted SE, gradient-echo, or echo-planar technique and applying strong gradients. The ability of these special pulse sequences to depict diffusion depends on the strength and duration of the diffusion gradients and on the direction in which they are applied. Figure 11-8 gives an example of how diffusion influences contrast and its dependence upon gradient direction.

The pulse sequences used are extremely sensitive to bulk motion within the patient, such as blood and CSF flow, and to movements of the patient himself. Thus, in some instances, special casts or mechanical devices have been produced to immobilize patients, but this restricts the application of the method. Faster imaging and additional software manipulation have been developed to overcome these problems.

Pathologically increased diffusion patterns in the brain have been observed in infarction, tumors, edema, multiple sclerosis, and cysts [6,14]. Diffusion changes indicate ischemia at a very early stage. This finding is helping MR imaging become the modality of choice in patients with suspected brain infarction.

Functional Imaging

'Functional imaging' is a misleading term because it is mainly used for changes of local blood supply in the brain activated by specific stimuli. Dynamic or cine imaging of other organs or, e.g., joints are not described as functional MR imaging (fMRI).

In 1990, Belliveau published the first observation of the stimulation of the human visual cortex by magnetic resonance imaging [1]. He used the observation of the first pass effect after bolus injection of a contrast agent to demonstrate changes in cortical perfusion upon activation with a photic stimulus.

The use of bolus tracking to study changes in perfusion was an exact analogue to previous experiments using the observation of tracers with PET or SPECT. It required the injection of contrast agent in two consecutive scans, one with and one without stimulus. The performance of such an experiment with MR, instead of nuclear-medicine techniques, offers the immediate advantage of vastly superior spatial and temporal resolution and the lack of radioactive tracers (see cerebral blood-volume studies and regional cerebral blood-volume described in Chapter 16). The need for dual contrast agent injection, nevertheless, poses a problem, especially for studies of brain activation in normal individuals.

This disadvantage was resolved by the demonstration of brain activation using the BOLD-contrast mechanism first described by Ogawa [16]. This very elegant technique has led to a fast proliferation of fMRI in various centers over the last few years.

BOLD-Contrast

The BOLD-contrast relies on the fact that paramagnetic deoxyhemoglobin — by comparison to diamagnetic oxyhemoglobin — possesses a strong magnetic moment. By interaction of the bulk magnetization of deoxygenated blood with the external field, local field variations in and around blood vessels are thus created. These susceptibility effects can be measured using appropriate MR imaging sequences.

The only source of energy of normal brain cells is the oxidation of glucose. Since the glucose storage capacity of brain cells is negligible, the brain very heavily depends on a constant supply of glucose and oxygen via the capillary bed. This increased demand leads to an increased amount of blood flowing to the activated

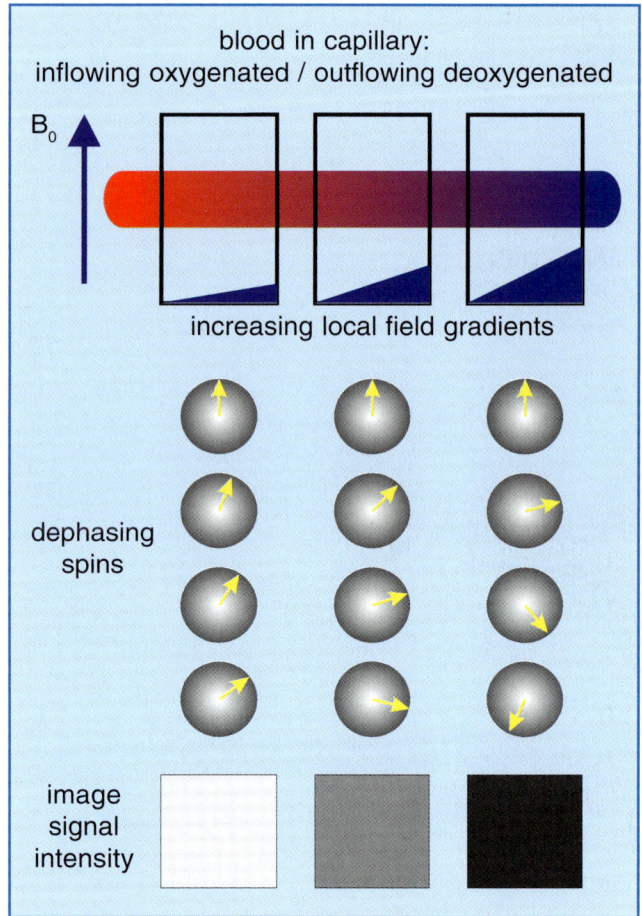

Figure 11-9:
BOLD-contrast. The presence of deoxyhemoglobin in a capillary causes a susceptibility difference between the blood vessel and the neighboring tissue. It induces a dephasing of the spins, thus a decrease in T2* and signal loss on T2-/T2*-weighted images.

- Not a direct measure of brain activity, but measurement of hemodynamic changes.
- Motion artifacts due to cardiac and respiratory motion.
- Physiological noise by arterial and cerebrospinal fluid pulsations.
- Spatial resolution limited to 3 to 4 mm,
- Partial volume effects.
- Temporal resolution limited by hemodynamics response; the MR imager is able to acquire images every 100 ms but the hemodynamic response is slower.

Table 11-2:
Limitations of monitoring of brain activity with BOLD imaging.

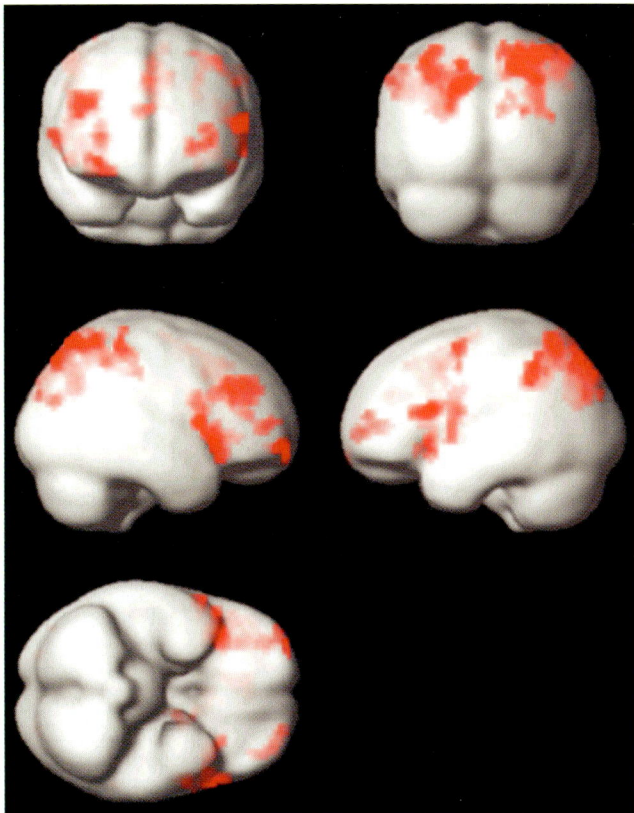

Figure 11-10:
Working memory test: typical activation pattern in the parietal cortex; cognitive/speech processing dorsolaterally.

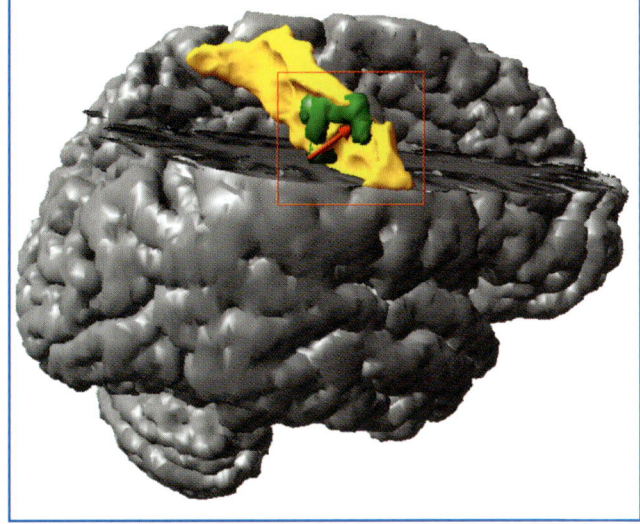

Figure 11-11:
BOLD-contrast. Activated brain regions (green) after electric stimulation of the median nerve. Yellow: sulcus centralis.

area. This in turn decreases the local susceptibility effect, which can be visualized using appropriate susceptibility-sensitive imaging techniques.

Susceptibility differences are greater at higher fields, and thus higher fields are desirable for this kind of studies.

The first brain activation studies by Kwong used gradient-echo-planar imaging (GRE-EPI)[11]. The EPI sequence uses multiple gradient refocusing to acquire all data necessary for image reconstruction after a single excitation pulse. In spite of its not very well defined signal behavior, EPI has turned out to be a very efficient technique for brain activation studies due to its short acquisition time.

Conventional gradient-echo imaging with long echo times (40-60 ms depending on field strength) has also turned out to be a useful technique for fMRI[7]. Its advantage over EPI lies in the fact that it allows the acquisition of high-resolution images, whereas the resolution in EPI is determined roughly by the number of echoes which can be acquired within the T2 of brain parenchyma.

Conventional gradient-echo imaging does, however, suffer from a number of severe drawbacks. The long acquisition time per image restricts the application to a single slice and thus requires prior knowledge about the area of activation. Partial-volume effects can lead to difficulties in data interpretation. Gradient-echo techniques also are very sensitive to inflow. Since vascular flow — especially in large veins — also changes upon stimulation, this can lead to the measurement of activation effects many centimeters from the area of activation[17]. These vascular signal changes can be much larger than the actual parenchymal effects, which seldom exceed 2-3% (Table 11-2).

The image quality of all susceptibility-sensitive techniques is strongly dependent on macroscopic susceptibility problems occurring especially at soft tissue/bone/air interfaces, leading to magnetic field inhomogeneities over several centimeters. These long-range effects will lead to image distortions when occurring in the direction of the readout gradient, which is normally of no practical consequence. Field inhomogeneities across the selected slice, however, will lead to signal attenuation and thus severely affect the image quality. The use of thin slices (or 3D data acquisition) is, therefore, to be preferred for fMRI. The strength of the stimulation effect will, of course, not be dependent on the slice thickness due to the small range of the BOLD effect.

Applications

The first experiments performed with fMRI used the well-known paradigm of photic stimulation with a flicker display or an alternating checkerboard pattern. This is known to lead to significant changes in perfusion and thus serves as a test tool for sequence development.

Meanwhile, quite a number of experiments have been performed, which led to new insights in neurocognitive research. Apart from activation in the primary visual cortex, activation of associated areas was demonstrated using a number of paradigms to test cognitive processing of motion, texture, color, object recognition, sound, memory, and others (Figures 11-10 and 11-11). Various paradigms using motor activation have been successfully examined. Numerous groups have investigated language processing using a number of well established paradigms. In addition to activation of the cerebral cortex, the involvement of the cerebellum in learning tasks has been demonstrated. Subcortical activation has been found, for example, in the nucleus geniculatus (upon visual stimulation).

References

1.   Belliveau JW, Rosen BR, Kantor HL, et al. Functional cerebral imaging by susceptibility-contrast NMR. Magn Res Med 1990; 14: 538-546.

2.   Buxton R, Kwong K, Brady T, Rosen B. Diffusion imaging of the human brain. J Comput Assist Tomogr 1990; 14: 514-520.

3.   Dixon WT. Simple proton spectroscopic imaging. Radiology 1984; 153: 189-194.

4.   Dixon WT and Lee JKT. Separate water and fat MR images (letter). Radiology 1985; 157: 552-553.

5.   Edzes HT, Samulski ET. The measurements of cross-relaxation effects in the proton NMR spin-lattice relaxation of water in biological systems: hydrated collagen and muscle. J Magn Reson 1978; 31: 207-208.

6.   Doran M, Hajnal JV, Van Bruggen N, King MD, Young IR, Bydder GM. Normal and abnormal white matter tracts shown by MR imaging using directional diffusion weighted sequences. J Comput Assist Tomogr 1990; 14: 865-873.

7.   Frahm J, Bruhn H, Merboldt KD, Hänicke W. Dynamic MR imaging of the human brain oxygenation during rest and photic stimulation. J Magn Reson Imag 1992; 2: 501-505.

8.   Jones RA, Haraldseth O, Schjøtt J, Brurok H, Jynge P, Øksendal AN, Rinck PA. Effect of Gd-DTPA-BMA on magnetization transfer: application to rapid imaging of cardiac ischemia. J Magn Reson Imaging 1993; 3: 31-39.

9.   Jones RA, Southon TE. A magnetization transfer preparation scheme for snapshot FLASH imaging. Magn Reson Med 1991; 19: 483-488.

10.   Jones RA, Southon TE. Improving the contrast in rapid imaging sequences with pulsed magnetization transfer contrast. J Magn Reson 1992; 97: 171-176.

11.   Kwong KK, Belliveau JW, Chesler DA, et al. Dynamic magnetic resonance imaging of human brain activity during primary sensory stimulation. Proc Natl Acad Sci USA 1992; 89: 5675-5679.

12.   Le Bihan D, Breton E, Lallemand D, Grenier P, Cabanis E, Laval-Jeantet M. MR imaging of intravoxel incoherent motions: applications to diffusion and perfusion in neurologic disorders. Radiology 1986; 161: 401-407.

13.   Lipton MJ, Sepponen RE, Tanttu JI, Kuusela T. Magnetization transfer technique for improved magnetic resonance imaging contrast enhancement in whole body imaging. Invest Radiol. 1991; 26 S1: S255-256; and S263-265.

14.   Moseley ME, Cohen Y, Mintorovitch J, et al. Early detection of regional cerebral ischemia in cats: comparison of diffusion weighted and T2-weighted MRI and spectroscopy. Magn Reson Med 1990; 14: 330-346.

15.   Muller RN, Marsh MJ, Bernardo ML, Lauterbur PC. True 3-D imaging of limbs by NMR zeugmatography with off-resonance irradiation. Eur J Radiol. 1983; 3, S1: 286-290.

16.   Ogawa S, Lee TM, Nayak AS, Glynn P. Oxygenation-sensitive contrast in magnetic resonance image of rodent brain at high magnetic fields. Magn Res Med 1990; 14: 68-78.

17.   Segebarth C, Belle V, Delon C, et al. Functional MRI of the human brain: predominance of signals from extracerebral veins. NeuroReport 1994; 5: 813-816.

18.   Szumowski J, Plewes DB. Fat suppression in the time domain in fast MR imaging. Magn Reson Med 1988; 13: 534-535.

19.   Tanttu JI, Sepponen RE, Lipton MJ, Kuusela T. Synergistic enhancement of MRI with Gd-DTPA and magnetization transfer. J Comput Assist Tomogr 1992; 16: 19-24.

20.   Wolff SD, Balaban RS. Magnetization transfer contrast (MTC) and tissue water relaxation *in vivo*. Magn Reson Med 1989; 10: 135-144.

Interlude Five

What is Normal ?

When Warren G. Harding became President of the United States in 1920, shortly after the First World War, his motto was *back to normalcy*. Unfortunately, Harding and his political friends never gave their definition of normalcy — their approach to normalcy in the domestic politics of the U.S.A. was rather stunning and dreadful.

In medicine, the concept of normalcy is different, but also rather unclear. In the early days of roentgenology, two standard books in radiology were published by German professors. The first one was written by Rudolf Grashey in 1905 [1], the second one by Alban Köhler in 1910 [2]. Since then, numerous reprints and new editions have described the borders of the normal range and the beginning of pathology in x-rays, and today there are many other books on the same topic.

Over the decades, tens of thousands of x-ray images of every part of the body have been taken to create a catalog of normal features and of varieties of the normal. The result is an overview of normalcy and the delineation of the borderlines to pathology.

With the appearance of MR imaging in routine clinical practice, an abundance of new insights arose in clinical imaging. Radiologists were confronted with tissues and tissue changes that previously had only been accessible to, and known by, pathologists. For instance, nobody in clinical diagnostic medicine had ever seen such accurate and distinct slices of the brain as MR imaging could now produce. Physicians have had to relearn anatomy and pathology.

Therefore, MR imaging has been a great boost for publishers of anatomy books, and the market for comparative books and CD-ROMs of anatomy with imaging techniques is still surging.

As new structures became visible, image reading was transformed into an even more delicate and difficult task, and contrast behavior was unpredictable, given the multitude of parameters influencing image contrast. Once again, the borderlines of medical normalcy and its variations, which should not be assessed as disease, were unclear.

The diagnosis of multiple sclerosis is the standard example of what can happen if there is no proper knowledge of the normal range. Multiple sclerosis plaques are easily visible in MR imaging, and thus the possibility of verifying the diagnosis proved attractive to an enormous number of physicians, as well as patients and their relatives.

MR imaging revealed white-matter lesions in many patients, who were therefore diagnosed as having multiple sclerosis. However, soon it became evident that such white-matter lesions could also be observed in control groups of normal volunteers. High-signal-intensity spots, called *unidentified bright objects* (UBOs) by some authors, were detected in both healthy subjects and patients with a variety of diseases or conditions [3].

The appearance of these spots is consistent with an increased water content and changes in myelin structure. A local lesion (e.g., a cerebral edema caused by circulatory changes or breakdown of the blood-brain barrier) may result in atrophic perivascular demyelination, myelin pallor, gliosis, infarction, and/or porencephalic changes, all of which can be seen as hyperintense spots in either intermediately or T2-weighted MR images.

Initially, these spots created some confusion, but soon this was removed. Joseph Durand, a French physician from Lyons, had described such lesions already in 1843 and given them the technical terms *état lacunaire* and *état criblé*. Now they could be seen *in vivo* in patients.

Several studies showed that the finding of UBOs was common in MR imaging, but that such changes are unusual in individuals less than 40 years of age. However, the frequency increases with age. It was also found that risk factors for cerebrovascular disease and a history of brain ischemia correlated positively with the number of lesions. Smoking, a known risk factor for atherosclerosis, correlates with the increased occurrence of changes [4]. However, many of these patients and volunteers had no neurological symptoms or psychometric changes. From a health point of view, they were normal.

Whereas previously it would be usual to read about findings *consistent with demyelinating disease*, today such descriptions are (hopefully) worded far more carefully, and they are only found if the clinical history suggests multiple sclerosis or another of the possible causes for white-matter lesions. In some reports, single bright spots in the white matter of the brain are still described in the findings section, but there will not be a pathology description in the impression section of such reports. The spots are not mentioned because they are not considered to be of clinical relevance, and the verdict is *normal*.

Then the question arises as to where is the border between normalcy and pathology and when such spots should be considered pathological. This question can become extremely important if there are far-reaching consequences for the patient or, as in the following case, for a group of people who might be potential patients.

As part of a larger study, the brains of a group of deep-sea divers were examined. These divers stay and work at great depths below sea level for several weeks. Before they can return to the surface, they have to undergo decompression. It has been postulated that decompression can induce minimal brain damage, which over the years leads to permanent damage.

As with all such studies, one needs a reference group with which to compare the results of the target population. This reference group should be similar but *normal*. In the case of the divers, police officers and offshore workers were chosen as occupational groups having to fulfil the same stringent medical selection criteria; but they were not diving.

The MR imaging results were striking. The white-matter changes for the divers and the control group were 33% and 43% respectively [5]. According to the literature, such changes should be expected in no more than 20% of the population.

Obviously, the control group did not resemble the normal population, but to make sure, another randomly chosen group was examined and they revealed less than 20% of such changes, as expected. It is unclear why the offshore workers/police group showed more white-matter changes than the divers and the normal population.

However, in epidemiological terms, choosing them as a control group was wrong.

What are the lessons of this study ?

First, despite the fact that millions of MR imaging brain examinations have been performed all over the world during the last two decades, normalcy ranges still have not been set. This is true not only for the brain but also for the spine, liver, etc.

Second, selecting a control group for clinical studies may be a more difficult task than is generally thought, particularly if such a group's normalcy has not been defined. This means that the results of such studies may have to be interpreted *cum grano salis*.

Some years later, functional MR imaging of the brain became possible and fashionable. Again, the question has to be asked: what is normal and was is pathological — and what is an artifact ?

Many blood flow alterations described in functional brain imaging rely on signal-intensity changes of less than 5%.

As Gustav von Schulthess once pointed out[6]:

»... a *caveat* for fMRI: it is a very interesting technique but signal changes are but a few percent. Hence, the method is technically demanding and 'the threshold of nonsense production is low'.«

References

1.   Grashey R. Atlas typischer Röntgenbilder vom normalen Menschen. Munich: Lehmann 1905.
2.   Köhler A. Grenzen des Normalen und Anfänge des Pathologischen im Röntgenbilde. Hamburg: Gräfe und Sillem 1910.
3.   Refer to list of references cited in the article in reference 4.
4.   Rinck PA, Svihus P, De Francisco P. MR imaging of the central nervous system in divers. J Magn Reson Imaging 1991; 1: 293-299.
5.   Todnem K, Skeidsvoll H, Svihus R, et al. Electroencephalography, evoked potentials and MRI brain scans in saturation divers. An epidemiological study. Electroencephalogr Clin Neuro physiol 1991; 79: 322-329.
6.   von Schulthess G. Clinical MR in the year 2010. MAGMA 1999; 8: 133-145.

Chapter Twelve — Contrast Agents: Fundamentals

More Magnetism

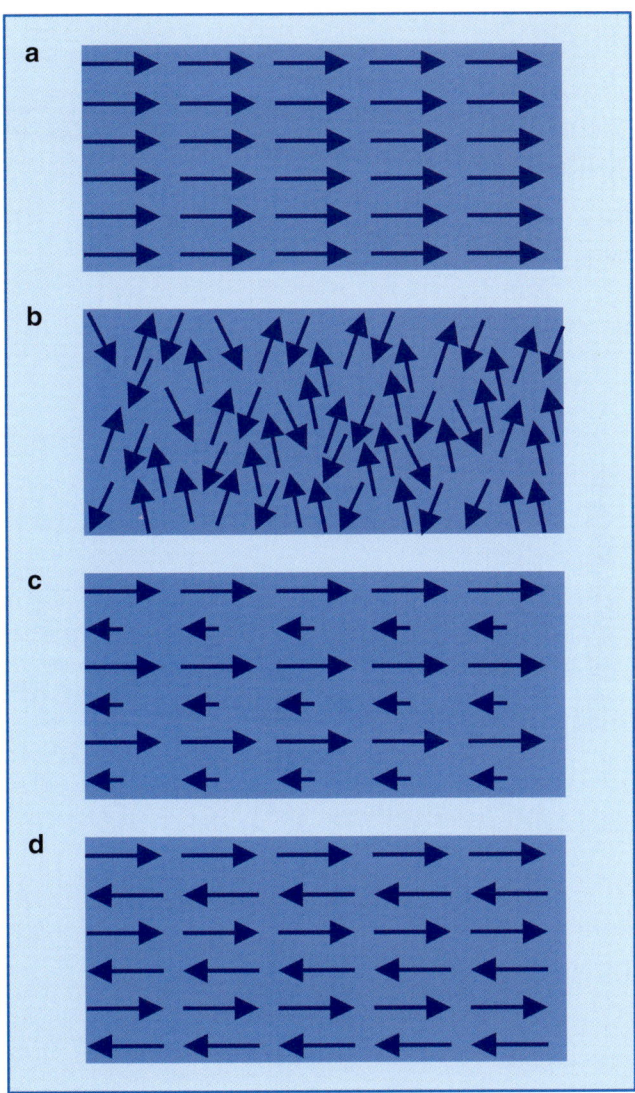

Figure 12-1:
(a) ferromagnetic material: atomic magnets are strongly coupled; (b) paramagnetic material in an external field: atomic magnets are weakly coupled; (c) ferrimagnetic material: weak overall magnetism; (d) antiferromagnetic material: atomic magnets coupled in an antiparallel manner, resulting in no magnetism.

Magnetic resonance contrast agents aim at changing signal intensity and thus image contrast. The main contrast parameters in MR imaging are proton density, the relaxation times, and magnetic susceptibility.

It is rather difficult to alter the water content of tissues. Therefore, the magnetic properties have been the major target for the development of contrast changing agents.

Magnetic *susceptibility* describes the ability of a material or substance to become magnetized by an external magnetic field.

All substances are *diamagnetic*. A strong external magnetic field speeds up or slows down the electrons orbiting in atoms in such a way as to oppose the action of the external field. These materials partly expel from their interior the magnet field in which they are placed. For diamagnetic materials, the value of susceptibility is always negative (Table 12-1).

The diamagnetism of some materials is masked either by a strong attraction (ferromagnetism) or by a weak attraction (paramagnetism).

Certain materials have *ferromagnetic* properties, among them iron, nickel, cobalt, and their alloys. Ferromagnetic materials are strongly attracted by magnets. In ferromagnetic materials, there is a strong coupling of the individual magnets, resulting in their lining up parallel to one another (Figure 12-1a).

They lose their magnetic properties when heated above a temperature known as the 'Curie point' (770° C for iron, 358° C for nickel, 1,120° C for cobalt).

Then they begin to show a kind of magnetic behavior which as called 'paramagnetic'. Many other elements and compounds are paramagnetic at all temperatures, among them oxygen, gadolinium, and manganese.

Paramagnetism is due to the presence of little colonies of atomic magnets, in which the individual magnets are weakly, if at all, bound to one another and therefore capable only of random orientation in the absence of an external field (Figure 12-1b). Paramagnetic substances are feeble in their response to an external magnetic field. *Superparamagnetic* substances have a substantially higher susceptibility.

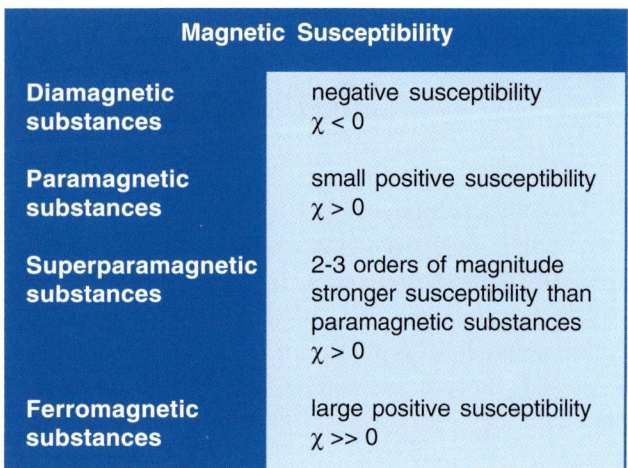

Table 12-1:
Fundamentals of magnetic susceptibility χ.

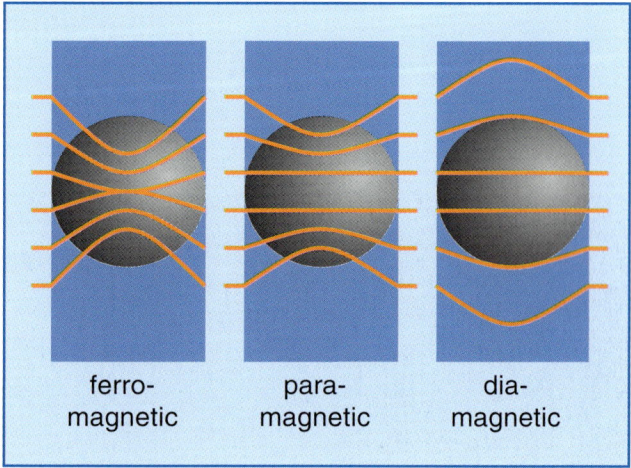

Figure 12-2:
Different magnetic compounds and outside magnetic fields.

Ferrimagnetic materials, such as ferrites, are also coupled in an antiparallel fashion, but the overall effect of the individual magnets pointing in one direction exceeds. Thus, the net effect is that of weak overall magnetism (Figure 12-1c).

Antiferromagnetic materials consist of elementary magnets coupled together in opposite directions, resulting in zero net magnetization (Figure 12-1d).

Figure 12-2 shows the response of diamagnetic, paramagnetic, and ferromagnetic substances to an outside magnetic field.

Table 12-2 summarizes the standard nomenclature of magnetic resonance contrast agent terms.

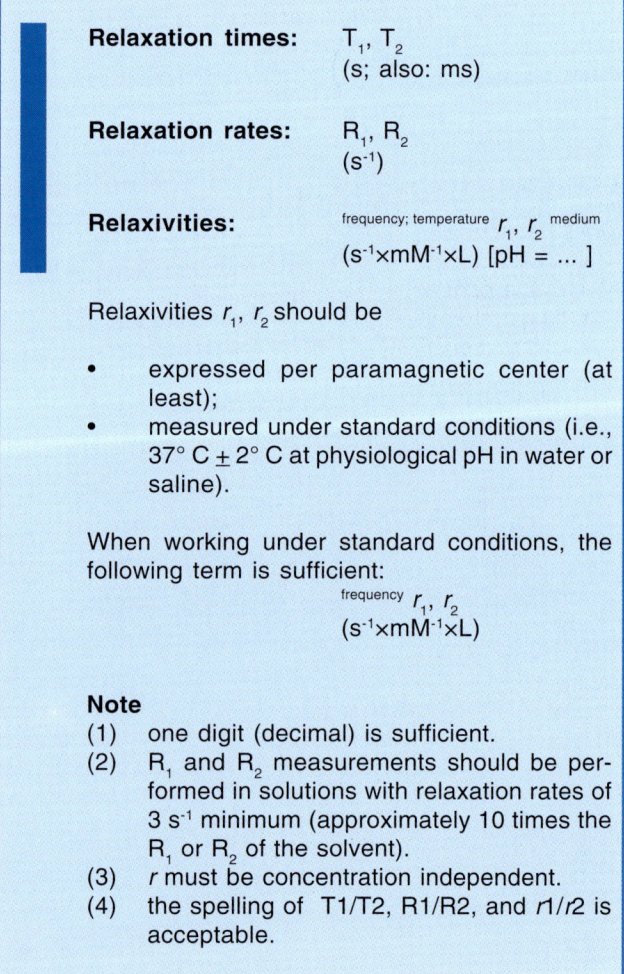

Table 12-2:
Recommendations for the Nomenclature of Magnetic Resonance Contrast Agent Terms. This standard was set in 1990 and revised in 1992 and 1996 (published in Acta Radiol 1997; 38, S1: 5).

Chapter Thirteen

Contrast Agents

General Remarks

Despite the fact that inherent contrast in MR imaging can be manipulated to a much greater extent than in other imaging techniques, certain diagnostic questions cannot be answered easily and require the application of contrast agents (Figure 13-1). In general, contrast manipulation in MR imaging by application of contrast agents is most useful when inherent contrast cannot be changed successfully.

The most important goals and requirements for the development and use of MR contrast agents are listed in Table 13-1.

Since it is nearly impossible to change the water content of tissues, contrast agents on the market or in clinical or pre-clinical trials focus upon relaxation time and susceptibility changes. Most of them are either para- or superparamagnetic. As early as 1946, in one of the first papers describing NMR, *paramagnetic catalysts* were mentioned to accelerate the T1 relaxation process[4], and were immediately translated into imaging studies after the invention of MR imaging[12].

Many efforts in contrast-agent development were channelled in certain directions more by the relative ease of chemical synthesis than by the goal of specific medical applications. Thus, the first-generation contrast agents available for clinical routine examinations today are relatively safe, good enhancers, but unspecific. This means that they do not highlight specific pathologies but rather unspecific pathological tissue changes. In many instances, the pattern of enhancement of paramagnetic contrast agents in MR imaging is very similar to that of contrast-enhanced x-ray CT. This allows the transfer and full use of the radiologist's longtime experience in the interpretation of contrast-enhanced CT scans. However, it should be taken into account that MR contrast agents behave differently from CT agents and do not in any case follow their enhancement patterns.

At present, paramagnetic contrast agents are the most frequently used. The most efficient elements are listed in Table 13-2. A different class is particulate agents. All of them come as positive or negative agents (Figure 13-2). Table 13-3 summarizes MR contrast agents currently in use or being developed.

Figure 13-1:
Nature likes mimicry; radiologists like to highlight lesions. With plain photography (or MR imaging), the object of the examination might be visible but not clearly delineated. Changing contrast with an extrinsic agent may help — for instance, painting the wall in the background or injecting a contrast agent.

Goals	Requirements
improvement of • tissue contrast • tissue characterization • overall sensitivity • overall specificity and thus improvement of • diagnosis and • therapy monitoring of function reduction of • artifacts • imaging time • overall costs	adequate • relaxivity • susceptibility • osmolality • biodistribution • tolerance • stability • elimination • metabolism • toxicity

Table 13-1:
Primary and secondary goals and requirements for the development of contrast agents for magnetic resonance imaging.

	Gd^{3+}	Mn^{2+}	Dy^{3+}	Fe^{3+}
Number of unpaired spins	7	5	5	4
Electron spin relaxation time	10^{-8} -10^{-9}	10^{-8} -10^{-9}	10^{-12} -10^{-13}	10^{-10} -10^{-12}

Table 13-2:
Some paramagnetic elements (transition metals) and their properties: gadolinium, manganese, dysprosium, and iron.

Positive Contrast Agents

The magnetic field produced by an electron is much stronger than that produced by a proton. However, in most substances the electrons are paired, resulting in a weak net magnetic field. As seen in Table 13-2, with its seven unpaired electrons and relatively long electron-spin relaxation time, gadolinium possesses the highest ability to alter the relaxation times of adjacent protons (highest relaxivity).

Paramagnetic contrast agents, with the exception of dysprosium, are called *positive* agents. The effect on T1 and T2 is similar, but since T1 of tissues is much higher than T2, the predominant effect at low doses is that of T1 shortening. Thus, tissues taking up such agents will become bright in a T1-weighted sequence (Figures 13-2 and 13-4). Figure 13-3 gives an example of a clinical case where only the application of such a positive contrast agent led to the diagnosis.

Figure 13-2:
Influence of positive (T1) and negative (T2, T2*) MR contrast agents upon signal intensity. Paramagnetic agents are mainly used to shorten T1 relaxation and thus brighten the region of interest, whereas ferro- and superparamagnetic shorten T2 and T2* and thus darken the image (gray arrows).

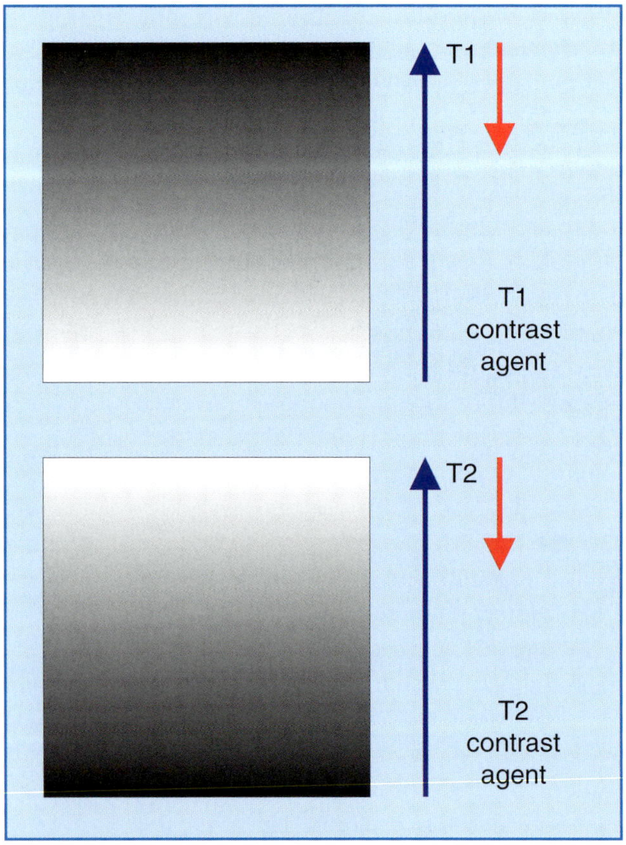

Figure 13-3:
Patient with breast cancer and recent neurological symptoms. Neither the CT images — which are not shown here — nor the MR images without contrast enhancement reliably revealed brain lesions. However, the contrast-enhanced MR imaging showed a large number of metastases. (a) and (c): precontrast T1-weighted images; (b) and (d): postcontrast T1-weighted images.

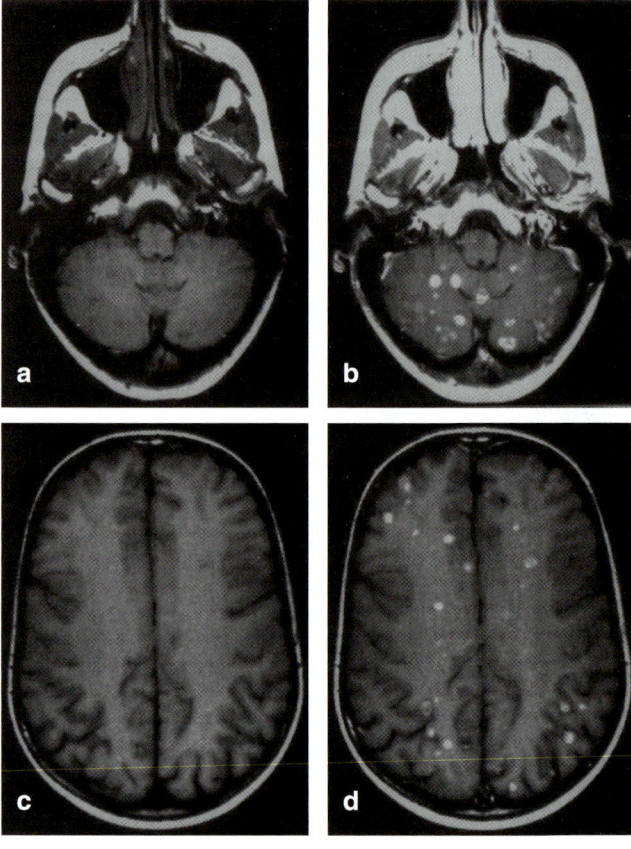

Negative Contrast Agents

Negative contrast agents influence signal intensity, usually by shortening T2* and T2. This darkens the region of interest (Figure 13-2).

Superparamagnetic and ferromagnetic agents belong to this group. Ferromagnetic agents consist of particles which show permanent magnetism. If one reduces their size, they lose their permanent magnetic characteristics and are then called *superparamagnetic* particles [31]. Depending on their particle size and coating, these compounds can also become T1 agents.

The ferro- and superparamagnetic contrast agents produce local magnetic-field gradients which disrupt the homogeneity of the local magnetic field. T2 is reduced due to the diffusion of water through these field gradients. However, their principal effect is a reduction in T2*. For this reason, the effects of such contrast agents are best observed using gradient-echo sequences where T2* effects are retained. This kind of effect is referred to as *susceptibility effect* and is field strength-dependent, with the effect increasing as the square of the field strength.

Magnetite, Fe_3O_4, is such a superparamagnetic particle. Coated with an inert resin, it can be used for oral or intravenous applications. Magnetites are subject to a specific uptake in the RES system after intravenous administration, the most simple example being uptake in Kupffer cells. It leads to a signal void in blood vessels. Superparamagnetic contrast agents can be used to enhance contrast in the liver, the spleen, lymph nodes, and other organs after intravenous application or in the abdomen and pelvis after peroral application (Figure 13-18).

Dysprosium can also be used as an intravenous negative contrast agent. Dy-DTPA-BMA is a paramagnetic bulk susceptibility perfusion agent. It possesses a large magnetic moment and a poor T1 relaxivity, despite its relationship with gadolinium and manganese [10].

Figure 13-4:
Influence of a positive (T1) and negative (T2, T2*) contrast agent upon signal intensity (SI).
T1 relaxation is accelerated by positive agents, and the spins recover faster. Therefore, at a given TR, signal intensity (yellow curve) is higher than in the same tissue without contrast agent (red curve). Only a T1-weighted SE sequence with short TE highlights this contrast enhancement.
With negative contrast agents, T2 relaxation is accelerated; signal intensity is lower (blue curve).

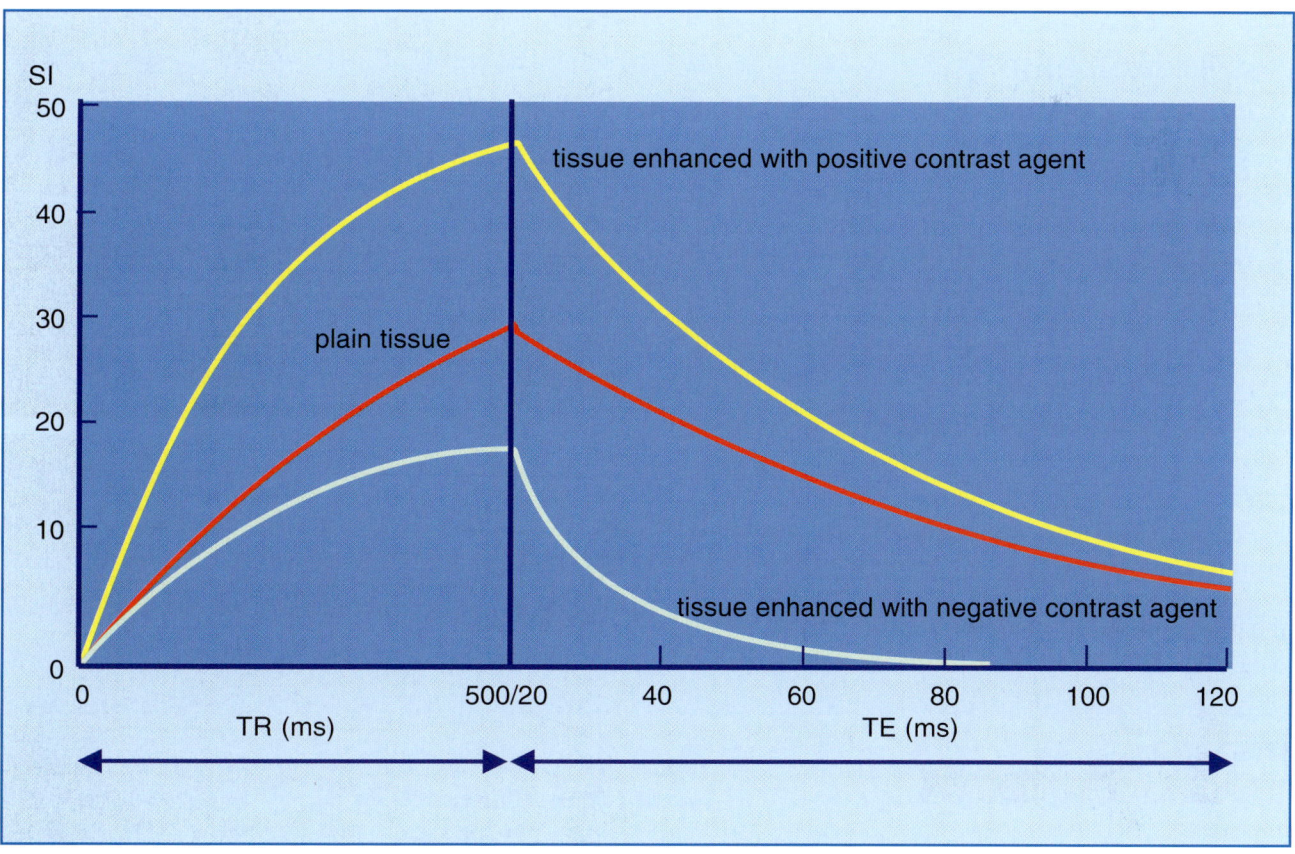

Short Name	Generic Name	Trade Name*	Enhancement Pattern
Extracellular Fluid (ECF) Space Agents**			
Gd-DTPA	gadopentetate dimeglumine	Magnevist	positive
Gd-DOTA	gadoterate meglumine	Dotarem	positive
Gd-DTPA-BMA	gadodiamide injection	Omniscan	positive
Gd-HP-DO3A	gadoteridol injection	ProHance	positive
Gd-DTPA-BMEA	gadoversetamide	Optimark	positive
Gd-DO3A-butriol	gadobutrol	Gadovist	positive
Gd-BOPTA/Dimeg	gadobenate dimeglumine	MultiHance	positive
Targeted / Organ-specific Agents***			
Liver			
Mn-DPDP	mangafodipir trisodium	Teslascan	positive
Gd-EOB-DTPA	gadoxetic acid	Eovist	positive
Gd-BOPTA/Dimeg	gadobenate dimeglumine	MultiHance	positive
AMI-25	ferumoxides (SPIO)	Endorem / Feridex	negative
SH U 555 A	ferrixan (SPIO)	Resovist	negative
Lymph Nodes			
—	gadofluorine-8		positive
AMI-227	ferumoxtran (USPIO)	Sinerem / Combidex	negative
AMI-25	ferumoxides (SPIO)	Endorem / Feridex	negative
Myocardium / Necrosis			
Gd-DTPA mesoporphyrin	gadophrin		positive
Blood Pool Agents			
NC-100150	PEG-feron	Clariscan	positive
MS-325		Angiomark	
—	gadomer-17		positive
—	gadofluorine-8		positive
MnHa/PEG			
AMI-227	ferumoxtran (USPIO)	Sinerem / Combidex	negative
Gd-BOPTA/Dimeg	gadobenate dimeglumine	MultiHance	positive
Enteral Agents (orally or rectally administered)			
Gd-DTPA	gadopentetate dimeglumine	Magnevist enteral	positive
—	ferric ammonium citrate	Ferriseltz	positive
—	manganese chloride	LumenHance	positive
—	gadolinium-loaded zeolite	Gadolite	positive
OMP	ferristene	Abdoscan	negative
AMI-121	ferumoxsil (SPIO)	Lumirem / Gastromark	negative
PFOB	perfluoro-octylbromide	Imagent-GI	negative
—	barium sulfate suspensions		negative
—	clays		negative
Ventilation Agents			
Perfluorinated gases			
Gadolinium-based aerosols			
Hyperpolarized gases (^{3}He, ^{129}Xe)			
Oxygen			

Table 13-3:
Classification of some magnetic resonance contrast agents approved, or to be approved, for clinical use. Some of the agents mentioned have been withdrawn from the market; there are numerous other agents in development. * ™ or ®; **with high local concentrations, negative contrast can be achieved (e.g., first-track bolus); ***all ECF space agents are also kidney-specific agents.

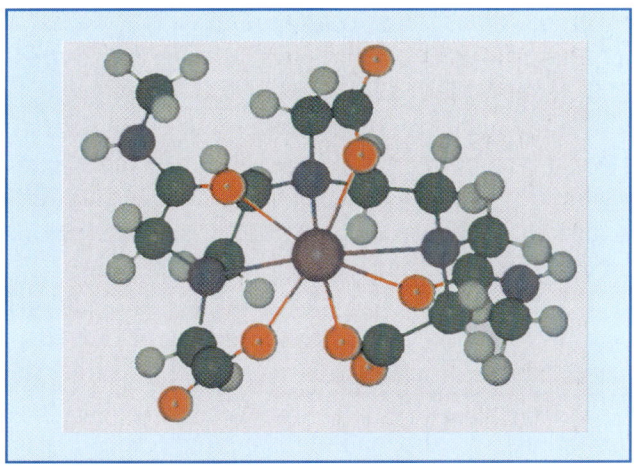

Figure 13-5:
The root of the word chelate means 'claw' in Greek. The gadolinium atom is kept by the rest of the molecule like a claw so that it cannot escape. This example shows a graphic depiction of the gadodiamide molecule, with the gadolinium atom drawn in dark magenta in the center. Many other contrast agents follow a similar pattern.

Gadolinium-based Extracellular Fluid Space Contrast Agents

Low molecular weight paramagnetic contrast agents distribute into the intravascular and extracellular fluid space of the body. Thus, they are also known as ECF space agents. Among the positive contrast agents, they are the most commonly used (Table 13-3). Their effect is caused by the metal ion in their center which contains unpaired electrons. Additional details on their safety are found in Chapter 18.

Chelates. Because the metals suitable as relaxation agents and their salts are rather toxic, they have to be bound in stable complexes in which they are usually kept until the contrast agent is excreted (Figure 13-5)[18]. Among these complexes are DTPA, DTPA-BMA, DTPA-BMEA, DOTA, and HP-DO3A. Bound to them, they form low molecular weight, water-soluble contrast agents which principally are excreted through the kidney. Gd-DTPA and Gd-DOTA are high-osmolality ionic agents, whereas Gd-DTPA-BMA and Gd-HP-DO3A possess lower osmolality and are nonionic. In this context, the terms 'ionic' and 'nonionic' are widely used; 'charged' and 'neutral' would be better terms.

Further developments include Gd or other paramagnetic metals bound to, e.g., BOPTA, EOB-DTPA, and similar ligands which are slightly more lipophilic and excreted via the kidney but also via the liver.

Gd-compounds are considered safe with few side effects. However, some severe anaphylactoid reactions have been reported. Proper supervision of the patient after injection, and access to an intensive care unit, must therefore be guaranteed.

Timing. The uptake of these agents is relatively fast (see Figure 16-6). Conventional imaging exploits the uptake after 4-5 minutes; often image acquisition already starts after two minutes. Depending on the tissue, the local enhancement peak in the extracellular space is reached after 5-30 minutes after injection.

Imaging parameters. Because paramagnetic contrast agents are T1-agents, their effect is most pronounced on T1-weighted MR images, for instance on spin-echo images with short repetition and echo times, and gradient-echo images with short repetition times and high flip angle (50°-90°).

Proton-density-weighted images will still show enhancement. However, the agents lose most of their efficiency on T2-weighted pictures; in many cases, even an isointense behavior can be observed (Figures 13-4 and 13-6).

In general, imaging protocols should include pre-contrast T1-weighted images to exclude or differentiate high signal intensity pathologies such as hematoma, and T2-weighted images to exclude pathologies, such as small edematous white-matter lesions. Pre- and post-contrast T1-weighted images only cannot be recommended in brain studies trying to rule out unknown pathology.

Uptake. The diagnostic development of low molecular weight paramagnetic contrast agents was primarily focused on their use in lesions of the central nervous system (CNS), an indication in which plain MR imaging had found its first major field of application.

These agents do not cross the intact blood-brain barrier (BBB). In the healthy CNS, clear enhancement is only seen in regions without this barrier, such as the choroid plexus. Normal enhancement in the brain is also seen in the pituitary gland and infundibulum, the cavernous sinus, dura mater, and nasal mucosa. Vessels may also enhance, particularly during the first pass of the contrast agent (Figure 13-7).

Pathological breakdown or absence of the BBB allow paramagnetic contrast agents to cross into the extracellular space and to alter T1 relaxation locally. This occurs in a variety of pathologies such as tumors, infarctions, infection, acute demyelination, etc.

Figure 13-8 (top row) is an example of the ability of an intravenous Gd contrast agent to enhance brain tumors. The application of a contrast agent allows the delineation and definition of tumor extent (with certain limitations, as explained in the bottom row of Figure 13-8 bottom row).

Dose. The regular clinical formulation of all Gd-based ECF-space agents contains gadolinium at a concentration of 500 mM. Increasing the dose of the agents above the recommended normal dose (i.e., higher than 0.1 mmol/kg body weight = 0.2 ml/kg body weight) may have both beneficial and unwanted effects (Figure 13-9). Depending on the field strength, it may facilitate the detection of small CNS lesions with mini-

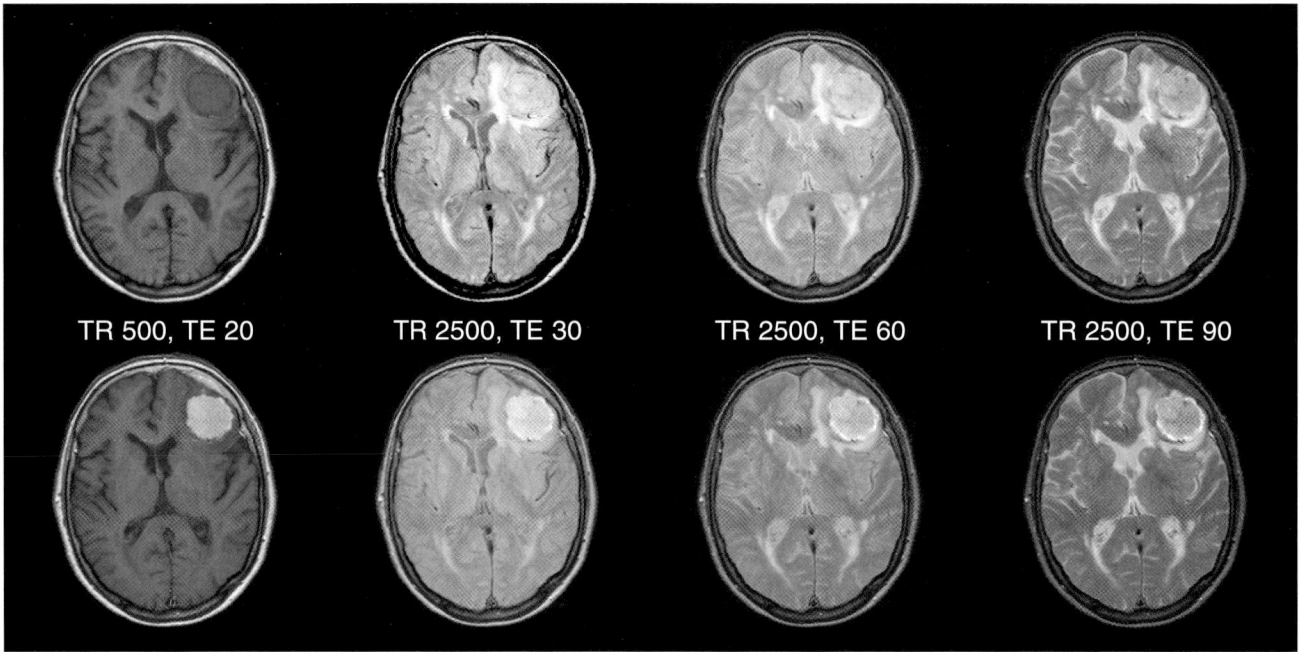

TR 500, TE 20   TR 2500, TE 30   TR 2500, TE 60   TR 2500, TE 90

Figure 13-6:
Multiple SE sequence before (top row) and after (bottom row) intravenous contrast application of a gadolinium compound. The images to the left are heavily T1-weighted, then ρ-weighted, and to the right increasingly T2-weighted.
The patient has a huge meningioma in the left frontal lobe, which is easily visible on the non-enhanced images, mostly because of its mass effect and the bright surrounding edema on T2-weighted images. However, this case is a good example of the enhancement pattern of Gd contrast agents. This kind of tumor enhances brightly on T1-weighted images; there is still enhancement on ρ-weighted images. T2-weighted images, however, show the same contrast pattern before and after injection of the agent.
If the meningioma or similar enhancing lesions are very small and no indirect signs of lesions can be found, only contrast enhancement will reveal the pathology (see Figure 13-3). You can simulate the case shown in this figure with ***MR Image Expert***.

Figure 13-7:
Top: Child with small ependymoma (a) before contrast application, (b) after contrast application. The contrast agent enhances the tumor, but also other highly vascularized parts of the head, among them the pituitary gland and infundibulum, bone marrow, nasal mucosa, and blood vessels. Bottom: Adult brain (c) before contrast application, (d) after contrast application; normal contrast enhancement in the choroid plexus and the superficial veins.

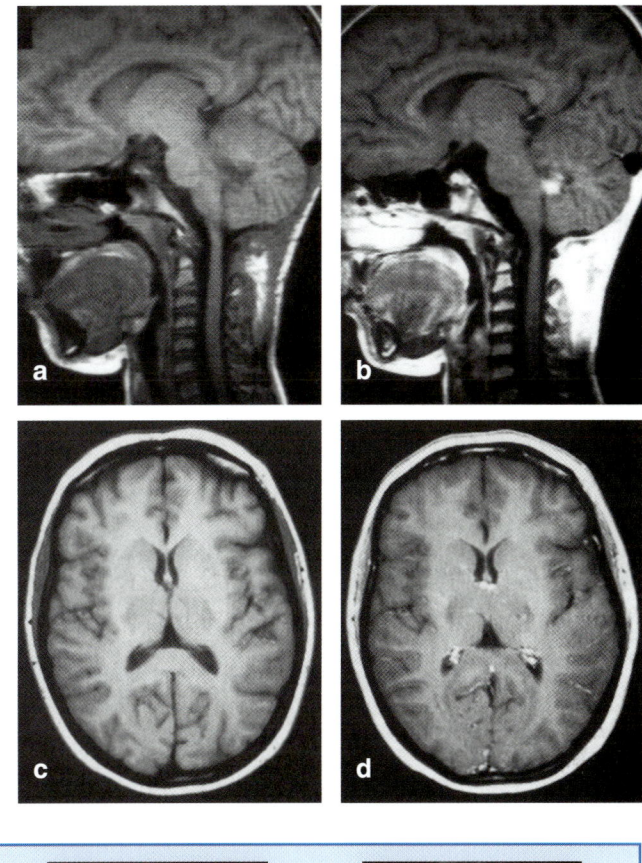

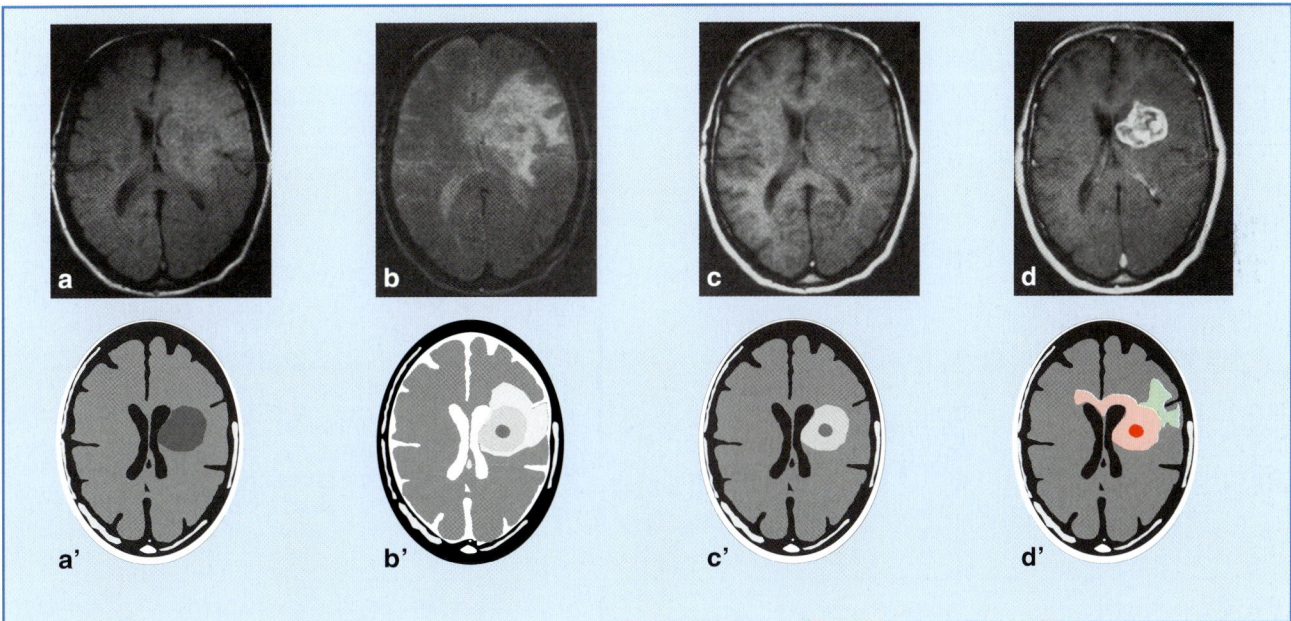

Figure 13-8:
Top row: Malignant brain tumor. (a) proton-density-weighted image; (b) T2-weighted image; (c) T1-weighted precontrast image; (d) T1-weighted postcontrast image. Although it is obvious that a large mass displaces the lateral ventricles, an exact delineation of the tumor is impossible on the first three images. Edema is well seen on the T2-weighted image, but the dark tumor areas are poorly delineated. Only after the administration of Gd-DTPA does the tumor become bright and its active parts are well delineated.
Bottom row: The contrast agent only enhances absence or breakdown of the blood-brain-barrier and high-vascularity lesions. This graphic depiction shows (a') and(b') T1- and T2-weighted precontrast images of a highly malignant tumor similar to the one seen in the top row; (c') the T1-weighted postcontrast image; and (d') the actual microscopic tumor extension through the corpus callosum, which is not depicted by the MR images. The enhanced image shows only the tumor region, with the effect on the blood-brain barrier.

Lesion Type (selected brain lesion)	Enhancement	Similarly behaving pathologies	Comments	Dose*
Intraparenchymal tumors **Metastases**	yes	Glioblastoma, abscess.	Large metastases require no CA; preoperative evaluation of number of metastases requires CA; patient with evidence of more than two metastases in separate locations usually requires no CA. For early detection of small metastases.	0.1 0.3 0.1
Gliomas (low grade) Astrocytoma I and II	no	Ependymoma and subependymoma (may enhance); encephalitis, multiple sclerosis, oligodendroglioma, infarction.		
Gliomas (high grade) Anaplastic astrocytomas, glioblastomas	yes	Hemangioblastoma, primary lymphoma, anaplastic oligodendroglioma, germinoma, medulloblastoma, metastases, abscess, teratoma.	Often not necessary for lesion detection, but for delineation and specificity. High dose 30-minute delayed images may be of advantage.	0.1 0.3
Meningeal tumors **Leptomeningeal carcinomatosis; carcinomatous meningitis**	yes	Meningitis, postoperative or posthemorrhagic changes, inflammation (sarcoidosis).	Enhancement focal/nodular (sometimes also patchy or linear), see *infection*. High dose if suspicion persists.	0.1 0.3
Meninigioma	yes	Meningeal metastases.		0.1
Pituitary microadenomas	no		Contrast agent may lead to loss of contrast; macroadenomas do enhance.	0.1
Cerebellopontine angle and internal auditory canal lesions **Acoustic neurinoma**	yes	Meningioma.	Very strong enhancement. Half dose may be sufficient.	0.1 0.05
Schwannoma	yes		Enhancement depends on predominant cell type of the tumor.	0.1
Inflammation **Demyelinating disease (e.g., multiple sclerosis)**	no yes		In case of BBB breakdown (= active plaques).	0.1
Acute disseminated encephalomyelitis	possible	Multiple sclerosis.		0.1
Infection	possible		Enhancement: in case of BBB breakdown; rim (capsule) in abscesses; homogenous linear in meningitis. Nodular or ring-shaped enhancement.	0.1
Toxoplasmosis **Tuberculoma**	yes yes			0.1 0.1
Ischemia and infarction	possible		Depending on stage, cortical patchy or linear; chronic infarcts do not enhance.	0.1
Cystic lesions	no	Epidermoid cysts.	Colloid cysts show variable enhancement (e.g., homogenous or ring).	0.1

Table 13-4:
Enhancement patterns of selected pathologies of the brain. Both plain T1- and T2-weighted images are necessary before the contrast-enhanced static T1-weighted study or studies, if different views are acquired. Modified from Rinck and Myhr [21]. For more details, check specified reference books or the appropriate handbooks.

BBB = blood-brain barrier; CA = contrast agent.

* Dose in mmol/kg body weight.

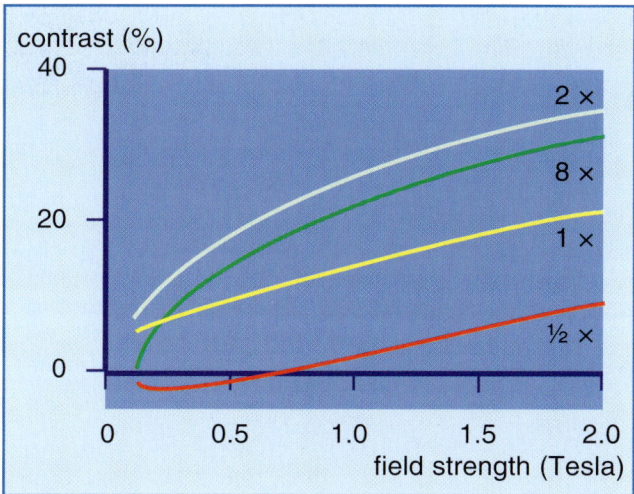

Figure 13-9:
Simulation of the influence of different tissue concentration of Gd (and thus injected dose) and magnetic field strength upon the signal. Before enhancement the contrast between glioblastoma and white matter is negative (not depicted).
The curves in color correspond to the contrast between glioblastoma and white matter after enhancement (red = half dose; yellow = regular dose of 0.1 mmol/kg body weight; light green = double dose; green = octuple dose). Only the regular and double dose give rise to sufficient contrast at all fields.

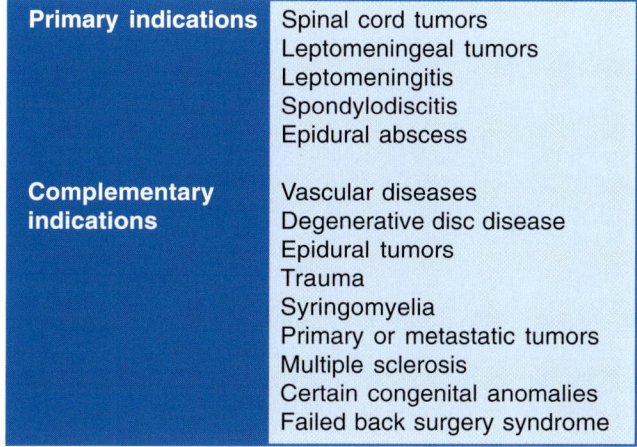

Table 13-5:
Indications for contrast applications in the spine.

mal blood-brain barrier breakdown. In other types of pathology, it may lead to loss of contrast. This is because a T2 shortening remains and can take over primary influence upon image contrast. The same holds for cutting the contrast dose, which in most cases is counterproductive.

Indications. Among other applications, Gd compounds have been found to be especially useful for increasing the detection rate of metastases and small tumors, and for improving tumor classification [5,6,8,21], the latter by allowing the differentiation of vital tumor tissue (well perfused versus impaired or absent blood-brain barrier) from central necrosis and from surrounding edema or macroscopically uninvolved tissue.

The contrast-enhancing effect of the contrast agent combined with the ease of demonstrating a lesion in different planes (sagittal, coronal, and transversal) with MR imaging, has proven to be of use in preoperative and pre-irradiation planning, as well as in follow-up during and after treatment.

Isointense benign tumors like meningiomas and hamartomas are among the major indications for paramagnetic contrast agents. Acoustic neurinomas, in particular small ones within the internal auditory canal, are clearly enhanced. In malignant tumors, contrast agents can help in delineating tumor and edema. Absence of enhancement is sometimes of as great a value as its presence, e.g., when distinguishing a low-grade astrocytoma at the cortical surface from a meningioma or small vascular white-matter lesions from metastases. The case depicted in the ***MR Image Expert*** Tutorial on the next page is an example of a non-enhancing lesion.

Tables 13-4 and 13-5 give an overview of the most common indications for contrast application in the brain and the spine, respectively.

Applications outside the CNS include the musculoskeletal system, ear-nose-throat diseases, the heart, kidneys and adrenals, gynecological diseases, lymphomas, joints, and the breast. Details can be found in numerous books on the clinical applications of MR imaging.

One of the approaches to increase specificity was the exploitation of faster imaging techniques. These techniques reduced imaging times from several hundred seconds per image first to below 100 seconds, then to under 10 seconds. Now the uptake of contrast agents in different organs could be followed. Details of such techniques are described in Chapter 16.

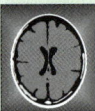 You can experience interactively the influence of contrast agents with **MR Image Expert**.

Lesion Enhancement with a Contrast Agent

Let's compare what happens if you inject a gadolinium-based contrast agent in a patient with a tumor.

- Select *File/Open* or the *Brain* icon in the **MR Image Expert** menu bar. Double-click on the file **1.5 T: Meningioma, plain, transversal slice**. Adjust the parameters of the image to TR = 450 ms, TE = 20 ms and click OK. A T1-weighted spin-echo image with these parameters will appear on the screen.

- Select *Draw/Spin Echo* or the *SE* icon three times, set TR = 2000 ms, and TE = 30, 60, and 90 ms, respectively. Each time you click OK.
  On these images, TE increases. The first one is intermediately (proton density) weighted, the other two are increasingly T2-weighted.

On all four images, you see a huge left frontal meningioma. Because it is so huge, the tumor is well visible and relatively well defined by the surrounding edema. The edema is nicely seen on the more T2-weighted images. If you just follow the signal-intensity changes of the meningioma, you see that its contrast versus gray and white matter is minimal.

Let's get some data on this observation. By using the *ROI* command, you easily can find out the T1, T2, and proton-density values of the tumor.

We have images of the same patient with a similar slice through the brain after the injection of the contrast agent:

- Select *File/Open* or the *Brain* icon and double-click on the file **1.5 T: Meningioma, postcontrast, transversal slice**. Create four images, as described above. Set for the first image TR = 450 ms, TE = 20 ms and for the following images TR = 2000 ms, and TE = 30, 60, and 90 ms, respectively.
  Adjust the window levels of the images until you like the contrast.
  You now see the tumor enhancing intensively on the T1-weighted image. Even the intermediately weighted image reveals minor enhancement, whereas the T2-weighted images do not enhance any more.

Now you can measure the T1, T2, and proton-density values with similar regions of interest as you did before. You will realize that T1 has dropped to less than one third of its original value, T2 has dropped by approximately 25%, and proton density has stayed the same.

Enhancement of an Artificial Lesion

Small meningiomas are isointense tumors. Whatever pulse sequence you choose, their signal intensity will be similar to that of the surrounding tissue. If they are very small, it is extremely difficult to discriminate them.
After the application of a contrast agent, they strongly enhance.

- Select *File/Open* or the *Brain* icon in the **MR Image Expert** menu bar. Double-click on the file **1.5 T: Normal brain, transversal slice**. Adjust the parameters of the image to TR = 450 ms, TE = 20 ms and click OK. Zoom the image by using the *Zoom* icon.

- Select *ROI/Freehand*. Position the cursor at the anterior pole of the right frontal lobe of the brain, press and hold left mouse button, draw a small meningioma, and release the button.
  When you have finished a dialog box is displayed. Select *Lesions*. Set ρ (rho) = 68%, T1 = 1350 ms, and T2 = 94 ms. Replace *Lesion 1* by typing in *Meningioma* and click OK.

MR Image Expert redraws the image. The meningioma you just defined is hardly visible on this T1-weighted image.

- Zoom out. Select *Draw/Spin Echo* three more times, set TR = 2000 ms, and TE = 30, 60, and 90 ms, respectively. Each time you click OK in the *Spin Echo* menu box, a new image is displayed.

TE increases and so does the T2-weighting of the images. The meningioma you just defined stays hardly visible.

What happens if we inject 0.1 mmol per kg body weight of a Gd-based contrast agent ? T1 and T2 relaxation times change.
In the table on page 159 you will find the values we need.

- Double-click, using the left mouse button, on the meningioma. The dialog box is displayed. Change T1 to 400 ms and T2 to 70 ms. Proton density does not change.

You will immediately see that the meningioma is highlighted on the T1-weighted image, but there is little change of contrast on the T2-weighted images.

Changing the Dose of a Gd-based Contrast Agent

Let's create another example: small metastases are extremely difficult to visualize.

- Clear your screen, create, as before, a T1-weighted and three increasingly T2-weighted images and draw a small lesion in the left deep white matter in the same way we have discussed before. Use the following values: T1 = 920 ms, T2 = 72 ms, rho = 72 %. This lesion is nearly invisible on all four images.

Now we inject a contrast agent at half the regular dose, i.e., 0.05 mmol/kg body weight. T1 is shortened to 412 ms, T2 to 60 ms. On the T1-weighted image, the lesion becomes faintly visible.

If we use the regular dose of 0.1 mmol/kg body weight, T1 drops to 205 ms, TE to 50 ms. Now the lesion is distinctly visible.

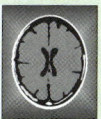

Let's double the dose: T1 = 102 ms, T2 = 44 ms. The lesion becomes slightly more visible, but the diagnostic information is the same.

Always be aware that these images are simulations. In reality, numerous factors add to the uptake of a contrast agent.

An increase of dose does not necessarily increase contrast proportionally; on the contrary, it may decrease contrast. This behavior depends on the field strength of the MR imaging equipment.

Do not increase or decrease the dosage below or above the dose recommended by the manufacturer for a certain indication.

Tissue/pathology	T1 (ms)	T2 (ms)	ρ (%)	T1 (ms)	T2 (ms)	Proton density (ρ) and T1 and T2 relaxation time values of some tissues at 1.5 Tesla.
White matter	582	73	72			
Gray matter	980	90	82	after 0.1 mmol/kg		Please note that the values are measured and adjusted for use in **MR Image Expert**.
Cerebrospinal fluid	4000*	1900*	100*	b.w.		
Brain edema	1150	140	88	i.v. Gd:		
Infarction	1000	95	88			
Multiple sclerosis plaque	1400	140	84			
Meningioma	1350	94	68	400	70	* due to partial volume effects, usually lower.
Acoustic neurinoma	1450	86	70	325	68	
Glioblastoma	1300	170	81	470	85	
Small brain metastasis	920	72	72	205	50	

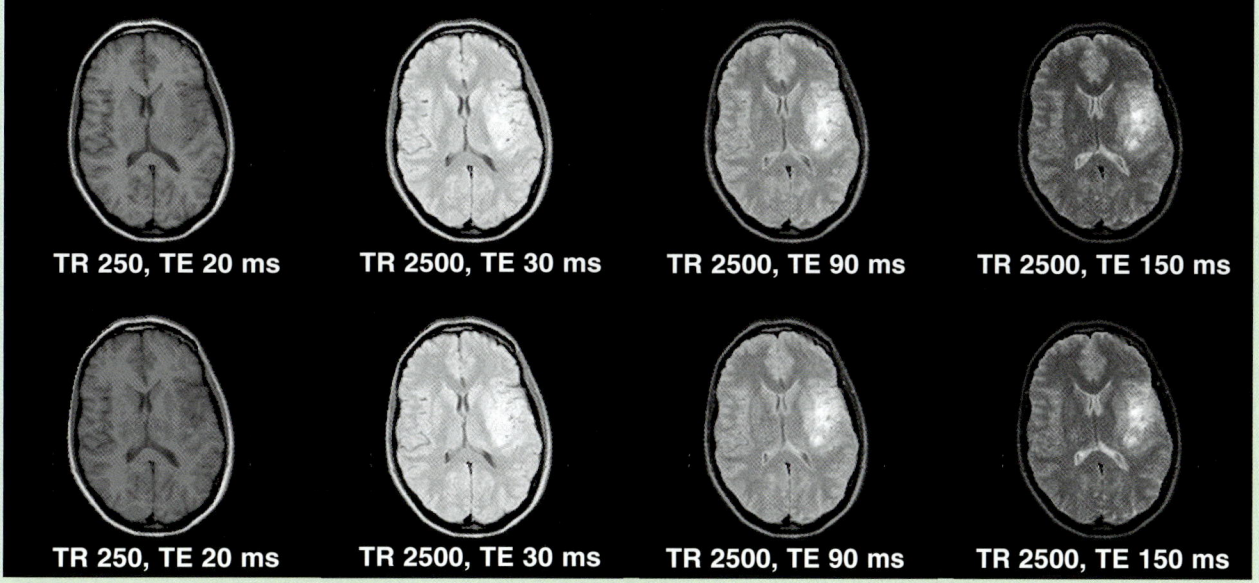

Figure 1:
Non-enhancing lesion (1.5 T: Old MCA infarction). Upper row before, lower row after contrast agent injection.

Targeted and Organ-specific Contrast Agents

As mentioned, the low molecular weight contrast agents available for clinical routine use today are unspecific agents. They can only be considered as targeted to the kidney because they are excreted by glomerular filtration. Targeting in this way is described as 'passive' targeting.

One of the ultimate goals of contrast agent development for magnetic resonance imaging is the identification of specific tissue- or pathology-seeking compounds, which would target pathological sites actively.

To optimize them, there are at least three parameters to be evaluated:
1. improvement of tolerance — although tolerance is already very good. This includes chemical and biological inertness, as well as complete elimination from the body;
2. improvement of the enhancing effect;
3. achievement of high local concentrations as a result of selective distribution in the body (organ- or pathology-specific tracers). This last point is very important.

For targeting organs and higher disease specificity, iron oxides and liposomes have attracted particular interest.

Iron Oxides

Iron-oxide particles are incorporated into cells of the reticuloendothelial system (RES) through phagocytosis. This opens a selective access route to the liver, spleen, lymph nodes, and bone marrow. They can also be applied for receptor and antibody imaging, as well as perfusion imaging of the heart and the brain. These agents can either be positive or negative enhancers, depending on particle size, composition and concentration, saturation magnetization of the material, and hardware of the imager and pulse sequences used. Their biodistribution is determined by size, shape, charge, hydrophilicity, chemical composition, and surface coating.

The majority of compounds are polydisperse (more than one size population of iron oxide crystals) and polycrystalline (particle aggregates). Actively targeted iron oxides preferably contain smaller superparamagnetic labels; they are monodisperse (only one size population of iron oxides) and monocrystalline (= MION, i.e., each particle consists of only one crystal).

Liposomes

Liposomes are another group of particulate contrast agents. Paramagnetic ions can either be encapsulated in the aqueous compartment of the liposomes or be linked to their lipid bilayers. Both manganese and gadolinium chelates have been attached to liposomes and studied preclinically. More sophisticated liposome compounds have been proposed, among them phospholipid spin-labelled and amphipathic chelate complexes.

Liver

The liver has been selected as the primary organ for developing passive targeting compounds. Table 13-6 gives an overview of liver contrast agents already on the market and in preclinical or clinical phases of development. Different possible enhancement patterns are depicted in Figure 13-10.

Aside from the vascular structures, either the hepatocytes or the RES can be targeted. Vascular structures and highly vascularized lesions are commonly highlighted by dynamic examinations with the conventional low molecular weight contrast agents.

Among the pharmaceuticals developed for hepatobiliary uptake are Mn-DPDP[25], Gd-BOPTA[26], and Gd-EOB-DTPA.

Mn-DPDP (Figures 13-12 and 13-13) is a positive liver-specific agent taken up preferentially by the hepatocytes. The contrast enhancement seems to be connected to a limited release of manganese ion. The enhancement is long-lasting and can be achieved with such low doses as 10 µmol/kg body weight.

Gd-EOB-DTPA and Gd-BOPTA/Dimeg (Figure 13-11) are two positive gadolinium-based agents with lipophilic side groups. Gd-EOB-DTPA is a targeted liver agent, whereas Gd-BOPTA/Dimeg is a multipurpose contrast agent.

Iron-oxide particles and liposomes are aimed at reticuloendothelial uptake[19,20]. Iron particles can influence both T1 and T2/T2* relaxation times. Used intravenously, iron-oxide particles such as SPIO (superparamagnetic iron oxides), and USPIO (ultrasmall superparamagnetic iron oxides) should possess a particle size smaller than 50 nm in diameter so that they are not entrapped during their passage through the lungs (Figure 13-14).

The receptor agents, among them the one targeted to asialoglycoprotein receptors, may allow the differ-

Vascular	
Gd-DTPA	Magnevist
Gd-DOTA	Dotarem
Gd-DTPA-BMA	Omniscan
Gd-HP-DO3A	ProHance
Gd-DTPA-BMEA	Optimark
Gd-DO3A-butriol	Gadovist
Gd-BOPTA/Dimeg	MultiHance

Hepatobiliary	
Mn-DPDP	Teslascan
Gd-EOB-DTPA	Eovist
(Gd-BOPTA/Dimeg	MultiHance)
Arabinogalactan-USPIO	—

Reticuloendothelial	
Iron-oxide particles	
SPIO, USPIO, MION	e.g. Endorem/Feridex, Resovist
Liposomes	

Table 13-6 (left):
MRI liver contrast agents on the market or in development.

Table 13-7 (bottom):
Pathologies of the liver and accumulation of MR contrast agents in target areas. Abbreviations: HBA, hepatobiliary agents; RES, agents targeting the reticuloendothelial system; + = accumulation in target area; - = no accumulation in target area.

Target Pathology	Accumulation of Contrast Agents
Metastases	HBA +; RES +
Hepatocellular carcinoma	HBA accumulate in well-defined tumors; RES -
Adenomatous hyperplasia	HBA +; RES +
Adenoma	HBA +; RES +
Focal nodular hyperplasia	HBA +; RES +
Cavernous hemangioma	Mn-DPDP -; AMI-25 + in late images
Hepatic cysts	HBA -; RES -

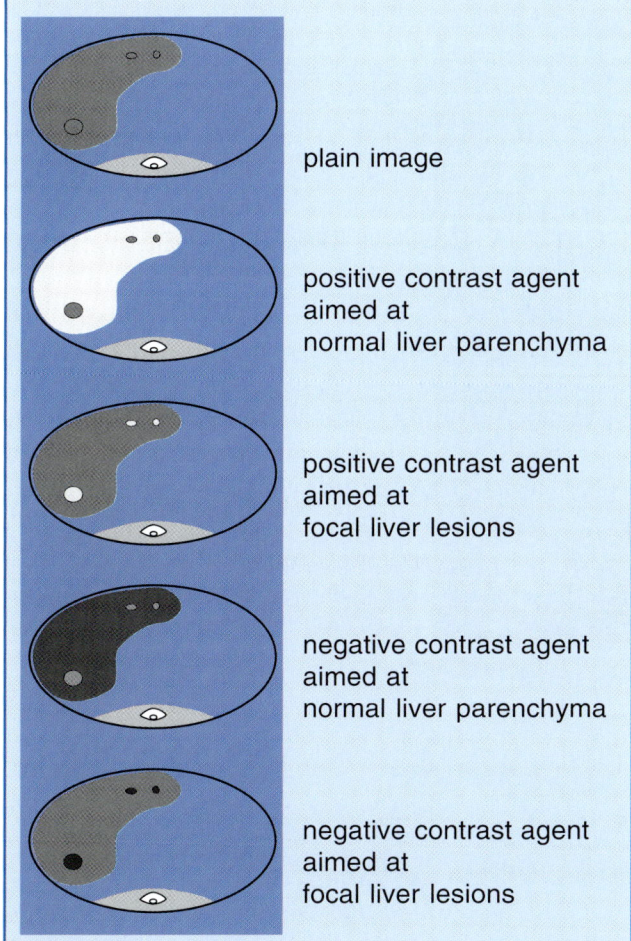

plain image

positive contrast agent aimed at normal liver parenchyma

positive contrast agent aimed at focal liver lesions

negative contrast agent aimed at normal liver parenchyma

negative contrast agent aimed at focal liver lesions

Figure 13-10:
Enhancement patterns of different liver contrast agents (modified from Leander[13]).

entiation between malignant liver tumors such as hepatocellular carcinoma and benign tumors such as focal nodular hyperplasia or adenoma.

Another group of possible liver agents is liposome-based compounds, such as liposomal gadolinium chelates [11,26]. These compounds reveal sustained intravascular contrast enhancement of vascular structures, the liver, and the spleen.

Table 13-7 summarizes accumulation sites of contrast agents targeted at the liver.

The spleen, pancreas, bone marrow, lymph nodes, adrenals, and heart, as well as inflammations and specific tumors, have been proposed as additional target regions for some of these agents.

Spleen

It is very unlikely that specific contrast agents will be developed for the spleen since the agents for liver imaging also enhance the spleen. Iron-oxide particles improve the detection of focal splenic metastases and the differentiation between benign and malignant splenomegaly (e.g., lymphoma or leukemia).

Pancreas

Mn-DPDP was primarily designed for imaging of the liver. The step to using Mn-DPDP as a selective contrast agent for the pancreas was taken after pancreatic enhancement, following infusion of Mn-DPDP, was noted. The results indicate that Mn-DPDP improves MR imaging of pancreatic tumors [7,24].

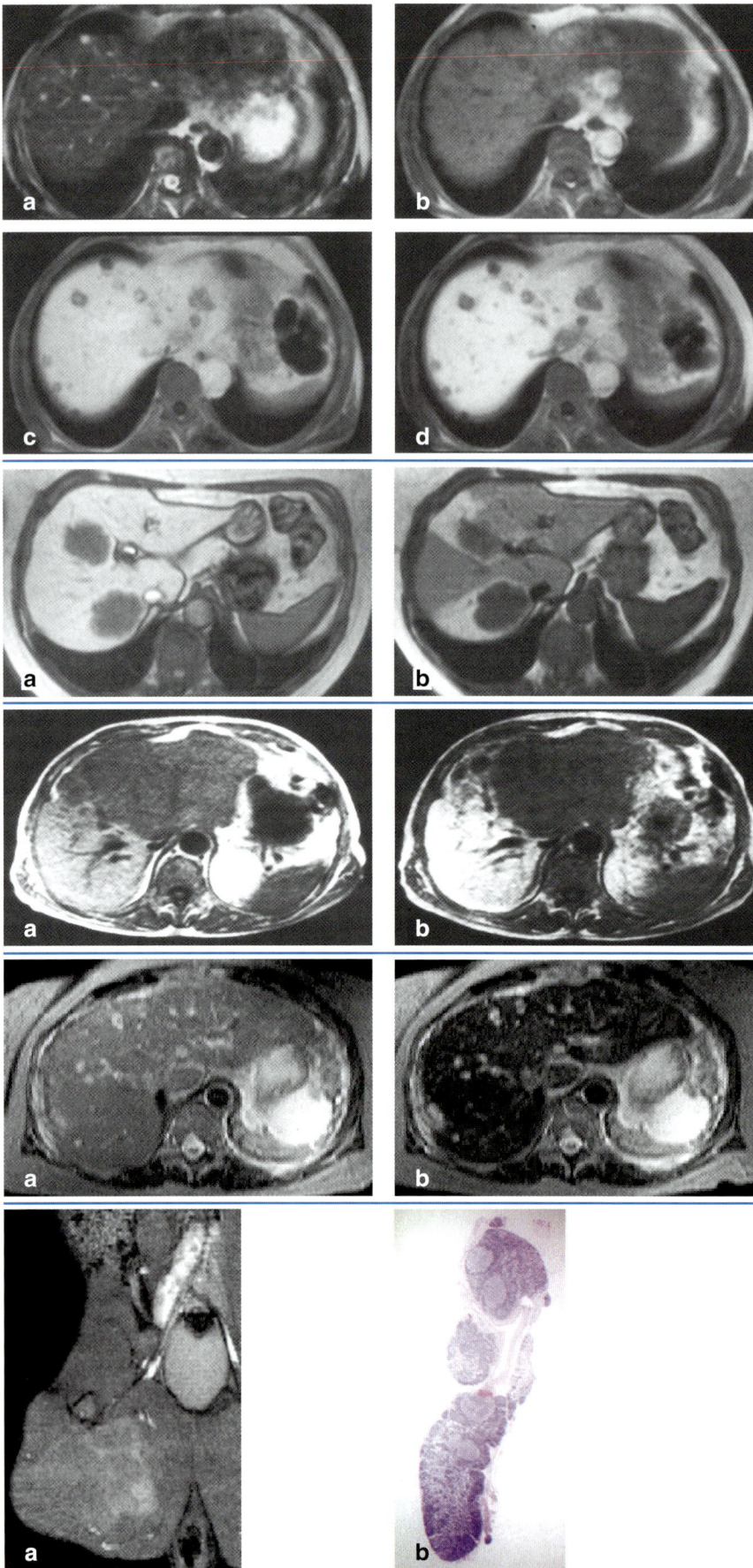

Figure 13-11:
Gadolinium-BOPTA in liver metastases of a pancreatic tumor. (a) plain T2-weighted GRE sequence; (b) plain T1-weighted GRE sequence; (c) enhanced T1-weighted GRE sequence 40 minutes after injection; (d) T1-weighted GRE sequence 90 minutes after injection.
Although the chemical composition of Gd-BOPTA sounds similar to the extracellular gadolinium agents, it combines both extracellular and liver-targeted properties, because some 5% of it is excreted through the liver, as is shown in this case of multiple metastases.
(Images courtesy of Drs. Caudana, Morana, and Pistolesi; Verona.)

Figures 13-12 (top) and 13-13:
Mn-DPDP uptake in the liver relies on the ability of hepatocytes to excrete metal ions. Manganese separates from the DPDP-complex and is taken up by the hepatocytes.
In the T1-weighted GRE images of Figure 13-13, the metastases are well delineated 15 minutes after the injection (a) and even 24 hours after administration some of the contrast agent remains (b).
In Figure 13-14a is a plain T1-weighted GRE image; the enhanced image (b) shows the large and smaller metastases well delineated.
(Images courtesy of Dr. Martí-Bonmatí; Valencia.)

Figure 13-14:
The particulate agents are taken up by endothelial and Kupffer's cells. They darken the liver tissue due to their effective shortening of the T2 relaxation time. In this case, the examination has been performed with ferixan; the liver metastases are well delineated on the post-contrast image; with this agent, the normal liver tissue becomes black.

Figure 13-15:
MR lymphography. (a) Gadofluorine-8 in a rabbit tumor model. Large tumor and brightly enhanced lymph nodes appear towards the top. Parts of the lymph nodes are darker which corresponds to the malignancy well seen on the histological cut (b). (Images courtesy of Dr. Misselwitz; Berlin.)

Lymph Nodes

MR lymphography has been the research topic of several groups. Here again, iron-oxide particles seem to have a high potential. Coated with dextran, such particles are transported to the lymph nodes. Intravenous injection will be the clinical technique of choice, although intralymphatic and interstitial injection of particulate agents have also shown positive results. Sinerem is the only lymph-node agent on the market today. Another approach is the use of positive agents such as gadofluorine-8 (Figure 13-15).

Adrenals

Derivatives of cholesterol, labelled with gadolinium-DO3A, have been used successfully to enhance the adrenals in animals [15]. The method has been developed for radioisotope tracers. Similarly, polylysine linked with Gd-DTPA was also found to be taken up by the adrenals. However, the mechanism of uptake is not yet completely understood [32].

Tumors - Myocardium

One of the main goals in contrast-agent research is targeting of tumors. Hard- and software development will allow fast cancer screening by combining rapid MR imaging of the whole body and specific tumor-targeting contrast agents.

Metalloporphyrins are compounds which are selectively retained by various tumors [16]. Water-soluble porphyrins, labelled with manganese, might be used as tumor-specific contrast agents.

New developments are encouraging; however, they have not yet reached the clinical stage.

Among them are gadolinium mesoporphyrin (Gd-MP) and manganese tetraphenylporphyrin (Mn-TPP) which showed nonspecific tumor enhancement during the early imaging phase and pronounced enhancement in nonviable tumor components during the delayed phase. These findings might elicit novel applications for such compounds, e.g., delineation of myocardial infarction and monitoring of tumor therapy [17].

Tumor-specific monoclonal antibodies labelled with e.g., gadolinium or MION particles, can be synthesized. They enhance, at least in animal studies, specific tumors and antibody sites, for instance, after infarction and in infectious diseases [29,30].

Details can be found in several overview articles, and annual or biannual publications [1,2].

Ventilation Imaging

Not all contrast agents are aimed at proton magnetic resonance imaging. One can also exploit magnetic properties of nuclei different from ^{1}H, for instance ^{19}F. The feasibility of *in vivo* application of perfluorinated compounds has been demonstrated for lung ventilation and perfusion studies (Figure on page 230) [22].

Recently gadolinium-based aerosols, hyperpolarized gases, and oxygen were introduced as possible ventilation agents. In the case of hyperpolarized gases helium-3 seems better suited than xenon-129 [3,14]. However, special hardware is necessary: devices for production, storage, and installation of the hyperpolarized gases, as well as special coils and receivers for imaging. This makes ventilation imaging with these gases more difficult than comparable methods.

Blood-Pool and Intravascular Agents

Another major area for contrast-agent development is angiography. Although MR can produce angiographic images without contrast agents, the quality becomes far better and more reliable with vascular and perhaps, in the future, blood-pool contrast agents. The applications of these agents are described in Chapters 14 and 16.

There are four different categories of possible angiographic agents. Figure 13-16 classifies them. Their categories are based on their global ability to cross the endothelium and to filter through the renal glomeruli.

The conventional gadolinium chelates are extracellular agents and not blood-pool agents. After the slow injection of an ECF-space agent, its concentration in the blood will rapidly decrease. Depending on the type, only 50% of the dose remains in the blood after 5-10 minutes. However, with bolus injections (injection time < 60 seconds), the initial first-pass concentration is high; it decreases rapidly immediately after the end of the injection. The contrast agent is diluted with the total blood-pool volume, it leaks from the capillaries into the extracellular space in many tissues (e.g., muscle), and it is excreted by the kidneys. Thus, for vascular imaging, these contrast agents can best be used for imaging the first pass of the applied bolus.

Blood-pool agents remain in the blood for a significantly longer time and their tissue uptake is limited.

Their imaging window is wider; examination can even be repeated if necessary. The ideal contrast agent for bright blood MRA would have a high *r*1 relaxivity to make T1 as short as possible and a low *r*2 relaxivity to keep T2*>2×TE to avoid spin dephasing effects.

Besides gadolinium-based agents, ultrasmall superparamagnetic iron-oxide particles also seem to be well suited for MR angiography, with efficient and long-lasting positive intravascular signal enhancement.

These compounds remain almost exclusively in the intravascular space and selectively display the blood vessels. Due to their prolonged plasma half-life, these compounds could also be used to enhance areas with vessels of varying permeability, and thereby reveal a certain tumor affinity. They also can help to define ischemia and reperfusion after treatment, for example of cerebral or myocardial infarction. With appropriate calculative algorithms, such agents can also be used to estimate tissue blood flow in myocardial and cerebral ischemia, pulmonary embolism, the vascularization of transplants, and perfusion of tumors (Chapter 16).

Enteral Contrast Agents

Problems in abdominal MR imaging are created by motion artifacts from respiration, cardiac motion, peristalsis, and blood flow, as well as a general lack of contrast between feces- and fluid-filled or collapsed bowel loops and adjacent organs or pathological structures. Cardiac and respiratory gating, special software, faster imaging sequences, and the use of contrast agents have solved some of these problems.

A number of substances were proposed and studied during the last years in phantom, animal, and clinical trials. Among them were positive and negative contrast agents. Gadolinium-containing contrast agents belonged to the first group, while fluorine-containing compounds and magnetic particles are found in the second group (Figure 13-17) [9,23].

Their use has been limited, and with today's imaging techniques, in particular custom-tailored pulse sequences, most clinical questions can be answered without the application of such agents.

Figure 13-16:
Pharmacokinetic classification of various categories of angiographic contrast agents. (a) ECF-space agents; (b) low-diffusion agents; (c) rapid-clearance blood-pool agents; (d) slow-clearance blood-pool agents.
Low-diffusion agents have an intermediate position between ECF-space and blood-pool agents, and their interstitial diffusion occurs at a lower rate than that of ECF-space agents. Rapid-clearance blood-pool agents are mainly confined to the vascular space, but are freely excreted by the kidneys, whereas the renal excretion of slow-clearance blood-pool agents is very restricted.

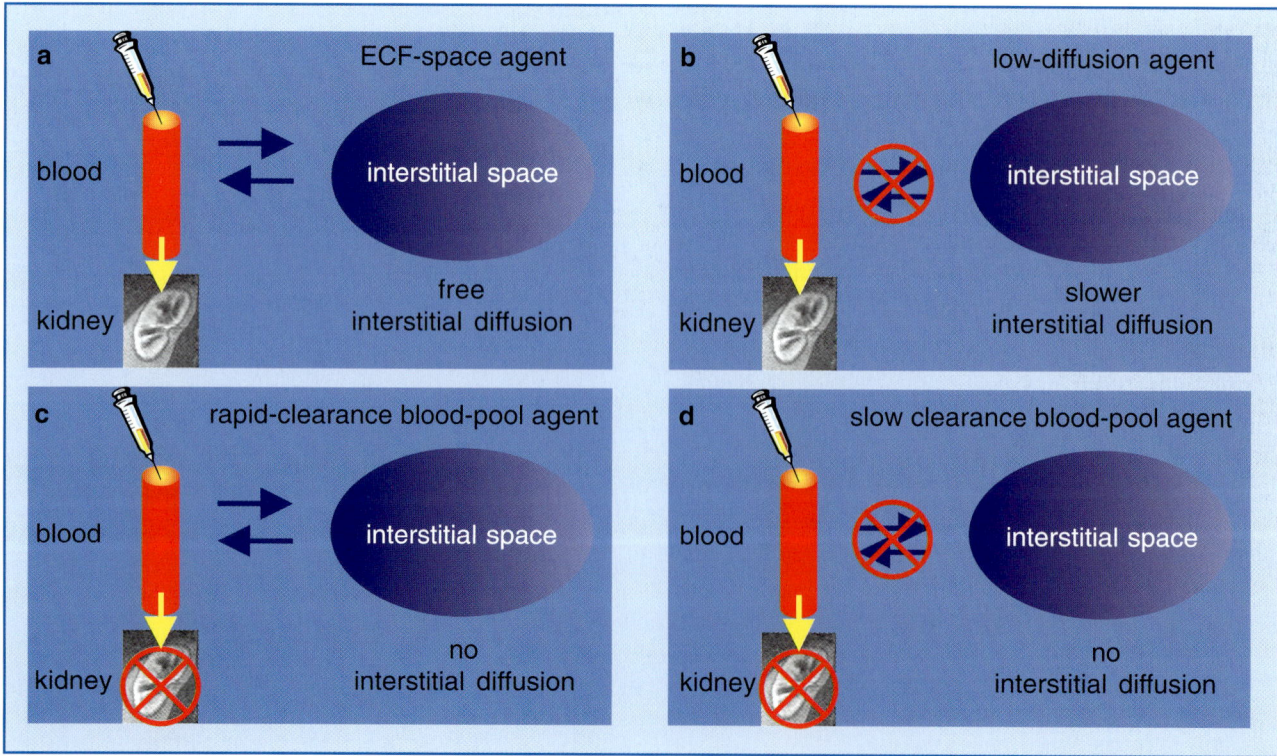

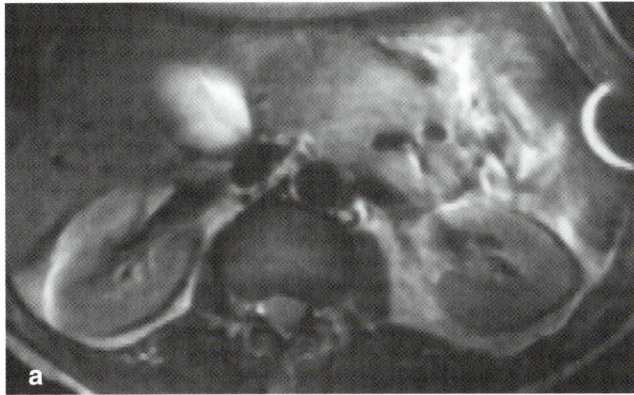

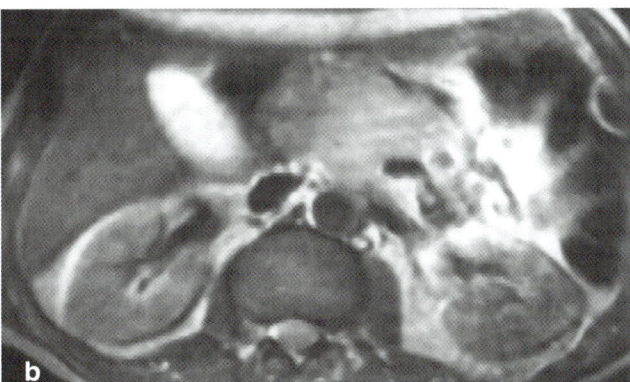

Figure 13-17:
Intermediately weighted images of the upper abdomen. Recurrent mesenteric tumor. (a) plain; (b) after ingestion of a negative oral contrast agent. On the enhanced image, the contour of the tumor is well delineated from the neighboring liver and intestines. Artifact created by ECG lead in the left abdominal wall.

Molecular Imaging

There are numerous other developments in contrast-agent research. A new approach is molecular imaging, using what is sometimes called 'intelligent agents'. Local thermal changes, for instance, can be targeted (see Chapter 9).

Given the increasing understanding of molecular mechanisms of disease and the development of innovative therapies at the genetic level, molecular imaging is aimed at the exploitation of specific molecules as the source of image contrast (Figure 13-18). This will allow researchers to leave some of the existing pathways in contrast-agent development. Instead of looking for relatively gross parameters of disease, they try to explore beyond the tissue level on the cell or even the molecular level. Imaging assessment of therapeutic effectiveness at the molecular level, long before phenotypic changes occur, is the final goal [28]. There are a lot of uncertainties in this kind of research. However, this will be one of the major innovations in imaging and contrast-agent research in the next twenty years.

Figure 13-18:
On the organ level, diagnostic imaging today is able to visualize gross parameters of disease and to describe, e.g, tumor burden. In the future, it will be possible to mark tumors and to target drugs as well as contrast agents, and on the genetic level to highlight genetic mutations and thus perform gene therapy.

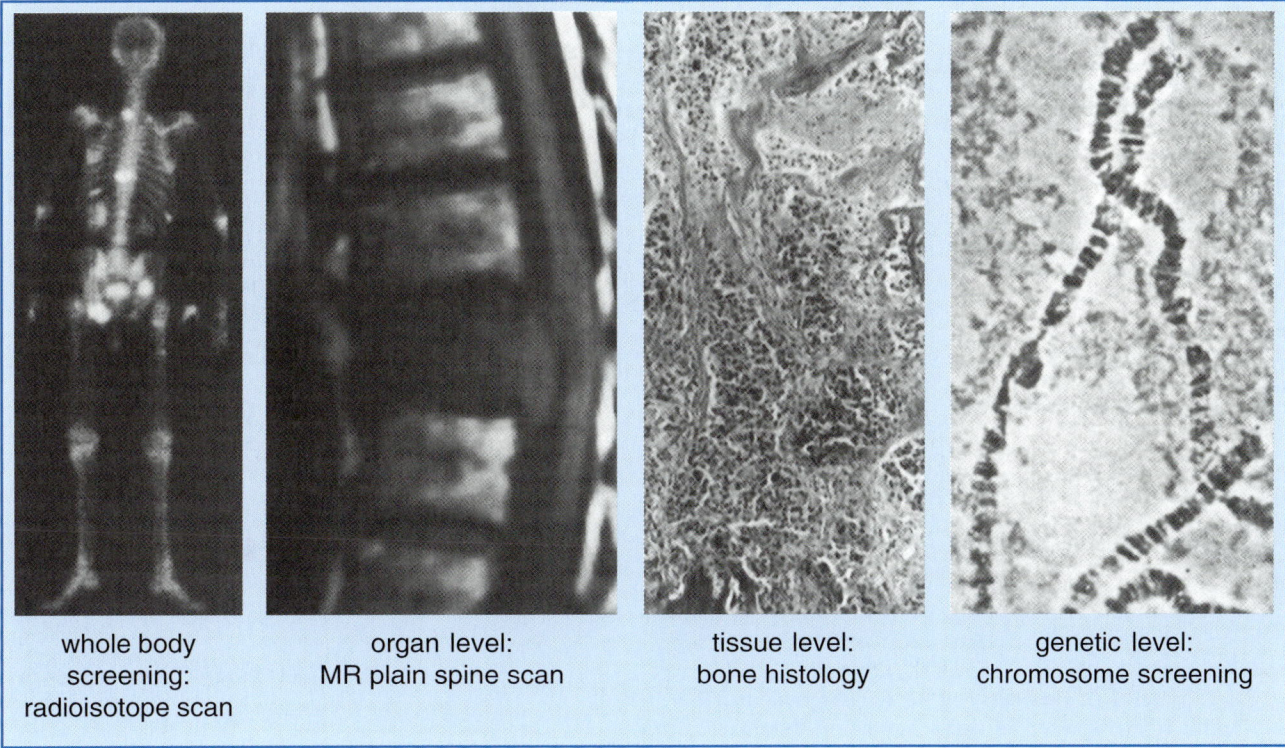

whole body screening: radioisotope scan

organ level: MR plain spine scan

tissue level: bone histology

genetic level: chromosome screening

References

EMRF publishes biannual proceedings on new developments in contrast-agent research. Details can be found at www.emrf.org.

The following two books contain dedicated reviews of magnetic resonance contrast agents:

1. Thomsen HS, Muller RN, Mattrey RF (eds). Trends in contrast media. Berlin, Heidelberg, New York: Springer 1999.
2. Dawson P, Cosgrove DO, Grainger RG (eds). Textbook of contrast media. Oxford: Isis Medial Media 1999.

3. Albert MS, Cates GD, Driehuys B, et al. Biological magnetic resonance imaging using laser polarized 129Xe. Nature 1994; 370: 199-201.
4. Bloch F, Hansen WW, Packard M. Nuclear induction. Phys Rev 1946; 69: 127.
5. Claussen C, Laniado M, Schörner W et al. Gadolinium-DTPA in MR imaging of glioblastomas and intracranial metastases. Am J Neuroradiol 1985; 6: 669-674.
6. Curati W, Graif M, Kingsley D et al. Acustic neuromas: Gd-DTPA enhancement in MR imaging. Radiology 1986; 158: 447-451.
7. Gehl H-B, Vorwerk D, Klose K-C, Günther RW. Pancreatic enhancement after low-dose infusion of Mn-DPDP. Radiology 1991; 180: 337-339.
8. Graif M, Bydder GM, Steiner RE et al. Contrast enhanced MR imaging of malignant brain tumors. Am J Neuroradiol 1985; 6: 855-862.
9. Hamm B. Contrast materials for cross-sectional imaging of the body. Current Opinion in Radiology 1992; 4 III: 93-104.
10. Haraldseth O, Rinck PA, Jynge, P, Jones RA. Perfusion imaging of pig hearts and rat brains with gadodiamide inj. (Gd-DTPA-BMA) and sprodiamide inj. (Dy-DTPA-BMA). in: Rinck PA, Muller RN (eds.). New developments in contrast agent research. Proceedings of the 3rd special topic seminar (Hamburg, Germany). European Magnetic Resonance Forum: Minusio, Switzerland 1993. 117-126.
11. Kabalka GW, Buonocore E, Hubner K, Davis M, Huang L. Gadolinium-labeled liposomes containing paramagnetic amphipathic agents: targeted MRI contrast agents for the liver. Magn Reson Med 1988; 8: 89-95.
12. Lauterbur PC, Mendonça-Dias MH, Rudin AM. Augmentation of tissue water proton spin-lattice relaxation rates by *in vivo* addition of paramagnetic ions. Frontiers of Biological Energetics 1978; 1: 752-759.
13. Leander P. Liver-specific contrast media for MRI and CT. Experimental studies. Acta Radiol Supplementum 1995; 36: S 396.
14. Middleton H, Black R, Saam B, et al. MR imaging with hyperpolarized H-3 gas. Magn Reson Med 1995; 33: 271-275.
15. Mühler A, Platzek J, Radüchel B, Frenzel T, Weinmann H-J. Characterization of a gadolinium-labeled cholesterol derivate as an organ-specific contrast agent for adrenal MR imaging. J Magn Reson Med 1995; 5: 7-10.
16. Nelson JA, Schmiedl U, Shankland EG. Metalloporphyrins as tumor-seeking MRI contrast media and as potential selective treatment sensitizer. Invest Radiol 1990; 25, S1: S71-S74.
17. Ni Y, Marchal G, Herijgers P, et al. Paramagnetic metalloporphyrins: from enhancers of malignant tumors to markers of myocardial infarct. Acad Radiol 1996; 3 Suppl 2: S395-397.
18. Niendorf HP, Haustein J, Cornelius I, Alhassan A, Clauss W. Safety of gadolinium-DTPA; extended clinical experience. Magn Reson Med 1991; 22: 222-228.
19. Paley MR, Mergo PJ, Torres GM, Ros PR. Characterization of focal hepatic lesions with ferumoxides-enhanced T2-weighted MR imaging. Am J Roentgenol. 2000; 175: 159-163.
20. Reimer P, Müller M, Marx C, et al. T1 effects of a bolus-injectable superparamagnetic iron oxide, SH U 555 A: dependence on field strength and plasma concentration—preliminary clinical experience with dynamic T1-weighted MR imaging. Radiology 1998; 209: 831-836.
21. Rinck PA, Myhr G. Gadolinium chelates: clinical applications. In ref. 2, 333-354.
22. Rinck PA, Petersen SB, Lauterbur PC. NMR-Imaging von fluorhaltigen Substanzen. 19-Fluor Ventilations- und Perfusionsdarstellungen. Fortschr Röntgenstr 1984; 140: 239-243.
23. Rinck PA, Smevik O, Nilsen G, Klepp O, Onsrud M, Øksendal A, Børseth A. Oral magnetic particles in MR imaging of the abdomen and pelvis. Radiology 1991; 178: 775-779.
24. Romijn MG, Stoker J, van Eijck CHJ, van Muiswinkel JM, Torres CG, Laméris JS. MRI with Mangafodipir trisodium in the detection and staging of pancreatic cancer. J Magn Reson Med 2000; 12: 261-268.
25. Rummeny EJ, Torres CG, Kurdziel JC, Nilsen G, Op de Beeck B, Lundby B. MnDPDP for MR imaging of the liver. Results of an independent image evaluation of the European phase III studies. Acta Radiol 1997; 38, Pt2: 638-642.
26. Spinazzi A, Lorusso V, Pirovano GP, Kirchin M. Safety, tolerability, biodistribution and MR enhancement of the liver with Gd-BOPTA: results of clinical pharmacology and pilot imaging studies in non-patient and patient volunteers. Acad Radiol 1999; 6: 282-291.
27. Unger E, Cardenas D, Zerella A, Fajardo LL, Tilcock C. Biodistribution and clearance of liposomal gadolinium-DTPA. Invest Radiol 1990; 25: 638-644.
28. Weissleder R. Molecular imaging: exploring the next frontier. Radiology 1999; 212: 609-614.
29. Weissleder R, Lee AS, Fishman AJ, et al. Polyclonal human immunoglobulin G labeled with polymeric iron oxide: antibody MR imaging. Radiology 1991; 181: 245-249.
30. Weissleder R, Lee AS, Khaw BA, et al. Antimyosin-labelled monocrystalline iron oxide allows detection of myocardial infarct: MR antibody imaging. Radiology 1992; 182: 381-385.
31. Weissleder R, Papisov M. Pharmaceutical iron oxides for MR imaging. Reviews Magn Reson Med 1992; 4: 1-20.
32. Weissleder R, Wang YM, Papisov M, et al. Polymeric contrast agents for MR imaging of adrenal glands. J Magn Reson Imaging 1993; 3: 93-97.

Chapter Fourteen

From Bulk Flow to MR Angiography and Cardiac MR

Figure 14-1:
Checking the vessels: Will there be laminar or plug flow ? Or do the plaques create turbulent flow ?

Figure 14-2:
(a) Laminar flow, (b) plug flow, and (c) laminar flow turning into vortex and turbulent flow after a vascular stenosis.
Laminar flow is relatively slow, whereas plug flow, as a special form of turbulent flow, is faster. All flow patterns comprise conceptual representations.

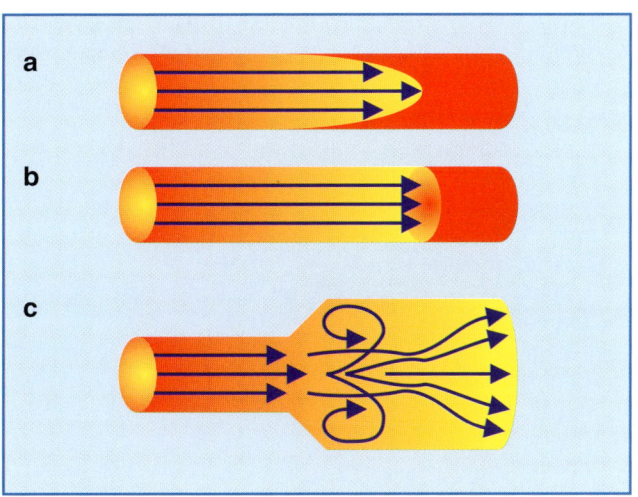

Some Fundamentals

Bulk or macroscopic flow of blood in vessels and of cerebrospinal fluid (CSF) adds still another parameter to image contrast in MR imaging. This kind of fluid motion is different from diffusion and perfusion and follows specific laws. A number of contrast features of flow in MR imaging and in magnetic resonance angiography (MRA) are rather complicated. The explanations in this chapter give a general overview, without attempting to cover the complexity of the topic in detail (Figure 14-1).

Flowing blood and CSF can appear bright or dark, depending on their velocity, direction and pattern of flow, and the pulse sequence used. In routine MR imaging normal flow effects can mimic pathology. Thus, understanding their influence upon image contrast is very important. This includes also knowledge of vascular anatomy and comprehension of vascular dynamics.

Blood flowing through a small caliber vessel usually exhibits *laminar flow* (Figure 14-2). Because of shearing forces, the blood closest to the vessel wall flows slowest. Blood velocity increases towards the center. Laminar flow is the predominant kind of flow in the human body.

Blood flowing faster in larger caliber vessels develops turbulence, particularly where vessel diameter changes, e.g., after stenoses or in vessels with irregular lumen, and moves more randomly, which produces phase shifts among blood cells. The spins dephase and the blood signal intensity decreases.

Plug flow is a special case of turbulent flow with a flat flow profile; all fluid elements possess the same velocity (Figure 14-2).

In general, flow velocity differs from arteries to veins.

Pulsatile flow in arteries, to a lesser extent also in veins, is cyclical and irregular, depending on systole and diastole of the heart. Thus, the appearance of the nature and velocity of the flow will depend on when during the cardiac cycle the image is taken.

This kind of flow is often turbulent during parts of the cycle.

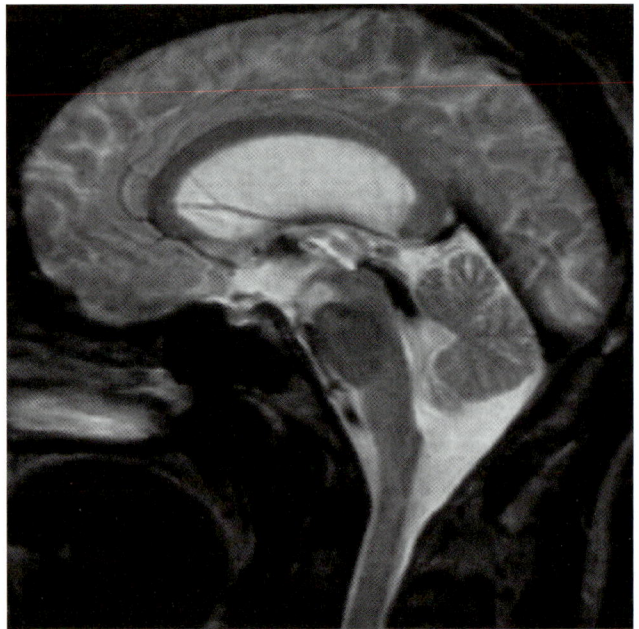

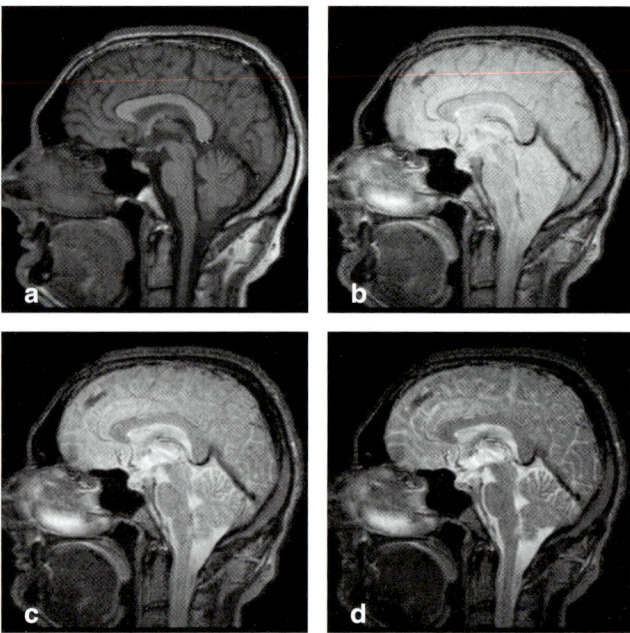

Figure 14-3:
Patient with hydrocephalus. On this T2-weighted SE image, cerebrospinal fluid should be bright. However, due to flow effects, the CSF in the aqueduct and the upper fourth ventricle appears dark.

Figure 14-4:
Midsagittal slice through a normal brain. SE pulse sequence, from T1-weighted (a) through intermediately weighted (b) and (c) to T2-weighted. The fluid signal of CSF changes accordingly, and the flowing blood in the straight sinus stays black.

Figure 14-5:
Intermediately weighted SE image of an infant's abdomen. The dark areas in the front are air-filled bowel loops. The dark structures in front of the spine represent rapidly moving blood in a porto-caval shunt.

Table 14-1:
Range of values of blood velocities in the human body. Different velocities in the aorta; from references [1, 2, 11]. a = artery.

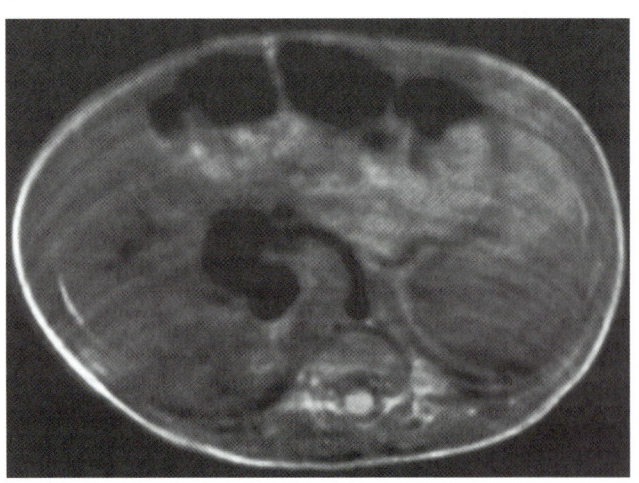

| | Velocity (cm/s) | |
	Peak	Mean Linear
Aorta diastole	65-80	20-80
Aorta systole	100-180	-
Inferior vena cava	40	10-20
Superior vena cava	20	-
External iliac a.	98-140	-
Common femoral a.	90-138	-
Superfic. femoral a.	77-103	-
Popliteal a.	56-82	-
Common carotid a.	80-120	-
Internal carotid a.	80-120	-
Vertebral a.	27-45	
Basilar a.	32-52	
Arteriole	-	1
Venole	-	1-10

Conventional Spin-Echo Pulse Sequences

In spin-echo images, fast-flowing blood appears dark, slow-moving blood appears relatively bright. Slow-flowing CSF appears dark on T1-weighted images and bright on T2-weighted images because the T1 relaxation time of blood (at 1.5 T) is approximately 1200 ms, while that of CSF is 3000 ms; the T2 relaxation times are 150 ms and 500-3000 ms, respectively.

Relatively fast-moving CSF, however, behaves like blood. Figure 14-3 shows an example of fast-flowing CSF and its influence upon contrast in a T2-weighted image. Fast-flowing blood appears dark in intermediately, T1- and T2-weighted spin-echo images (Figure 14-4). This can be very helpful for the differential diagnosis of aneurysms, angiomas, vascular malformations or similar diseases (Figure 14-5). Bright signal intensity where flow voids are to be expected, supports the diagnosis of slow flow or thrombosis.

Figure 14-6 explains this phenomenon for blood. In spin-echo sequences, one main effect contributes to signal intensity behavior, the *time-of-flight* (TOF) effect of inflowing blood. This effect originates from the movement of the blood during the time between the application of the excitation and the refocusing RF pulse. Blood can move so fast (Table 14-1), that it is not subject to both the 90° and 180° pulses of an SE experiment, but only to one of them. Whichever pulse it is, no signal will be received from the moving blood; there will be a signal void. Turbulent flow contributes to this effect.

Figure 14-6:
Basic effects of time-of-flight flow phenomena upon signal intensities in spin-echo images. The same imaging plane is first exposed to a 90°, then to a 180° pulse. If there is no flow, a bright signal will be visible. Slow flow creates a signal intensity at the bright end of the gray scale, whereas fast flow leads to low signal or no signal at all (*signal void*). Now all excited spins have left the imaging plane by the time the refocusing 180° pulse is applied. v = velocity.

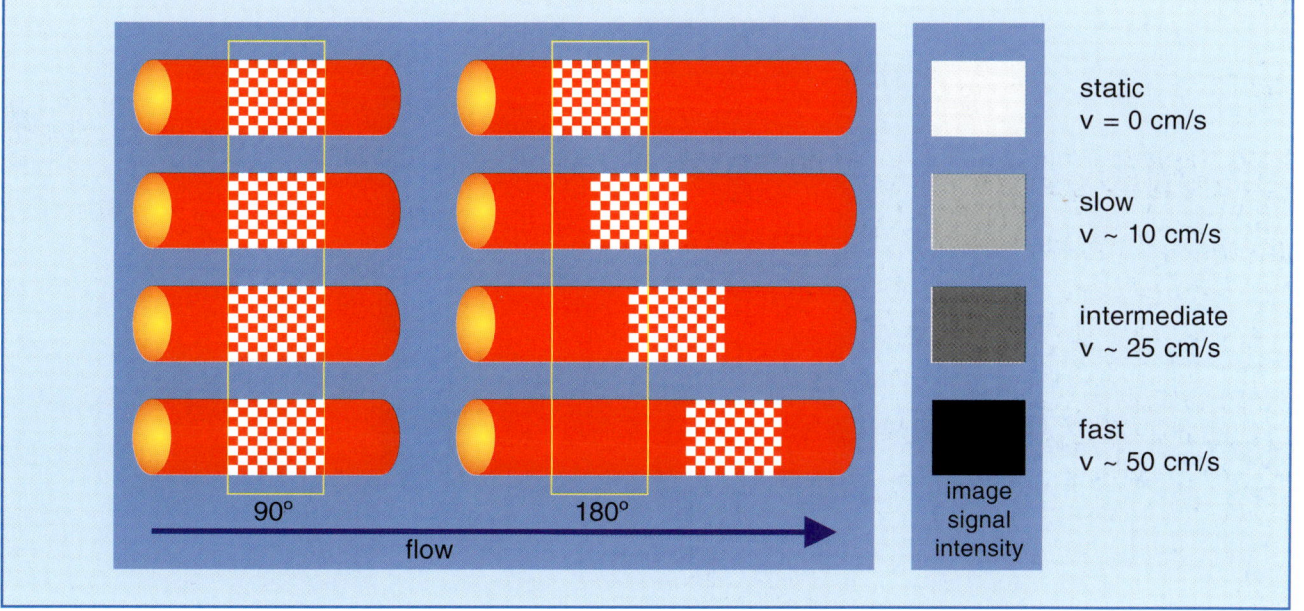

Gradient-Echo Sequences

In two-dimensional rapid imaging sequences using gradient echoes, the signal-intensity behavior is more straightforward than in SE images. Here, stationary material experiences the effect of all applied RF pulses, resulting in a signal which is only a small percentage of the equilibrium signal.

In the presence of flow, the spins in the slice are replaced by spins which have not experienced any of the preceding RF pulses and therefore give a much greater signal, providing that the flow is not turbulent.

With increasing blood velocity, signal intensity increases until it reaches a steady state (Figures 14-7). Blood velocity is not the only factor influencing signal intensity. Slice thickness and profile, T1 (and thus field strength), the repetition and echo times, and other intrinsic and extrinsic factors add to the extreme complexity of flow signals in MR images. Figure 14-8 explains this.

Figure 14-9 shows an example of high-signal blood in an examination of the pelvis.

This behavior has been exploited for cardiac studies (Figure 14-10) and is fundamental to magnetic resonance angiography, which uses gradient-echo sequences.

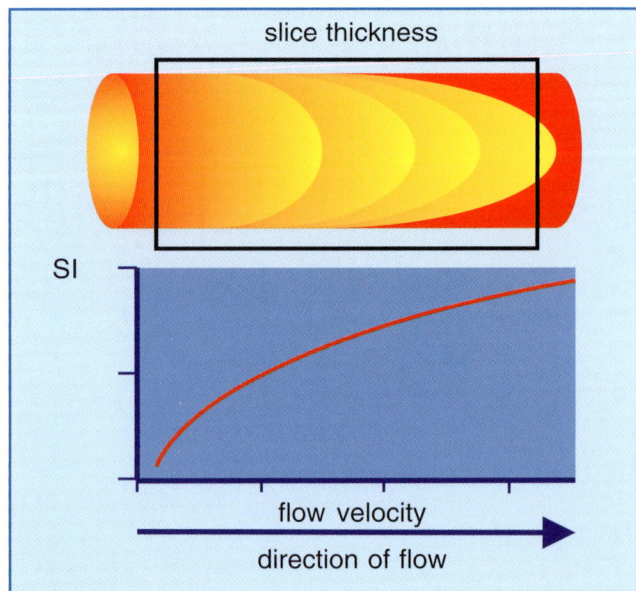

Figure 14-8:
Two components contribute to the flow signal of a gradient-echo sequence: those spins which have entered the slice before the current excitation pulse and those which have been excited by the preceding pulse and remained in the slice. The fresh spins travelling fastest carry full magnetization and contribute most to the signal; the partially saturated spins travelling slower contribute less. The signal intensity (SI) is velocity-dependent: the faster the flow, the brighter the signal, as shown in Figure 14-7.

Figure 14-7:
Conceptualized flow signal intensity patterns of a GRE sequence in the steady state. v = velocity.

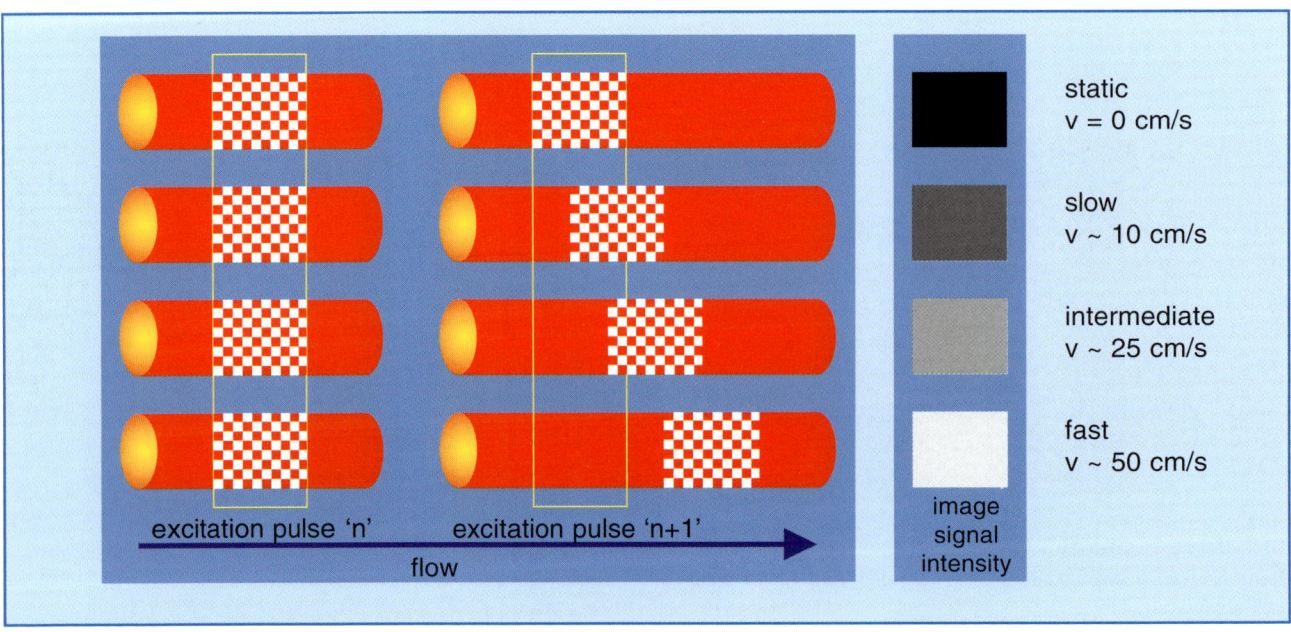

Angiography

MR angiography (MRA) is a further development of flow-related MR methods. In contrast to x-ray angiography, MRA does not require the use of contrast agents, rather the blood itself is used as an intrinsic contrast agent.

However, there are general problems with all MRA techniques which are difficult to overcome. They include flow voids or regions of low signal, where turbulent flow prevails, and difficulties in depicting smaller vessels.

Two groups of techniques are used for directly imaging flow in arteries, veins, and CSF-containing spaces:

- time-of-flight (TOF), and
- phase-contrast (PC).

Although both techniques are fundamentally different, they are markedly affected by normal and abnormal blood-flow patterns.

TOF and PC are both bright blood methods, i.e., the blood appears bright on images (Figure 14-11). Both are available in two-dimensional or three-dimensional versions, and PC also as a cine-technique.

The choice of methods depends on flow velocity, imaging time available, and a number of other conditions. Both techniques have advantages and disadvantages, which are summarized in Table 14-2 later in the chapter.

Black blood techniques are derived from TOF methods. They depict flowing blood dark and are preferably used in regions with high turbulence, e.g., in the exact assessment of stenotic lesions.

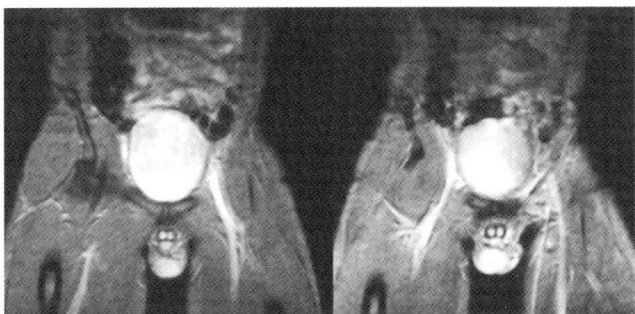

Figure 14-9:
In gradient-echo images, flowing blood appears bright (0.5 T; TR = 400 ms, TE = 28 ms, FA = 20°).
Note: These images are plain GRE images; this is not (yet) an MR angiogram.

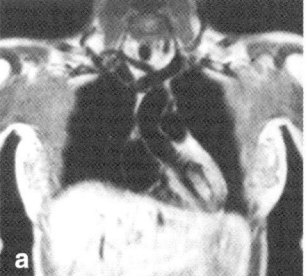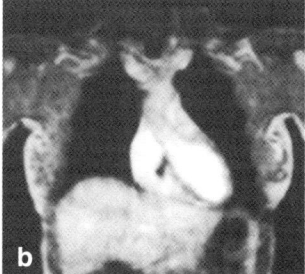

Figure 14-10:
Effects of blood flow upon signal intensity in a spin-echo and a gradient-echo pulse sequence. In the spin-echo image (a), flowing blood within the heart chambers and the ascending aorta is black, whereas in the gradient-echo image, flowing blood appears bright (b).

Time-of-Flight Angiography

Time-of-flight techniques were first described in 1959 by Jerome R. Singer[9]. These techniques are also known as *inflow* or *wash-in/wash-out* techniques.

As described, they take advantage of the contrast between inflowing fully magnetized blood and the saturated surrounding tissue (Figure 14-7). In this case, the flowing blood is bright, whereas the surrounding tissue is dark[4].

However, this holds for a single-slice experiment only. If we move to multiple slices, flow effects become more complex. To understand them, we have to consider the flow pattern of blood in vessels, as described on page 167. Flow signal intensity and image contrast also depend on some of the principal extrinsic contrast parameters of MR imaging. For instance, in multiple-slice images, flow signal intensity is influenced by the position of the particular slice and the direction of flow relative to slice excitation.

At the time of inflow, one slice of the flowing blood is excited. This excited blood continues travelling in laminar flow and leads to different signal patterns in the neighboring slices (*concurrent flow*: Figure 14-11).

During the image-acquisition period, the volume-of-interest receives multiple RF pulses saturating the non-moving spins within the volume. Fully magnetized flowing spins enter the volume, presenting greater signal intensity than the stationary tissue.

Reversing the slice-excitation direction changes this signal pattern (*countercurrent flow*). In this case, the central signal void is usually less pronounced but still visible.

The most common implementation is to acquire a series of parallel thin slices using a rapid GRE sequence, usually with flow-compensated gradients to minimize the dephasing effects.

Flow-compensation methods include such techniques as gradient moment nulling (GMN), motion-artifact suppression technique (MAST), and field even-echo rephasing (FEER). They return spins moving at a constant velocity along a magnetic gradient into phase at the same time as stationary tissue. This enhances blood (or CSF) signal intensity (Figure 14-12).

The blood flowing perpendicular, or with a perpendicular component, to the slices gives a strong signal. The series of slices are viewed on screen as a 3D stack forming a 3D picture of the flow perpendicular to the slice direction or as a single-projection image.

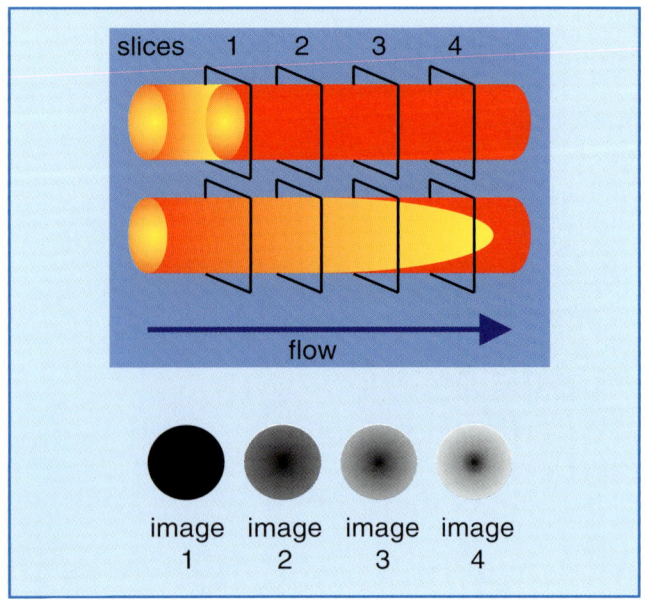

Figure 14-11:
Flow-related contrast enhancement in multiple-slice SE images. The dark gray depicting the spins in slice 1 illustrates the fate of these spins while flowing through the imaged region. Because the parabolic laminar flow has a higher velocity in the center, doughnut-shaped patterns develop. This kind of flow enhancement is not pathological. If flow direction and slice excitation direction are the same, the flow is called *concurrent*. Flow enhancement changes according to the direction of flow and sequence of slice excitation.
Countercurrent flow, which is not shown in this figure, has a different enhancement pattern from concurrent flow.

Figure 14-12:
Time-of-flight angiography.

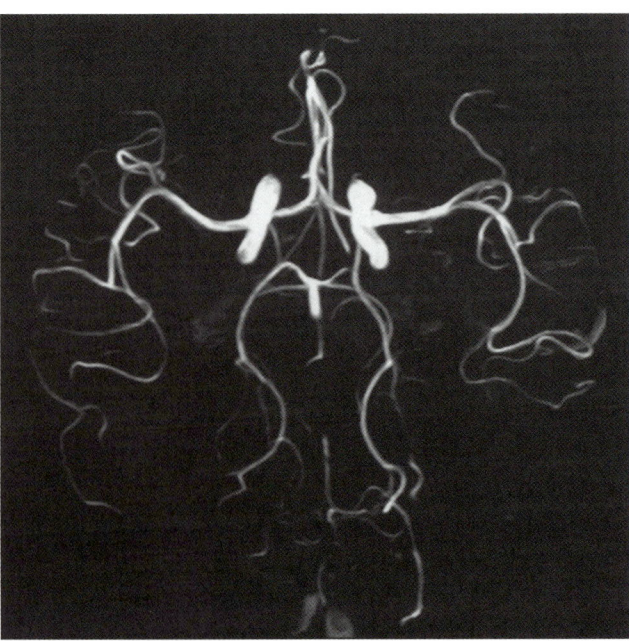

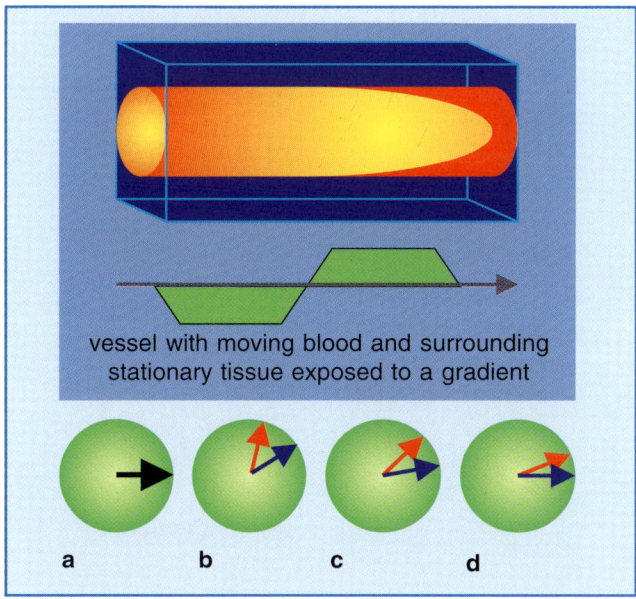

Figure 14-13:
Phase contrast. (a) Spin system at time 0; (b) spins are dephasing; (c) after switching of gradient spins are rephasing; (d) stationary spins are rephased.

Because of the balanced gradients, stationary spins in the surrounding tissue completely rephase. The flowing spins will stay out-of-phase. This phenomenon is used for phase-contrast angiography.

The phase angle depends on the flow velocity. The red dephasing arrows represent flowing spins; the blue dephasing arrows represent stationary spins.

Figure 14-14:
Phase-contrast angiography.

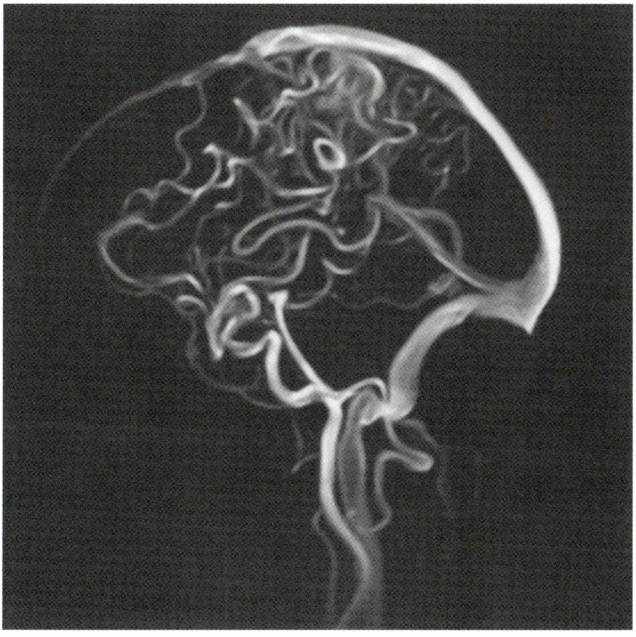

Phase-Contrast Angiography

Regular MR images are magnitude (modulus) images, with signal intensity being the basis for image reconstruction. Phase-contrast MR angiography exploits the shift in the phase that occurs when spins move in the presence of an imaging gradient[3].

As discussed in Chapter 6, spins dephase in the presence of a field gradient. For flow perpendicular to the gradient, the motion will cause the spins to experience different gradient strengths during their application. To counteract this force, for *stationary* spins the gradients are balanced and thus have no influence upon these spins (see Figures 6-9 and 6-10). However, they influence *moving* spins, leading to a net phase shift (Figure 14-13 and 14-14). The phase angle represents the flow velocity.

This method has the advantage that its sensitivity can be adjusted to the velocity of moving blood or CSF. Weak gradients allow the detection of fast-moving flow, whereas strong gradients are more sensitive to slow flow. Because the velocity of blood flow differs in the human body, as seen in Table 14-1, this is a great advantage of the PC method. The size of the phase shift depends on the gradients, the time separating them, and the flow rate.

Thus, for known gradient parameters, the flow rate can be calculated. In this way, PC angiography can produce quantitative velocity-encoded images or, in general, velocity mapping.

Because PC techniques depend on encoding of flow in all three spatial directions, the data acquisition takes longer than for TOF methods — after acquisition of a reference phase image, up to three images sensitized to flow in the three directions have to be collected; to eliminate the stationary background signal, the reference phase image is then subtracted from each of the three sensitized images. Similar considerations hold for 2D versus 3D: 2D acquisition is faster, 3D acquisition has better signal-to-noise ratio and spatial resolution.

When acquiring PC-MRA images, one has to know the approximate velocity of the flowing blood, the velocity-encoding (VENC) value. This value represents the maximum velocity present in the imaging volume.

The faster the spins in the vessel are moving, the greater their phase shift. Spins with a higher velocity than the VENC value will cause aliasing on the final image (wrap around; see Chapter 17).

An overview of TOF and PC characteristics is given in Table 14-2.

Maximum-Intensity Projection

With both time-of-flight and phase-contrast methods we now have isolated vascular structures by acquiring individual two-dimensional or three-dimensional images. However, an angiogram reflects entire vascular structures in a two-dimensional or pseudo three-dimensional overview.

In conventional x-ray and digital subtraction angiography, this is created by subtraction of a mask from the contrast-enhanced angiography images.

In MRA, a special computer algorithm is used for this purpose: the maximum-intensity projection.

Vascular structures on the MR angiography images we have acquired show bright signal intensity. The *maximum-intensity projection* (MIP) algorithm allows the selection of bright pixels in all parallel 2D slices or in the 3D slab or volume and their projection into one image. In 2D imaging, this is easily understandable as depicted in Figure 14-15.

The projection method is similar to a shadowgram, with the exception that only high-signal-intensity pixels are projected into the final image. The corresponding pixels in each original image finally form the projection angiogram.

In 3D imaging, this algorithm can be used to create images in any projection wanted. On screen, one can visualize a rotating 3D angiogram.

MIP is a relatively simple and useful technique for processing angiographic MR data. It has some drawbacks such as the lack of discrimination between arteries and veins and high-signal-intensity non-vascular structures such as fat. New methods are being developed to overcome these problems, among them vessel tracking and volume rendering.

For *black blood* angiography, the contrary of MIP is done: those pixels with the lowest signal intensity are chosen, and with *minimum-intensity projection* (mIP) a black blood angiogram is created.

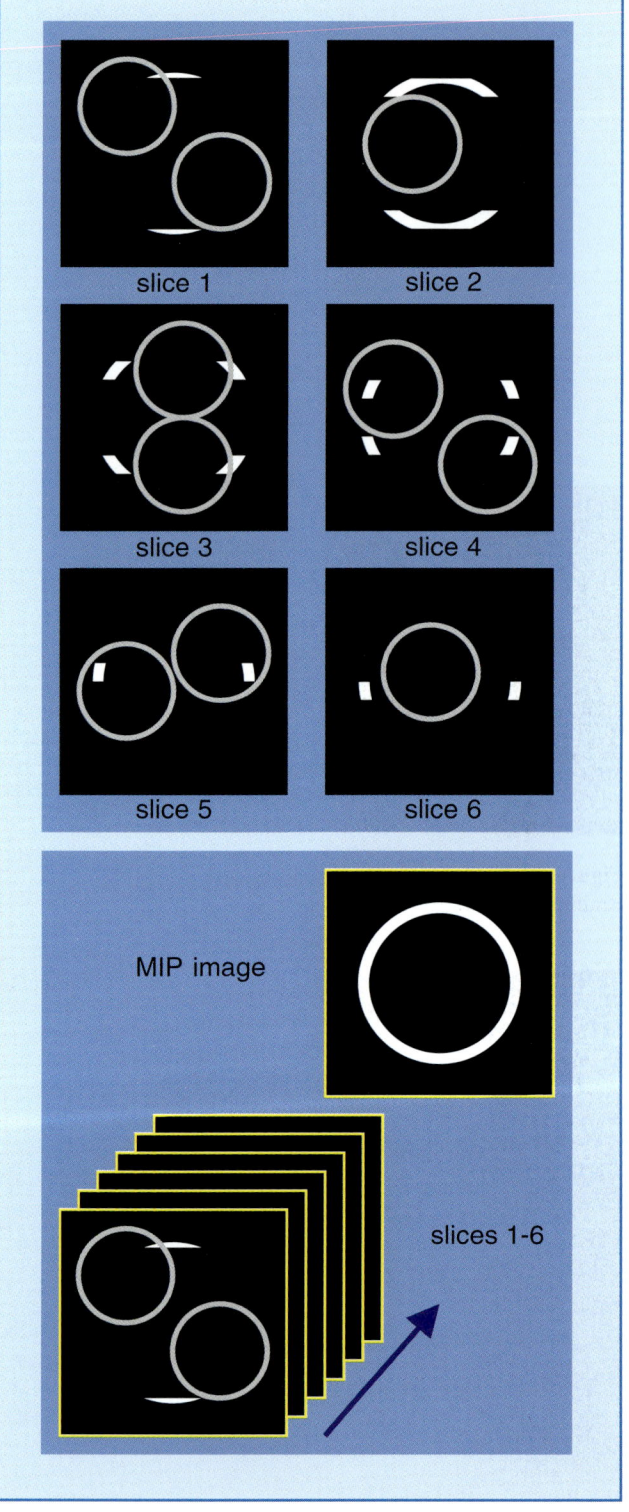

Figure 14-15:
The MIP operation. In this case, six slices have been acquired. Flow-related structures give the highest signal intensity (the big white circle on the final reconstructed image); however, there are other structures visible with intermediate signal intensity (small gray circles on the original images). Only the highest valued pixel is represented on the final image. Thus, other structures will disappear.

One disadvantage of MIP is that in routine clinical angiography bright non-vascular tissues (e.g., fatty tissues) can represent highest signal intensity on the original pictures and thus be depicted on the angiogram. Such tissues can only be discriminated from vascular structures by their anatomy.

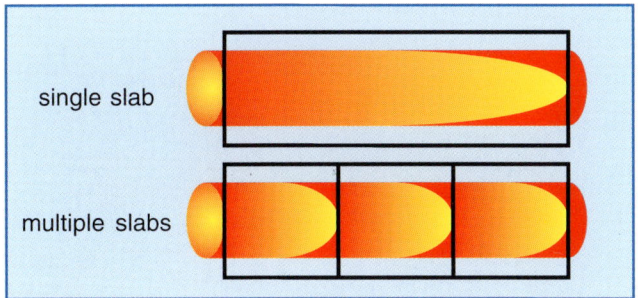

Figure 14-16:
MOTSA (multiple overlapping thin acquisition). Instead of one thick slab, multiple thinner slabs are acquired. MOTSA reduces overall saturation effects, but introduces the 'Venetian blind' arti-fact.

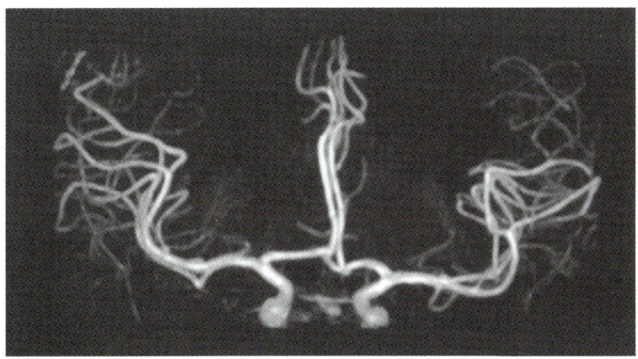

Figure 14-17:
Brain angiogram using TONE and MTC.

Reduction of Saturation Effects

Saturation effects are the gradual loss of T1 signal intensity by repeated excitation pulses. Too short TR leads to progressive loss of M_z; the same happens when increasing the flip angle.

Saturation effects associated with thick slabs can be partly overcome with a technique called 'multiple overlapping thin slabs acquisition' (MOTSA). Instead of one thick slab, several smaller slabs are acquired. MOTSA creates the 'Venetian blind' artifact (Figures 14-16 and 14-20).

Such artifacts do not exist when one uses 'tilted optimized non-saturation excitation' (TONE). TONE applies ramped flip angles to the different slices of the slab. Increasing the flip angle counteracts saturation effects, in this case of slow flowing blood in deeper slices.

Combined with magnetization transfer contrast (MTC), which suppresses background signal from brain parenchyma, TONE boosts the visibility of small vessels (Figure 14-17).

Table 14-2:
Advantages and disadvantages of various TOF and PC imaging techniques used for MRA.

	Advantages	**Disadvantages**
TOF	presaturation works well; multiple projections, including subvolumes are possible.	thrombus or other compounds with short T1 may simulate flow; tortuous vessels give less contrast; motion artifacts.
2D-TOF	sensitivity to slow flow; relatively long data acquisition times; lack of saturation effects.	thick slices and large voxels; long TE; insensitivity to in-plane flow.
3D-TOF	short data-acquisition times; high spatial resolution; short TE; good signal-to-noise ratio.	insensitivity to slow flow; saturation effects; field distortion artifacts, such as air-bone susceptibility gradients.
PC	variable velocity-encoding, allowing imaging of slow or fast flow; excellent background suppression; minimized saturation effects; differentiation between stationary and flowing blood; directional flow information.	relatively long TE; presaturation less effective
2D-PC	short data-acquisition times.	no reprojection images; large voxel size; low signal-to-noise ratio; loss of signal intensity with overlapping vessels.
3D-PC	small voxel size; reprojection and subvolume images possible.	long acquisition time; relatively long TE; field distortion artifacts.
Cine-PC	variable velocity-encoding, allowing imaging of CSF, venous, or arterial flow; quantitative flow measurement; time-resolved information; hemodynamic flow information.	loss of signal intensity with overlapping vessels; large voxel size; requires ECG-triggering.

Contrast-Enhanced MR Angiography

Both TOF-MRA and PC-MRA have limitations. Enhancement of blood can be erratic, mostly due to the influence of flow irregularities. In some body regions, motion of the surrounding organs by breathing, peristalsis or pulsation affects angiographic depiction of vessels negatively, and saturation effects influence signal intensity and contrast of blood vessels.

There are numerous other inherent MR properties which can easily deteriorate both TOF or PC images.

Thus, the dream of finally having a completely noninvasive imaging method was shattered once again. If MR angiography was to compete with x-ray angiographic methods, higher spatial and temporal resolution and more reliable enhancement would be necessary — with the application of contrast agents.

If one combines rapid T1-weighted GRE-imaging with the injection of an ECF-space contrast agent, one can get an excellent angiogram (contrast-enhanced MR angiography, or CE-MRA) [5,6]. In the future, they may be replaced by blood pool contrast agents.

CE-MRA depends mainly on T1 effects less on TOF- or PC-imaging techniques.

If (during and immediately after injection) blood has the shortest T1 of all tissues, it will show the brightest signal and thus the blood vessels will be visible in the MIP image. Furthermore, even in periods of slow flow (diastole for most vessels), there still is good signal from the blood which reduces ghosting and/or eliminates the need for cardiac synchronization. This makes CE-MRA much easier to perform.

However, even with contrast agents, the scan time can still be relatively long (20 seconds to 2 minutes). Therefore, it is necessary to keep the arterial concentration high continuously by injecting during the entire scan. As a rule of thumb, the duration of the injection is equal to or slightly shorter than the scan time (Figure 14-18). The delay between the start of the injection and the start of the scan depends on the delay between the start of the intravenous injection and arrival of the bolus in the arteries of interest. This delay depends on the distance of the arteries of interest to the heart, the cardiac output, and the quality of the veins in which the agent is injected (Figure 14-19). The injected dose volume depends on the maximum allowed dose, as well as on price. Usually single or double doses are used (single dose: 0.1 mmol/kg body weight).

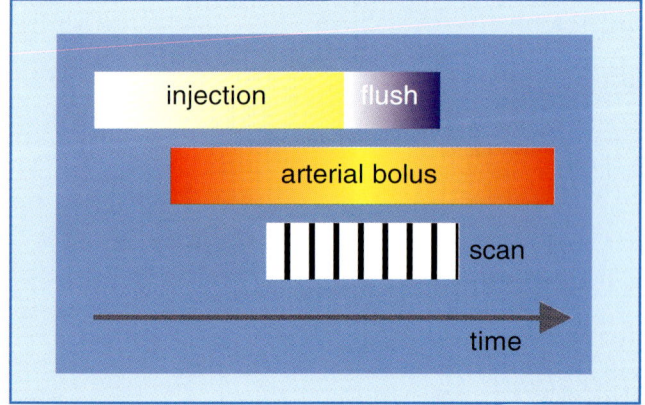

Figure 14-18:
Schematic drawing of a bolus injection for contrast-enhanced MR angiography.

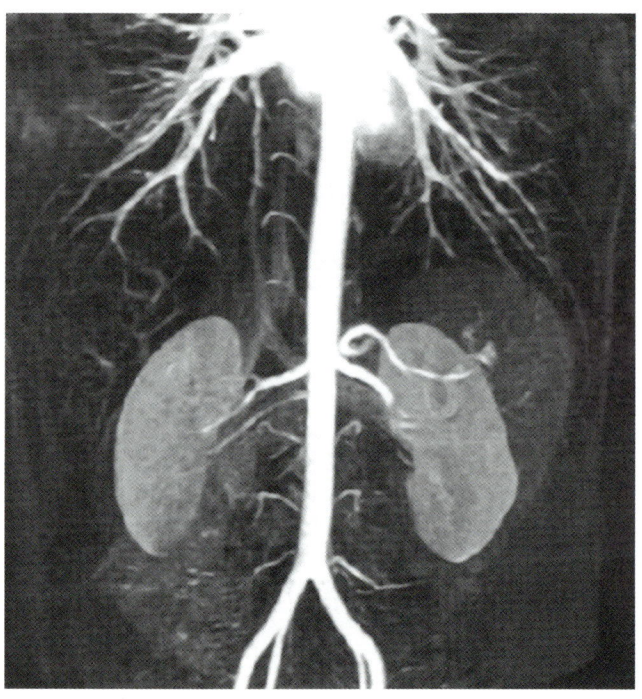

Figure 14-19:
Contrast-enhanced MR angiography of the abdominal aorta.

Figure 14-20:
Contrast-enhanced time-of-flight angiogram of a hand.

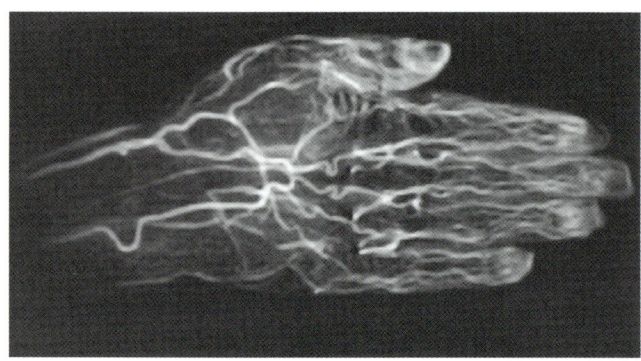

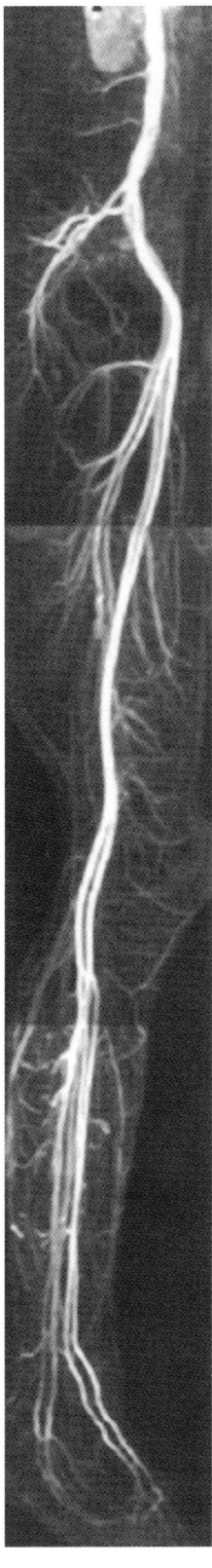
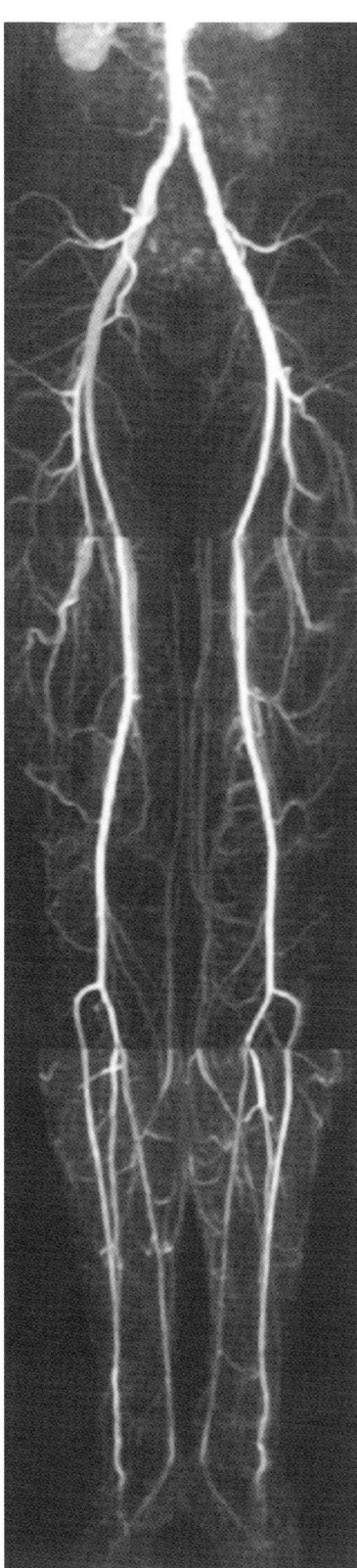

Figure 14-21:
Moving-bed contrast-enhanced MR angiogram of the abdominal aorta and lower extremities. The combination of rapid automatic table movement and automatic injection and follow-up of the bolus allows multiple successive coronal acquisitions.

In order to make sure that the arterial bolus is at its peak during imaging, several techniques can be used.

The delay between the start of the intravenous injection and arterial arrival of the bolus can be determined by a small test injection of one or two ml. The slice orientation of the test injection scan can be chosen in any direction, but if it is chosen perpendicular to the flow, presaturation slabs on both sides of the slice have to be used to suppress inflow effects so that only T1 effects will be visible. Another approach is prospective bolus detection, where the acquisition is triggered by the arrival of the arterial bolus. Because of the time needed for breath-hold instructions and the unknown delay between injection and arterial bolus arrival, both prospective and retrospective bolus detection can be problematic when combined with breath hold.

The protocol used for strong T1-weighting is relatively simple and is similar to that used in 3D inflow. The main difference is the flip angle and the freedom of slice orientation. Usually, a 3D gradient-echo sequence is applied. To suppress background tissues, a short TR (typically between 5 and 15 ms, depending on gradient system and sequence) and a large flip angle (between 40° and 70°) are used. Such a large flip angle cannot be used for a normal 3D inflow protocol (without contrast agent) because blood will become saturated too fast.

In combination with mechanical devices the entire peripheral vascular system can thus be examined after a single contrast agent injection (Figure 14-20).

The differentiation between arteries and veins is still problematic. The easiest differentiation is by morphology or, if a contrast agent is injected, by following the first pass.

Different approaches have been applied to distinguish between arteries and veins, both during image acquisition and by postprocessing image data. None of these approaches has been found to be sufficiently reliable. Among them is the presaturation method described in Figure 17-10. However, if presaturation slabs are used for selective demonstration of veins, the venous signal intensity in retrograde pathways may inadvertently be suppressed.

The use of gadolinium-based contrast agents obviates the dependence on inflow and allows imaging with large fields-of-view in the coronal or sagittal plane, despite substantial in-plane venous flow. Subtraction techniques offer a selective demonstration of veins, but a vein-free arterial study must be obtained first[8].

Cardiac MR Imaging

One of the biggest challenges for MR imaging is the heart[7]. It not only contracts, it also moves due to respiration; it contains flowing blood, and its axes are not orthogonal to the rest of the body (Figure 14-22).

The latter point is not a restriction in MR imaging because it is basically a three-dimensional technique; the former points limited cardiac MR imaging strongly, because it used to be a slow modality and the time resolution required for cardiac imaging is less than 50 ms.

Thus, cardiac imaging requires synchronization of data acquisition and the different kinds of motion of the heart, otherwise images are degraded.

Three types of synchronization are possible: gating, triggering, and slice following.

Gating means opening a gate or a time window during which the data acquisition can run freely. The acquisition is stopped when the gate is closed, and continued as the gate opens again. The time window during which the gate is open is not necessarily the same from gate to gate. Opening and closing of the gate is controlled by a physiological monitor, such as a chest elevation monitor from respiratory gating.

Triggering means receiving a signal that starts one or several pulses. The number of pulses is the same every time. For good timing, the trigger pulse should stem from an easily definable and exact physiological incident, like the peak of the R-wave in the QRS-complex of an electrocardiogram.

Distortions of the detected QRS-complex are readily introduced by gradient pulsing or by material in ECG electrodes.

Some problems cannot usually be avoided, such as the elevation of the T-wave due to the flow of blood through a magnetic field (see Chapter 18). This effect increases with field strength and with dobutamine stimulation of the heart in stress examinations. Since the elevation of the T-wave can give falsely detected R-waves, the triggering parameter should be set well ahead of the QRS-complex. There are some advanced ECG systems that use an out-of-magnet recorded vector ECG as a mask to get an undisturbed ECG inside the magnet during imaging.

Slice following is a technique where both the movement of the diaphragm during respiration and the heart during contraction is monitored by the system. Through checking the positions of anatomical landmarks imme-

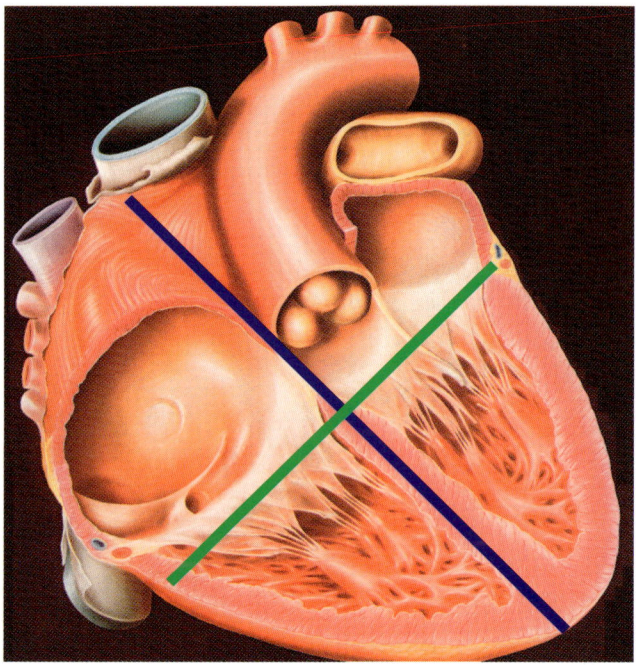

Figure 14-22:
Long axis (blue) and short axis (green) through the heart.

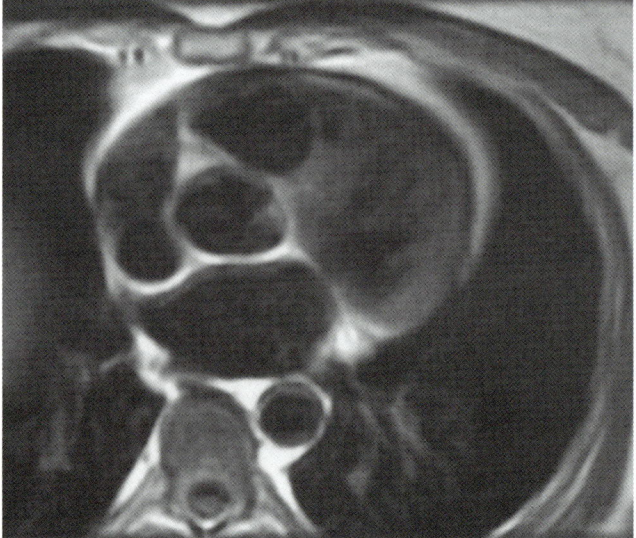

Figure 14-23:
Heart morphology: Transverse cut through the heart at the level of the aortic valve.

diately before and after the acquisition of an imaging profile, the displacement of the acquired profile can be checked (and discarded if outside given borders) and the displacement of the following profile can be predicted. Slice following is often used in combination with triggering to achieve better temporal resolution.

Measurement of myocardial wall motion becomes possible by myocardial *tagging*. Tagging entails labeling a strip in the myocardium by magnetic saturation. Thus wall motion can be directly monitored [12].

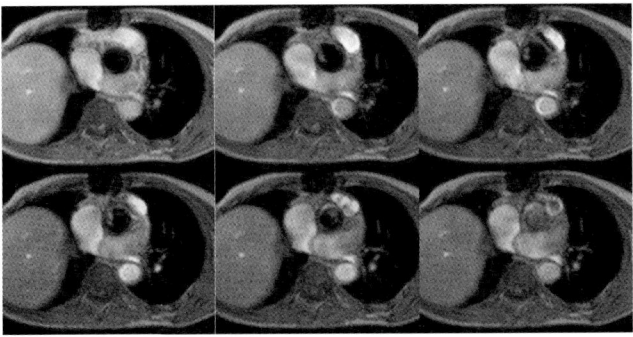

Figure 14-24:
Patient with several transient ischemic attacks. Gradient-echo images four millimeters above the aortic valve during different phases of the cardiac cycle show a pendulating thrombus.

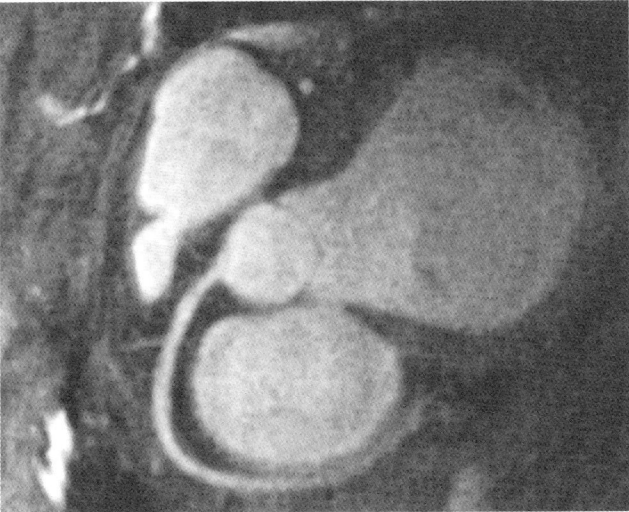

Figure 14-25:
Depiction of the coronary arteries without contrast agent application by spiral 3D rapid gradient-echo using navigator techniques and diastolic gating.

Static Studies

For the depiction of cardiac and great vessel morphology, static studies in several phases of the heart cycle are performed, usually as SE or RSE studies; primarily as black blood images (Figure 14-23). They should be performed as a multi-slice, multi-phase, double-oblique angulated acquisition, where special attention must be paid to the patient's heart frequency. Care should be taken about the chosen imaging planes; four-chamber views are generally coronal, but long-axis views can be either sagittal or transverse, whereas short-axis views can be either transverse or sagittal.

Flow Studies

Bulk flow either from shunts across the septum, regurgitant jets through valves closing insufficiently, or just through lumina and vessels can be visualized by gradient-echo techniques.

These studies must have reasonable temporal resolution to describe the different phases of the heart cycle, typically 16 or more, depending on the clinical question (Figure 14-24). The results will yield an image with muscular tissue in gray, static liquid in white and high-velocity jets in black (signal void). Consequently, both ordinary flow and regurgitant jets are seen, but cannot be quantified immediately.

One quantification method being employed is to measure the area (or volume) of the regurgitant jet (signal void) and compare it to the area (or volume) of the chamber.

Another technique tracks the signal intensity of the blood in the chamber during the cardiac cycle. The total signal intensity increases in normal patients during systole, but decreases markedly in patients with regurgitation. The percentage of decrease is found to be dependent on the severity of the regurgitation. Angiographic techniques such as flow quantification can also be utilized and then net flow through an orifice can be accurately quantified.

Clinical Application

To date, the combination of SE/RSE and fast gradient-echo imaging of the heart has been found to be the most efficient method to image the heart by magnetic resonance. The goal of such an examination is to combine the evaluation of morphology with functional features. In clinical routine, imaging time should not exceed thirty minutes; image processing, and particularly interpretation, will take longer. Great care has to be put into planning and optimizing a heart examination, and certain trade-offs should be realized. The key problem will always be to chose an imaging plane for an SE multi-slice, multi-phase series.

In the spin-echo images, regional abnormal wall motion or abnormal wall thickness is seen, as well as pericardial disturbances. Fatty deposits in the myocardium of the right ventricle and intra- or extracardiac tumors, together with crypts and ducts, are also generally found with ease.

Gradient-echo images tend to give somewhat poorer edge description of the endocardium, but a good overview of different hypertrophies. Specially designed gradient-echo sequences like 'True FISP' and 'Balanced FFE' do, however, give excellent blood-tissue contrast. Furthermore, both restricted and dilated cavities, insufficient valves and tracts are easily seen. The general flow pattern and the total overview of the heart add to the general understanding of cardiac performance.

New Techniques

Both first-pass contrast uptake and late-enhancement imaging are gaining ground in cardiac diagnostic imaging. These kinds of contrast studies are used to evaluate cardiac perfusion and perfusion reserve and to qualify the possible viability of cardiac tissue. Coronary artery imaging is also progressing using multiple different 2D- and 3D-imaging sequences (RSE, rapid GRE, SE-EPI, rapid GRE-EPI). The resolution reached has been good enough to persistently describe 7-10 cm of the main coronary arteries (RAD, LAD, and circumflex), but not the collaterals (Figure 14-25). This resolution is sufficient to evaluate the patency of grafts and MR coronary artery imaging can be used in patients with severe anaphylactic reactions to contrast media, but it is still not a sufficiently robust screening tool.

The discussion has not yet finished about MR imaging's indications compared with clinical methods and other para-clinical techniques such as x-ray CT, thallium imaging or echocardiography. A task force from the European Society of Cardiology, in collaboration with the Association of European Paediatric Cardiologists, published some guidelines on the issue in 1998 and 1999 [10].

MR imaging was considered the method-of-choice for the examination of congenital heart disease, acquired diseases of the great vessels and tumors infiltrating or close to the heart. Cardiac MR imaging was considered valuable in a number of other clinical areas such as cardiomyopathies, pericardial diseases and post-transplantation examinations.

References

1. Bradley WG, Waluch V, Lai KS, et al. The appearance of rapidly flowing blood on magnetic resonance images. Am J Roentgenol 1984; 143: 1167-1174.
2. Bradley WG. Flow phenomena. In: Stark DD and Bradley WG (eds.). Magnetic resonance imaging. 2nd edition. Vol. 1. St. Louis (USA): Mosby Year Book Inc. 1992, 253-298.
3. Gedroyc W. Phase-contrast magnetic resonance angiography in the abdomen. In: Aichner FT, Felber SR, Muller RN, Rinck PA (eds). Three-dimensional magnetic resonance imaging. Oxford: Blackwell 1994, 259-270.
4. Hausmann R. Time-of-flight MR angiography: physical principles and clinical applications. In: Aichner FT, Felber SR, Muller RN, Rinck PA (eds). Three-dimensional magnetic resonance imaging. Oxford: Blackwell 1994, 247-258.
5. Marchal G, Bosmans H, Van Hecke P, Jiang YB, Aerts P, Bauer H. Experimental Gd-DTPA polylysine enhanced MR angiography: sequence optimization. J Comput Assist Tomogr 1991; 15: 711-715.
6. Marchal G, Michiels J, Bosmans H, Van Hecke P. Contrast-enhanced MRA of the brain. J Comput Assist Tomogr 1992; 16: 25-29.
7. Pettigrew RI, Oshinski JN, Chatzimavroudis G, Dixon WT. MRI techniques for cardiovascular imaging. J Magn Reson Imaging 1999; 10: 590-601.
8. Shinde TS, Lee VS, Rofsky NM, Krinsky GA, Weinreb JC. Three-dimensional gadolinium-enhanced MR venographic evaluation of patency of central veins in the thorax: initial experience. Radiology. 1999; 213: 555-560.
9. Singer JR. Blood flow rates by nuclear magnetic resonance measurements. Science 1959; 130: 1652-1653.
10. Task Force of the European Society of Cardiology, in Collaboration with the Association of European Paediatric Cardiologists. The clinical role of magnetic resonance in cardiovascular disease. Europ Heart J 1998; 19: 19-39.
11. van As H, Schaafsma TJ. Flow in nuclear magnetic resonance imaging. In: Petersen SB, Muller RN, Rinck PA (eds.) An introduction to biomedical nuclear magnetic resonance. Thieme: Stuttgart, New York 1985, 68-96.
12. Zerhouni EA, Parish DM, Rogers WJ, Yang A, Shapiro EP. Human heart: tagging with MR imaging — a method for noninvasive assessment of myocardial motion. Radiology 1988; 169: 59-63.

Chapter Fifteen

Image-Processing and Visualization

Introduction

Basically all digital images are processed in one kind or another. Even analog images may be processed nowadays (Figure 15-1).

Image-processing was a well-established research field in its own right when clinical MR scanners became available in the early 1980s. The digital nature of MR imaging, coupled with a wide range of applications, spurred an enormous activity in image-processing of MR imaging data in the 1980s and 1990s.

In this chapter we give a brief description of image-processing techniques that have been applied to MR imaging. We also describe the somewhat related field of visualization techniques, with a special focus on 3D visualization methods. Some of the methods mentioned in this chapter relate directly to the next chapter on dynamic imaging.

Some Fundamentals

The main objective of medical image-processing is to facilitate the gathering of or provide diagnostic information not easily seen, or not seen at all, on non-processed images. In general, the processing of digitally acquired images is aimed at improving pictorial information for human interpretation and/or processing data for autonomous machine perception. Both aims have been targeted in magnetic resonance imaging and in both areas successful applications have been found.

All computed image-processing requires digitized imaging. In digitized radiology, the equivalent of a regular x-ray is taken and digitized directly by a specialized x-ray system. In nuclear medicine, CT, and MR, imaging slices through the human body are acquired and subdivided into volume elements. The numerical signal from each voxel, in turn, can be translated into a distinct shade of the gray scale and be represented as a picture element in the final image (Figure 15-2).

Both single images or a series of similar images can be manipulated, e.g., by noise reduction, edge or contrast enhancement[5]. In *multichannel imaging*, several channels representing n different parameters can

Figure 15-1:
As a radiologist, you always should be prepared for the unexpected. Digitized imaging brings more of the unexpected into your life. Still there are some easily recognizable features in most images. You recognize them immediately:
(a) When, where, and from where was this picture taken and what does it show ? (b) When and where was this picture taken ? Answers appear on the next page.

be acquired simultaneously or by consecutive procedures, leading eventually to n images of exactly the same object (Figure 15-3). Image-processing allows the connection of picture element data of the same location in different images with changed parameters; known connections can be computed, e.g., by using appropriate equations. Such procedures may extract additional information or allow quantification of data and thus an objective definition of structures, tissues, or metabolic processes. Several single images can, e.g., be added to a new multispectral image (synthetic image) which does not necessarily add useful information or even depict reality (Figure 15-4). Details on image-processing can be found in a number of monographs, such as [6, 12].

There are different ways of classifying image-processing techniques, for instance, they can be defined by what they are supposed to achieve. Types of techniques include noise reduction, image segmentation, feature extraction, and classification.

Whereas noise reduction is of vital importance for more noisy modalities like ultrasound, MR imaging has, due to a rapid development of MR hardware and software, not the same need for such techniques, perhaps with the exception of dynamic imaging (Chapter 16).

However, image segmentation and classification have found much more widespread use in MR imaging, partly due to the possibility of acquiring multichannel data suitable for such processing.

Another important group of image-processing technique is image registration or image alignment, which sometimes employs image-segmentation techniques to align images. Image registration is important for aligning multimodality data (for instance, nuclear medicine data and MR imaging data from the same patient) or registration of time series.

A specific type of time series (dynamic contrast-enhanced MR imaging) is described in the next chapter. Another important type of MR imaging time series is BOLD functional MR imaging (fMRI) of the brain. Image registration is now routinely performed on fMRI data. Furthermore, time series can also be used for monitoring tumor growth or growth of bones in children.

Improvement of diagnostic performance to reduce the ever existing level of uncertainty is one of the main propelling forces in diagnostic imaging research.

Since the human visual-perception system is unable to perform multichannel analysis in order to

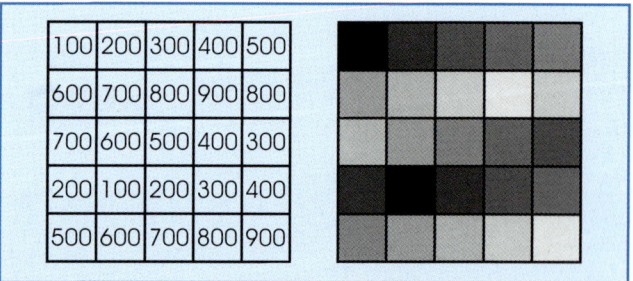

Figure 15-2:
Numerical image date output (left) turned into gray-scale image (right). Typically, one finds medical images with an image matrix of 256×256 or 512×512 and 256 gray levels.

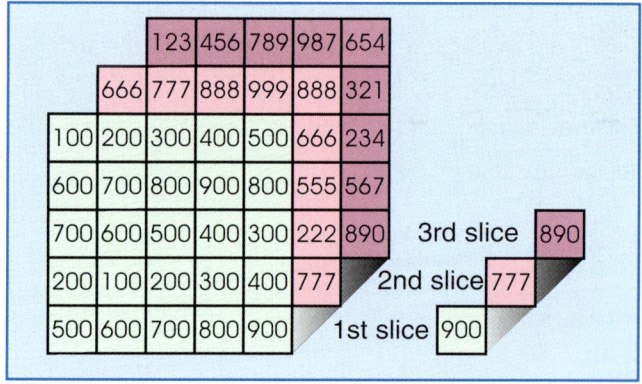

Figure 15-3:
If multichannel images of the same object are properly aligned, it is easy to compare or compute their signal intensities.

Answers to the questions in Figure 15-1:
(a) These questions were just asked to confuse you. You thought the answers were easy - a bird's-eye view of Central Park in Manhattan. Wrong: this is a vodka commercial with a vodka bottle which looks like Central Park. This picture has been image-processed. Even if you believe that you know what you are seeing — think twice.
(b) When: you are right — before World War II (in 1928). Where: you are wrong — not in Chicago, but in Berlin. Always check the patient's history before you read your images and make a diagnosis. Not all diagnostic questions can be answered by image-processing.

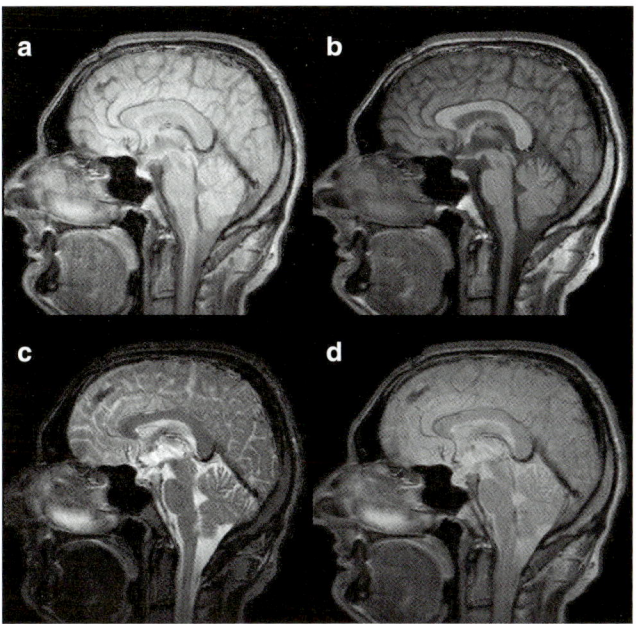

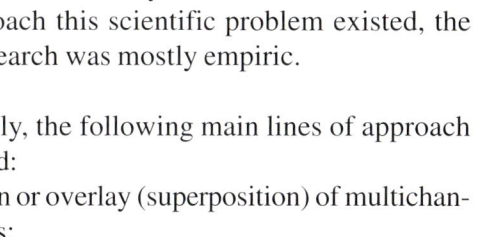

achieve a new dimension, image-processing developed in parallel to the introduction of MR imaging as a clinical tool. Researchers wanted to detect any message, possibly hidden, in a single MR image or a series of MR images. Given that only minimal information about how to approach this scientific problem existed, the course of research was mostly empiric.

Historically, the following main lines of approach were followed:
- subtraction or overlay (superposition) of multichannel images;
- quantification of MR parameters, i.e., T1, T2 and proton density;
- image segmentation and multispectral analysis;
- 3D visualization.

Subtraction or Overlay Images

Multispectral images of the same body region can simply be overlaid to give an impression of the exact location of certain contrast-enhanced structures. Usually, a high spatial resolution T1-weighted MR image is used as a background image to show the anatomical structures and the contrast-enhanced image is projected onto this picture.

This has been first shown with perfluorinated ventilation images [10, 11], and today it is commonly used in fMRI studies or in intermodality comparison between, e.g., CT or MR and PET images. Here, the information obtained from PET is overlaid or imprinted onto the more detailed anatomic images acquired with MR imaging or CT.

Practical Applications. The method is useful since it allows a better visualization to locate certain processes. The implementation is relatively simple. It is mainly used as an auxiliary tool to facilitate visualization of enhancement visible on postcontrast images (Figure 15-5), but also in MR angiography to highlight veins after subtraction of the CE-MRA images of the arterial phase, and in MRSI (Figure 5-8).

Figure 15-4:
Example of multichannel images: (a) proton-density-weighted, (b) T1-weighted, and (c) T2-weighted images of a slice through the brain. The anatomic location of the pixels is exactly the same; according to image-weighting the pixel representation is different. (d) is a pixel-by-pixel compilation of images a-c. This synthetic image does not reveal any additional diagnostic information.

Figure 15-5:
Example of multichannel images: (a) plain image of a liver, (b) contrast-enhanced liver, with depiction of focal nodular hyperplasia, part of a dynamic series, and (c) overlay of enhanced lesion on plain liver image.

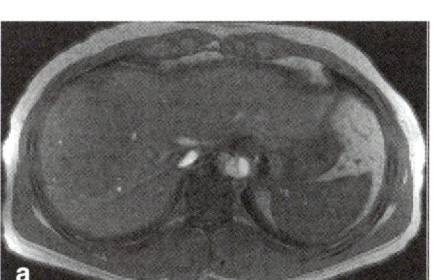
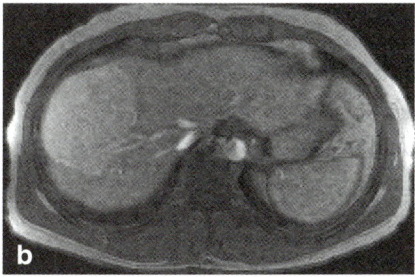
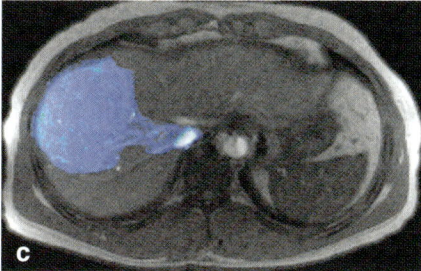

Quantification of MR Parameters

With numerous tissue parameters, MR imaging has substantial theoretical potential for tissue discrimination in different organs. The most important intrinsic contrast factors are proton density, T1 and T2 relaxation, and bulk flow.

The use of relaxation times for medical applications was first proposed in 1971[4]. Voxel-by-voxel *in vivo* relaxation-time measurements, partly turned into T1- and T2-maps, have been tried out over the years by a large number of researchers[13]. However, parametric T1 and T2 images did not enter into clinical routine. They were restricted to a single parameter only and revealed less information than images representing several parameters combined with different parameter-weighting.

This was one of the first major lessons to be learnt in MR image-processing: if you have more than one known factor influencing the contrast of an image, and if the change in contrast is perceivable by the human eye, it is not worthwhile to extract such a factor to create a parametric image. This holds in particular if this factor cannot be quantified exactly. In the case of relaxation times, only an estimation is possible *in vivo*. In 1985, it was realized that even carefully performed *in vivo* T2 measurements cannot be used as a diagnostic method in cancer detection, characterization, or typing[9].

Practical applications. For specific applications, pure relaxation-time maps can be used to simulate image contrast behavior. Such recalculated images can be created with special software programs, e.g., *MR Image Expert*[15,16]. The simulated images have substantially less noise than images acquired directly on an MR imager and can be used when looking for specific anatomical or pathological features or to evaluate best pulse-sequence parameters for contrast-agent enhancement. The drawbacks of such image-processing programs are their dependence on specific data-acquisition sequences and their time-consuming nature.

T1 maps are also used as the basis for calculating tissue concentrations of contrast agents in dynamic imaging. Here, two measurements are necessary, one before injection of the contrast agent, a second one after injection together with drawing a blood sample to determine the blood concentration of the contrast agent. This makes this technique cumbersome, and it is rarely used in clinical routine. A different non-invasive approach is described on page 191.

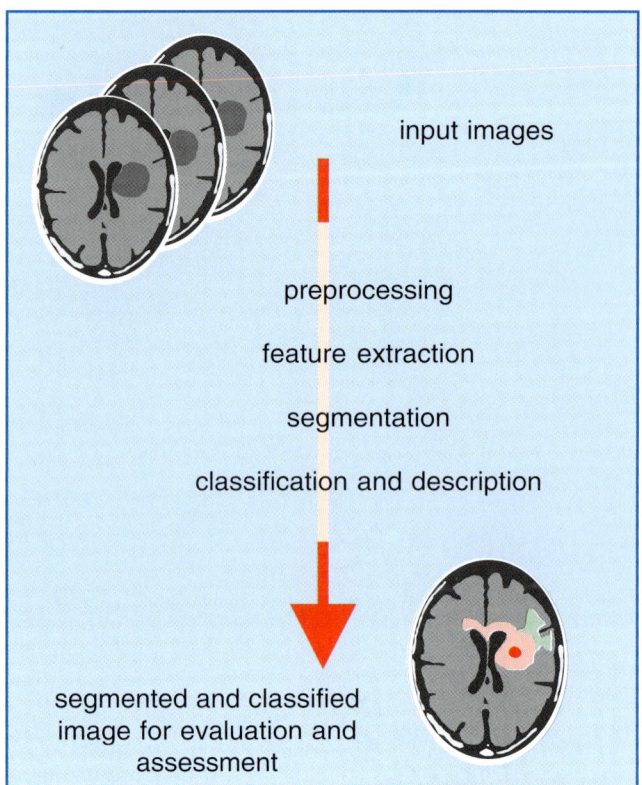

Figure 15-6:
Steps in image analysis: preprocessing improves the quality of the image by reducing artifacts; feature extraction and selection provide the measurement vectors on which segmentation is based. After segmentation, classification and description allow pattern recognition.

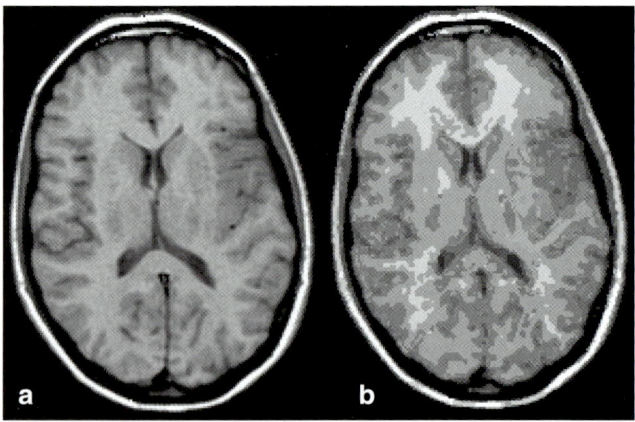

Figure 15-7:
(a) original brain imaging, and (b) segmented image presenting 90 different tissues.

Image Segmentation and Multispectral Analysis

Space and military technologies were the forerunners of many image-processing applications which later found their way into medical imaging. One of the most important was the Landsat program of Nasa. Landsat created sets of images of the earth consisting of four or more images of different spectral windows (usually, two within the visible spectrum and two within the infrared spectrum). Similar approaches are used today in medical image-processing. Image segmentation is one of the most important tools in automated image analysis [2, 7].

Plain and postcontrast T1-weighted, T2-weighted, and diffusion images can be used as multispectral images. Reducing the representation of an image to a small number of components was one of the image-processing projects based on such pictures, a process called feature extraction. It permits the separation of the basic parts of an image by sets of features that can be extracted from the image and, in turn, can be used to calculate other features such as edges and textures.

Segmentation is also applied in preprocessing of images for multimodality image registration. Image segmentation can be used in static images and, quite important for the use of contrast agents, in dynamic time-varying images. The detection of gray-level discontinuity allows the highlighting of points, lines, and edges in an image.

Similarity techniques reveal areas of similar signal intensities using thresholding, region growing, as well as region splitting and merging [6]. An overview of the components of an image-analysis system is given in Figure 15-6. A detailed description of segmentation is beyond the limits of this chapter, but is available in other reports [3, 6, 12].

Multispectral models can be divided into supervised and unsupervised models. An unsupervised classification (like cluster analysis) into connected regions is generally sufficient to provide good partition of an image into relevant component structures. Supervised pattern recognition is mainly successful where a reliable classification can be expected on the basis of *a priori* knowledge of the tissue parameters.

An example is gray- and white-matter separation on the basis of relaxation time data (Figure 15-7).

Practical Applications. In medicine, segmentation is applied for the division of images into components

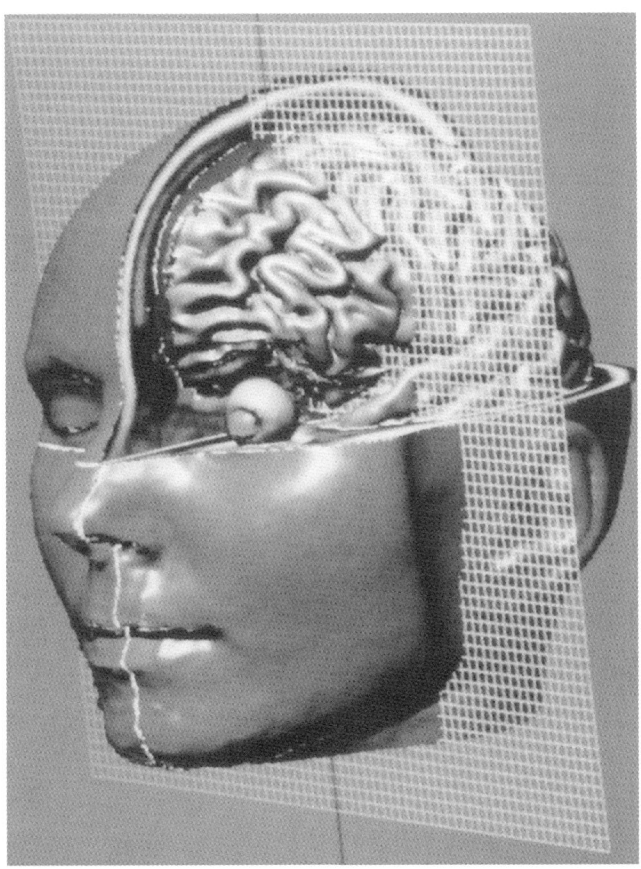

Figure 15-8:
Dissection of an MR imaging-based head model. A wire mesh is used to define cut planes [14]. Similar reconstructions can be used in surgery and radiation therapy planning. One of the main problems in 3D image-processing is that objects within the 3D domain may obscure each other. Therefore any visualization must be preceded by a segmentation step in which 3D regions belonging to an organ must be identified.

reflecting the same or similar tissues. Today, the concept of segmentation and its application of volumetry have become fast and clinically usable. Segmentation allows identification of anatomical areas of interest for diagnosis and therapy, for instance for planning of surgery. Measurement of tumor volume before and after treatment has become a relatively easy task with image segmentation. Among other applications are quantitative measurements of brain atrophy in patients with Alzheimer's disease or alcoholic brain damage.

In cardiac MR imaging segmentation methods and contour-detection methods have been successfully applied to detect the borders of myocardium in order to calculate parameters like ejection fraction and myocardial mass. Automatic contour detection can be used for the three-dimensional depiction of bone or soft tissue structures, e.g., to produce prostheses.

3D Visualization

As with most imaging modalities, MR imaging data are normally presented as two-dimensional gray-scale images. However, MR imaging is essentially a three-dimensional method and can produce three-dimensional data sets of virtually any body organ[1].

The simplest way of visualizing such data sets is by letting the radiologist flip through the data set displayed as slices of 2D images, leaving it up to the radiologist to visualize the structures. Whereas this approach is often suitable for some purposes, like diagnosis, it is less suitable for other purposes, like surgical planning or radiotherapy planning. Thus, there is a need for 3D visualization techniques.

By performing segmentation, surface- or volume-rendering techniques can be applied[8]. The advantage of surface-rendering techniques is that they are easy and fast to visualize and manipulate (by rotation, zooming, etc.).

Since a segmentation has been done, it is possible to manipulate the 3D data set by removing tissues, request volumes and sizes, etc. (Figure 15-8). The disadvantage is that segmentation of the data is required before visualization can be performed, and that some information is lost in the segmentation step.

An alternative technique is volume-rendering. Volume-rendering does not require segmentation. However, the method requires more powerful computers to be fully interactive, and normally requires some interaction to visualize structures of interest.

References

1. Aichner FT, Felber SR, Muller RN, Rinck PA (eds.): Three-dimensional magnetic resonance imaging. An integrated clinical update on 3D-imaging and 3D-postprocessing. Oxford: Blackwell Scientific Publications 1994.
2. Bezdek JC, Hall LO, Clarke LP. Review of MR segmentation techniques using pattern recognition. Med Phys 1993; 20: 1033.
3. Clarke LP, Velthuizen RP, Phuphanich S, Schellenberg JD, Arrington JA, Silbiger M. MRI: Stability of three supervised segmentation techniques. Magn Res Imag 1993; 11: 95.
4. Damadian R. Tumor detection by nuclear magnetic resonance. Science 1971; 171: 1151.
5. Godtliebsen F. A study of image improvement techniques applied to NMR images. Doctoral thesis. Trondheim: The Norwegian Institute of Technology, Division of Mathematical Sciences 1989.
6. Gonzalez RC, Wintz P. Digital image-processing. 2nd ed. Reading (U.S.A.): Addison-Wesley 1987.
7. Lundervold A, Myhr G, Bosnes V, Myrheim J. Automatic recognition of pathological tissues in the central nervous system using MRI contrast agents and pattern recognition techniques. Oslo: Norwegian Computing Center, Report 858, 1992.
8. Maintz JB, Viergever MA: A survey of medical image registration. Med Image Anal 1998; 2: 1-36.
9. Rinck PA, Meindl S, Higer HP, Bieler EU, Pfannenstiel P. MRI of brain tumors: discrimination and attempt of typing by CPMG sequences and in vivo T2-measurements. Radiology 1985; 157: 103.
10. Rinck PA, Petersen SB, Heidelberger E, et al. NMR ventilation imaging of the lungs using perfluorinated gases. Proceedings. The Society of Magnetic Resonance in Medicine. Second Annual Meeting. San Francisco 1983, 302-303, and: Magn Reson Med 1984; 1: 237.
11. Rinck PA, Petersen SB, Lauterbur PC. NMR-Imaging von fluorhaltigen Substanzen. 19-Fluor Ventilations- und -Perfusionsdarstellungen. Fortschr Röntgenstr 1984; 140: 239.
12. Russ JC. The image-processing handbook. 2nd ed. Boca Raton (U.S.A.): CRC Press 1995.
13. Skalej M, Higer HP, Meves M, Brückner A, Bielke G, Meindl S, Rinck PA, Pfannenstiel P. T2-Analyse normaler und pathologischer Strukturen des Kopfes. Digit Bilddiag 1985; 5: 112.
14. Tiede U, Bomans M, Höhne KH, Pommert A, Riemer M, Schiemann T, Schubert R. A computerized three-dimensional atlas of the human skull and brain. In: Aichner FT, Felber SR, Muller RN, Rinck PA (eds.): Three-dimensional magnetic resonance imaging. An integrated clinical update on 3D-imaging and 3D-postprocessing. Oxford: Blackwell Scientific Publications 1994, 61-74.
15. Torheim G, Rinck PA, Jones RA, Kværness J. A simulator for teaching MR image contrast behavior. Magn Res Materials 1994; 2: 515.
16. Torheim G, Rinck PA. MR Image Expert - interactively teaching contrast behavior in magnetic resonance imaging. In: Lemke HU, Vannier MW, Inamura K, Farman AG (eds.): Computer Assisted Radiology, CAR '96. Amsterdam: Excerpta Medica 1996, 619.

Chapter Sixteen

Dynamic Imaging

Introduction

The combination of MR imaging with extrinsic contrast-changing agents have led many researchers to try to highlight anatomical and pathological structures and processes which, *per se*, are invisible on single plain images [18]. Image series of the same anatomical structure during a time period before and after the application of a contrast agent add another dimension, commonly described as perfusion imaging.

Perfusion describes blood delivery to tissues, usually flow on the capillary level.

Perfusion imaging can be categorized into two classes: those methods monitoring target-tissue signal changes after the application of an extrinsic contrast agent, and those relying upon intrinsic factors such as increased blood volume or blood oxygenation and flow in the microvasculature. The former can be visualized by dynamic imaging, the latter by functional imaging.

However, even the individual images of a dynamic time series will, in some cases, not yield all information contained in them. Thus, the combination of dynamic MR and mathematical image manipulation has been proposed (Figure 16-1).

This chapter gives an overview of some of the techniques. It cannot be exhaustive considering the wide spectrum of the field [15,27].

Traditionally, the following methods have been applied to analyze dynamic contrast-enhanced images:

- Visual inspection of time-intensity curves;
- Visual inspection of the time series by running the images in a movie;
- Subtraction images.

However, the most intriguing and interesting way of processing the data is by the creation and subsequent visualization of parametric maps: images combining parametric images derived from the information present in the image series with anatomical information (Table 16-1). To be clinically useful, such a postprocessing method must be robust, reliable and automatic or semi-automatic.

Figure 16-1:
Combining image-processing, time series images, and contrast agents can give surprising results. Always check if the outcome makes sense.

1	Acquisition of a time series of images before and after the contrast agent is administered.
2	Noise-filtering of the acquired images (optional).
3	Motion correction (optional).
4	Definition of time-intensity curves.
5	Processing of time-intensity curves.
6	Presentation of the data.

Table 16-1:
Steps of the entire dynamic acquisition and image-processing procedure. The noise-filtering and motion-correction steps are optional, but in many instances necessary and fairly complicated.

Inherent Problems

In order to process the time series (= the series of images starting with one or several precontrast images followed by the contrast-enhanced images) on a pixel-by-pixel basis, the spatial location of picture elements within the image matrix must be exactly the same.

In many instances, this is not the case due to movement (Figures 16-2 and 16-3).

This hampers the calculation and evaluation of time intensity curves and parametric images[5]. Pixels or regions-of-interest have to be reallocated manually or, preferably, automatically, e.g., by contour recognition. The realignment of the images is often referred to as 'image registration'.

This problem is even more complicated in contracting organs such as the heart[26]. Here artifacts can be corrected by highlighting the edges of the organ by drawing the boundaries either manually, semiautomatically or automatically if there is some *a priori* knowledge, e.g., previous knowledge of the outline of the organ and the ratio between its length and width (Figure 16-4)[28]. A more sophisticated approach is the creation of auto-ROIs[26].

Image registration of contrast-enhanced images of non-rigid organs is a difficult problem due to the changes not only in contrast but also in the structures visible. Whereas rigid organs like the brain can be aligned by performing translations and rotations, non-rigid organs require non-rigid reformations, which potentially can change the structures to be observed.

An additional problem is the propagation of mistakes, e.g., through artifacts which are not recognizable any more on processed images. They can be caused by a change of relative or absolute signal intensity on the image influenced by outside factors such as surface coils.

Furthermore, due to the frequent presence of time constraints, the signal-to-noise ratio is often quite low in dynamic contrast-enhanced MR imaging. This necessitates the need for noise-reduction techniques[21] which are commonly performed by applying so-called 'filters' either in the spatial (image) or in the temporal domain.

There are some very sophisticated filters which can remove noise in a way which was previously only possible by increasing the magnetic field strength or improvements in the amplifiers or other hardware.

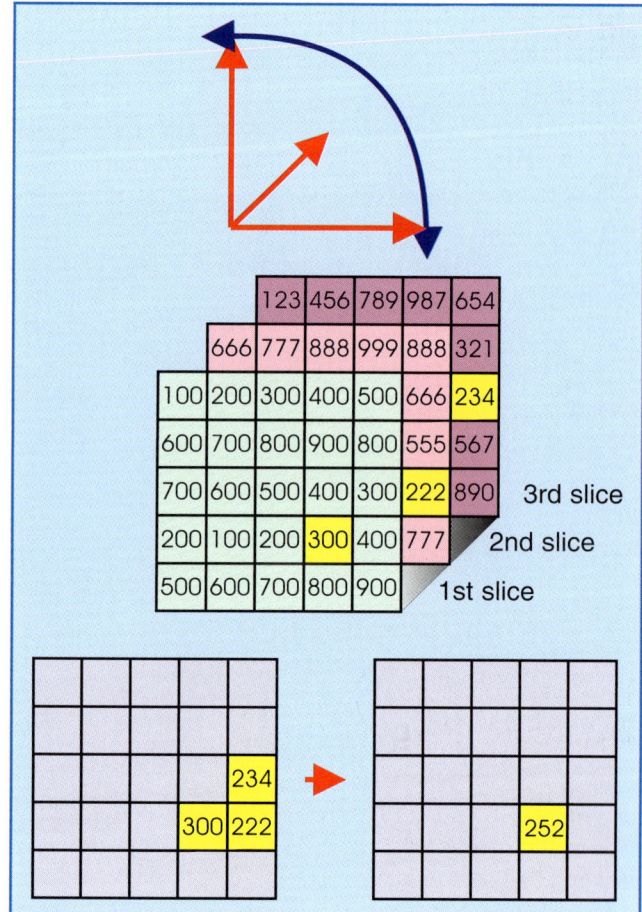

Figure 16-2:
Organs in the human body may move in all three dimensions and rotate at the same time. Thus, multichannel images can easily be out of alignment due to motion and rotation. Pixels have to be realigned for image-processing and analysis.

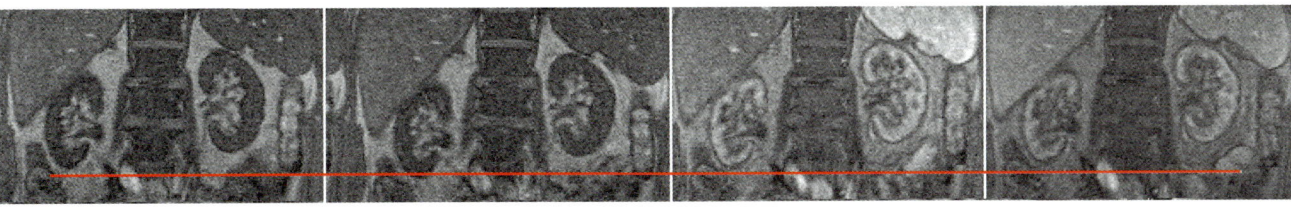

Figure 16-3 (top):
Three images taken out of a dynamic study of the kidneys. During the examination, both kidneys move in all three dimensions and rotate. In this example, the right kidney moves several pixels up and down during the time series (red line). Mathematical processing of the images can help eliminate some of the movement and facilitate processing the time series.

Figure 16-4 (bottom):
Dynamic study of the kidneys. (a) a reference image is selected; (b) the boundaries of the kindey can be manually drawn and adjusted in each image of the examinations series; or (c) an edge-detection program can define the boundaries through contour-enhancement and automatically or semiautomatically adjust them by comparing the ratios between fixed parameters.

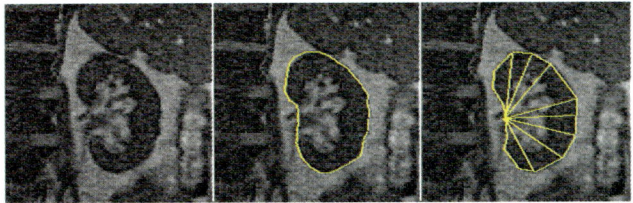

Dynamic Image-Processing

For a long time, one of the main objectives researchers wanted to achieve was the improvement of MR image contrast by means of possible electronic contrast agents without the application of any exogenous medium. Relaxation-time images and, at a later stage, segmented images were the first research targets of this kind of MR image-processing.

However, this notion of the existence of electronic contrast agents that can be employed to highlight pathologies through image-processing derives from a wrong hypothesis, namely that there is hidden information about tissue structures or processes in the original images or raw data. The additional information given by a pharmaceutical contrast agent, i.e., vascularity, membrane permeability, etc., is not given by any plain imaging modality and therefore cannot be electronically enhanced; in other words, if the contrast-to-noise ratio is zero in the raw data, no contrast enhancement by image-processing will be possible.

On the other hand, it might be possible to enhance minimum quantities of contrast-agent uptake, which create only minor contrast changes if appropriate image-processing techniques are applied.

Since the temporal uptake pattern of contrast agents *in vivo* can vary between healthy and diseased tissue, differences in uptake can be of diagnostic value. In signal intensity-versus-time curves, some lesions reveal steeper slope, higher maximal signal intensity, and

faster wash-out of the contrast agent than the surrounding tissue [7,11,25].

A factor contributing to the uncertainty of quantitative data acquired with dynamic images is the wide scatter of blood volumes and transit times in different organs, or even within similar organ structures. In the brain, blood volumes and transit times tend to be higher in the pons, cerebellum, and medulla than in the midbrain and forebrain, which suggest some general regional differences in microvascularity [16]. These differences can be substantial and may overlap with the values of pathologies. Proper knowledge of the normal range of anatomy and physiology is essential for the final assessment of data or calculated 'parametric' images.

Dynamic imaging to highlight these additional contrast parameters was developed for nuclear medicine and CT.

Dynamic PET scans for regional cerebral blood flow measurements, for instance, can be performed at intervals of several seconds. However, radioisotope methods lack good temporal and spatial resolution, whereas CT is fast and shows anatomical structures in detail.

Conventional MR imaging is a slow imaging technique. All of the classical imaging sequences have long scan times. Only in the late 1980s did rapid imaging pulse sequences become available. They made it possible to follow the dynamic signal-intensity changes after contrast injection (Figure 16-5).

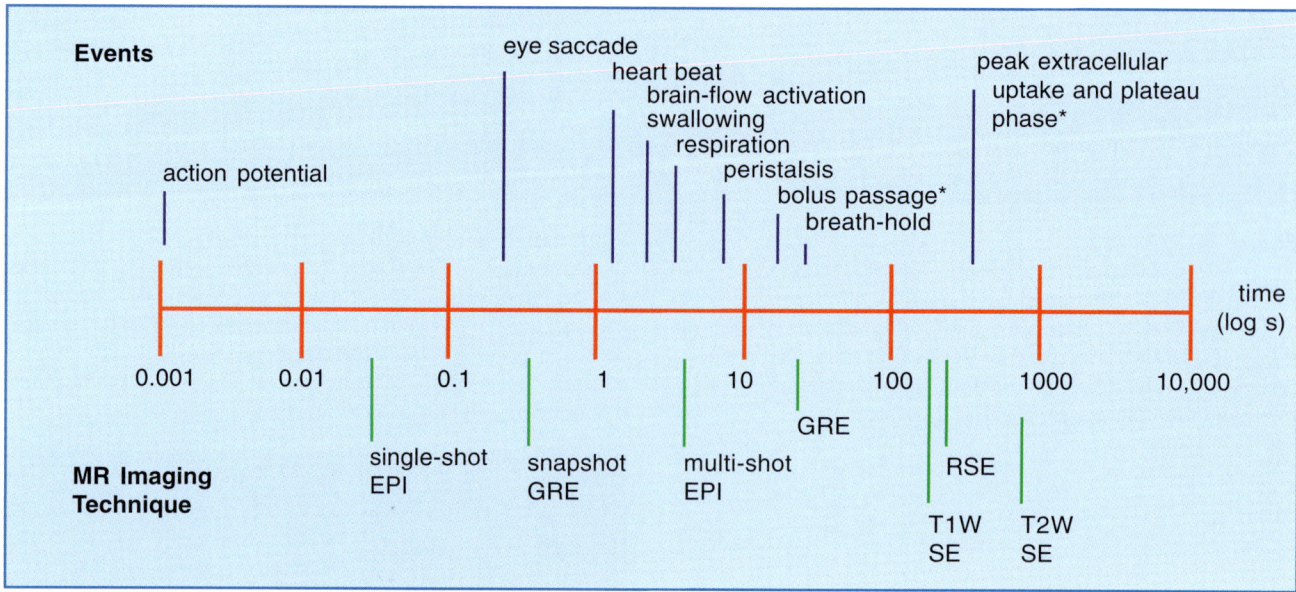

The prerequisite of sufficient spatial and time resolution was given with rapid imaging, and the uptake of contrast agents could be readily monitored by MR imaging.

Maximal vascular concentration of an ECF-space contrast agent will occur during the first pass through the body in the vascular phase. Maximal tissue concentration, as well as enhancement rate and onset, depend on a number of not yet completely understood factors such as vascularity, membrane permeability, and leakage of the blood-brain or the tumor barrier, and venous outflow.

Thus, maximal tissue concentration may occur between several seconds and minutes after the injection of the contrast agent (Figure 16-6). The agent then diffuses into the extracellular space until an equilibrium is reached during the plateau phase. Concentration decreases during the wash-out phase which follows. The enhancement processes can be monitored by time-intensity curves.

Although both the vascular and the extracellular phase of uptake go hand-in-hand, one must distinguish monitoring the dynamic signal-intensity changes of the first pass of a contrast agent bolus such as in brain and heart imaging and the slower tissue uptake as it is commonly done in breast lesions.

The time resolution required to study the dynamics of the contrast agent uptake in an organ depends on the requirements of the individual investigation. Better time resolution will improve the characterization of the bolus while passing through the body region-of-interest, in particular if these images are to be treated by mathematical techniques to analyze bolus behavior[10].

Figure 16-5:
Graphic depiction of the time scale of different events in man versus MR imaging methods. Imaging speed must be 3-10 times faster than the event to be monitored or avoided.
* = of contrast agent.

Parameter
Onset of enhancement (onset time): time delay between bolus injection and point of, e.g., 10% of maximum enhancement.
Slope of enhancement within a certain time period.
Maximum relative signal intensity (also: peak enhancement, maximum enhancement).
Time to peak.
Maximum wash-in slope, i.e., steepest slope of enhancement.
Mean gradient: average rate of change in SI between 10% and 90% of maximum enhancement.
Relative signal intensity/contrast enhancement.
Wash-out slope.
Mean transit time.
Cross-correlation coefficient between different non-pathological and pathological ROIs.
Area under the curve: the relative regional blood volume is taken to be proportional to the area under the time-concentration curve, determined either by numerical integration or after fitting a gamma-variate curve to the data.

Table 16-2:
Some parameters in dynamic imaging. Note: there is no general agreement on terms and nomenclature — and there is no agreement on the diagnostic implication of these parameters.

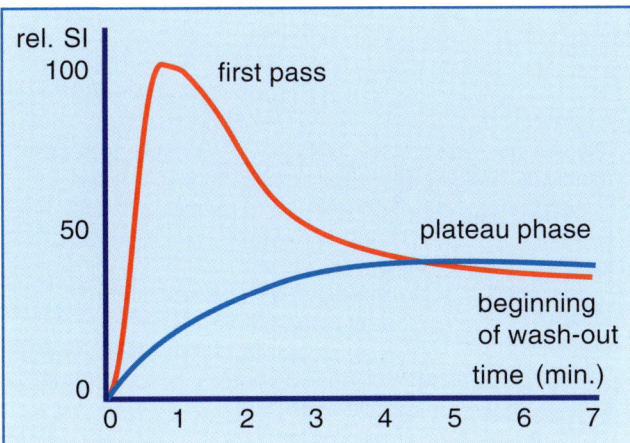

Figure 16-6:
Bolus injection of contrast agent: rapid intravascular uptake after bolus injection leading to first-pass phenomenon versus relatively slow extracellular uptake in easily accessible regions. rel. SI = relative signal intensity (%).

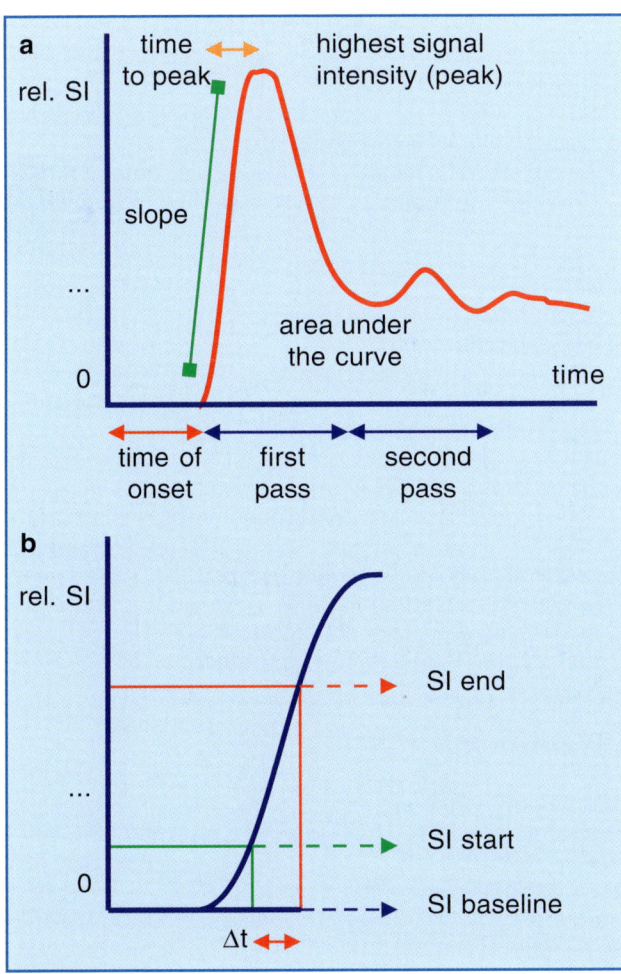

Figure 16-7:
Parameters of bolus injection of a T1 contrast agent. The maximum wash-in slope is considered one of the most important parameters, indicative of, e.g., tumor vascularity. It can be calculated as follows: (SIend - SIstart / SIbaseline × DT) × 100 (%/s).

Such mathematical treatment which can lead to parametric images is beyond the capability of the human brain, which is unable to make a correlation if the number of features exceeds three. The differences can be distinguished either by a qualitative description of the change in signal intensity observed over time or by quantification of certain parameters by fitting the concentration-time course to a pharmacokinetic model [13,23,25]. Explanations of the parameters are shown in Figure 16-7.

In this context, it is important to remember that MR imaging has no standard units for signal intensity. It cannot be measured in units comparable to Hounsfield units in CT. Several attempts have been made to create a quantitative scale, but none of these methods creates an absolute standard.

Thus, signal intensity-versus-time curves remain only qualitative representations of contrast passage through tissues. To acquire quantitative data, concentration-versus-time curves have to be fitted by the principles of tracer dilution kinetics. Time-intensity curves of bolus injections have to be corrected, i.e., cleaned of the influence of recirculation to achieve better raw data for parameter calculation, e.g., by gamma-variate fitting [6,21,22].

An overview of qualitative or semi-quantitative parameters commonly determined by time-intensity curves is given in Table 16-2.

The concentration (C) of the contrast agent in each pixel, at each time point, can be calculated from the known relaxivity of the contrast agent and the respective relaxation times, T1 and $T1_0$:

$$C = (R1 - R1_0) / r1$$

where $R1 = 1/T1$ at the measured point during the dynamic study, $R1_0 = 1/T1_0$ prior to injection, and $r1$ is the longitudinal relaxivity of the contrast agent at the given magnetic field strength and 37° C.

The resulting concentration-time curves can be fitted to pharmacokinetic models [22]. From this, values of permeability and leakage volume are derived. For extended image-processing and data analysis, independent software systems running on personal computers have been developed. Software systems such as **Dynalize** offer an integrated, standardized way of performing both image-processing and image analysis on all kinds of dynamic images independent of the imaging equipment on which the images were acquired in a standardized way [26,27].

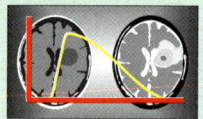

You can experience interactively dynamic image processing with **Dynalize**.

Contrast-enhanced Dynamic Imaging

This tutorial gives you a quick overview of how to process and analyze data using **Dynalize**.

Install **Dynalize** following the instructions given in the Software Appendix of this book.

Start **Dynalize**. When you want to begin working with **Dynalize** clinical data, insert the EMRF CD-ROM into the CD drive.

Select *File/Open*, or click on this *File Open* icon

- Select the file Breast T1.xv in the *Dynalize Data* directory on the CD. **Dynalize** will ask you to specify which slices of the data set to load. Mark slice 22, and click *OK*.
  This data set is of a patient with breast carcinoma. A total of 44 slices covering the breast were acquired before, during, and after injection of a gadolinium-based ECF-space contrast agent. Nine time frames were acquired with a temporal resolution of 58 seconds. The scan sequence was a T1-weighted RF-spoiled gradient-echo sequence.

- Click with the right button inside the image, and select *Window Setting...* in the menu. Adjust window level and width according to your wishes.

- To appreciate the dynamic nature of the data set, select *View/Cine Loop...* Increase the speed of the animation by dragging the speed bar followed by *OK*. The contrast agent is entering this slice in image frame number 4. Close the animation by clicking *Close*.

- Click on *Select Frame* and go to frame 4. Draw an elliptical ROI by positioning the mouse, pushing and holding the left mouse button while dragging the mouse, and releasing the button when the ROI has the desired size and shape. Position the ROI as shown in Figure 1.

- When you release the button, a dialog box will appear on screen. Click *OK*. Dynalize will now automatically define the ROI in the image, create identical ROIs in all frames, and create and display a time-intensity (TI) curve.
  Each point on the TI-curve is the average intensity value from the corresponding ROI. The curve should

look similar to the one in Figure 3. A large and rapid increase in signal intensity has been correlated to malignancy, although also some benign tumors can exhibit similar contrast behavior.

- Now, create a parametric image by selecting *Analysis/ Parametric Image/Enhancement/Max(abs)*. Select *Whole image* when prompted for it. Then, select frame 2 as baseline. This will automatically create a parametric map by calculating the maximum increase in intensity from the baseline.

- Then, unless you are already looking at frame 4, click *Select Frame* and select frame 4.

- Adjust *Window level and width* to get an image similar to the one shown in Figure 2.

Let's look at brain perfusion and another kind of parametric maps now.

- Load the data set *brain_perf.xv* from the *Dynalize Data* directory on the EMRF CD-ROM.
  Select slice 7. This data set is a T2*-weighted perfusion study of a patient with recent onset of brain ischemia (stroke). A rapid EPI sequence was used. In contrast to the T1-weighted breast data, the intensity from this data set decreases at the time the contrast arrives in its target region.

- Create a parametric image by selecting *Analysis/ Parametric Image/Enhancement/Min(%)*. This will give you an image as shown in Figure 4a.

- Since non-perfused areas are black in this image, select *View/Invert Images* to invert the gray scale and create a negative image. This will give you the image in Figure 4b. Now, the non-perfused areas are bright. However, due to noise and other factors, it is not easy to detect the non-perfused areas for the untrained eye — or even for the trained.

Definitions

ROI: region-of-interest.
Frame: an image series along the time axis ('time series').
Slice: an image along the z (spatial) axis.
Time intensity (TI) curve: a curve created by calculating the average intensity in ROIs in a frame (time series of images).
Parametric image: an image created by calculating a parameter value for each pixel in a slice and plotting the values as an image, typically a gray-scale image with the intensities proportional to the parameter value.

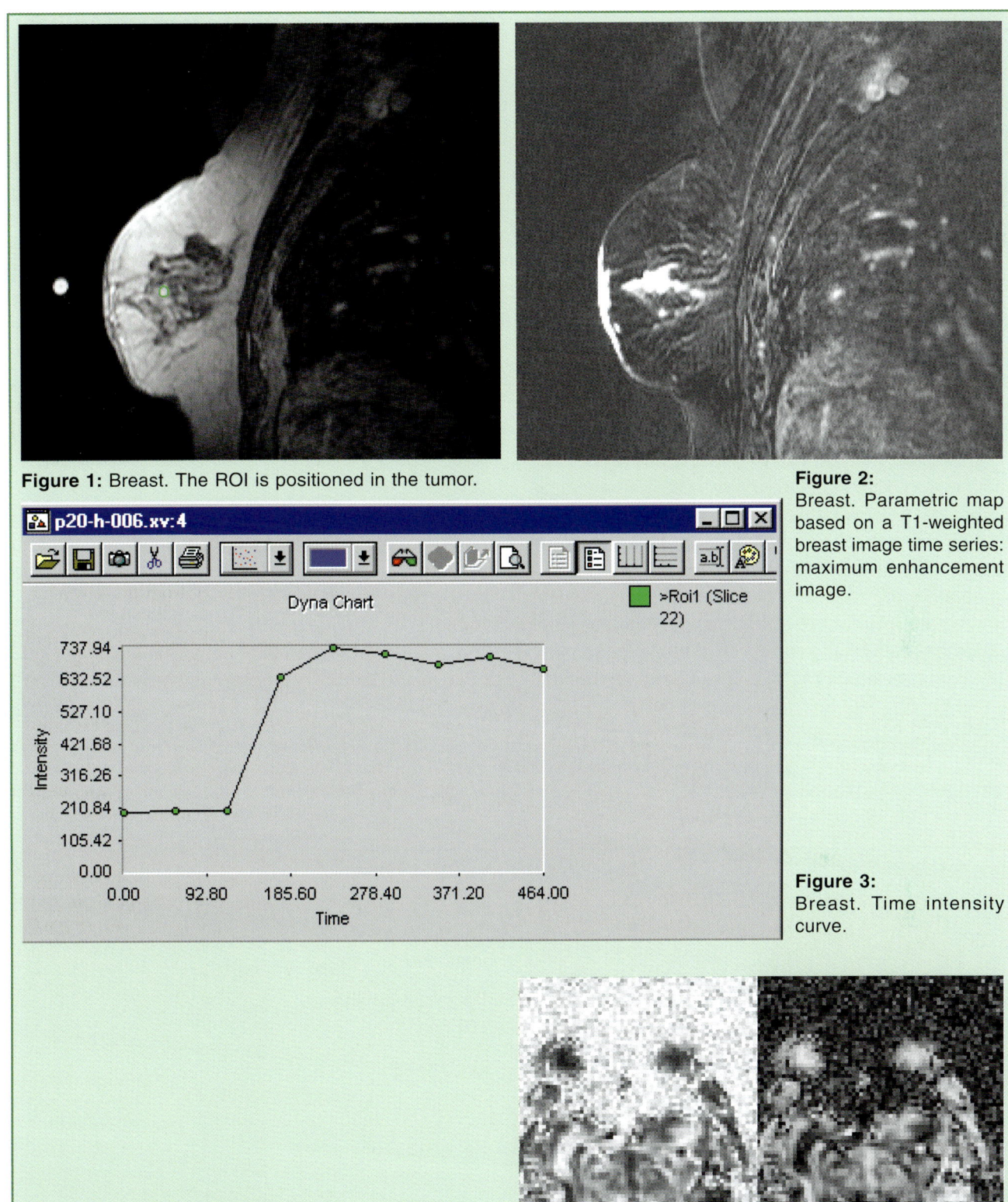

Figure 1: Breast. The ROI is positioned in the tumor.

Figure 2: Breast. Parametric map based on a T1-weighted breast image time series: maximum enhancement image.

Figure 3: Breast. Time intensity curve.

Figure 4: Brain. Minimum enhancement image; (a) positive, (b) inverted = negative image.

Clinical Examples

Breast Imaging. Dynamic imaging of the breast became the first major application of dynamic MR imaging with gradient-echo pulse sequences used. The combination of rapid imaging and contrast-agent application increased both sensitivity and specificity of breast MR imaging (magnetic resonance mammography = MRM) and allowed the differentiation between benign and malignant lesions.

Signal intensity-versus-time uptake curves showed that malignant lesions take up contrast agent faster than benign lesions, although there remained a certain overlap. Since these measurements were done manually, it was difficult to find the pixel or region of highest uptake of contrast agent within the breast, in particular on gray-scale images. In these cases, postprocessing became very valuable. Originally, subtraction images were used; however, this approach highlights all pixels with contrast enhancement without any differentiation between fast and slow enhancement over time.

Then, mathematical approaches to enhancement curves were introduced. Pixel-by-pixel calculation of enhancement intensity and speed or slope led to parametric images which can be color-coded in a way that regions of fast and high enhancement are highlighted in a specific color. For instance, enhancement of more than 90% in less then 90 seconds on T1-weighted images, or signal intensity loss of > 20% during the first 30 seconds after contrast material injection on T2*-weighted images, is considered typical for malignant breast lesions, although not all malignant lesions follow this pattern (Figure 16-8) [3, 4,7, 8,11,12].

Processing of dynamic imaging with color coding can also visualize the enhancement pattern over time. When large enough, fibroadenomas usually demonstrate initial peak enhancement in the center of the tumor, whereas carcinomas tend to enhance in the periphery.

However, since the enhancement pattern depends on the vascularity of the lesion, no direct histological tumor-typing by dynamic MR imaging is possible.

Brain Imaging. Patients with acute stroke make up the group of most interest for dynamic imaging of the brain [17]. In clinical practice, perfusion imaging has already proven to be an early and reliable predictor of prognosis in stroke patients (Figure 16-9).

A number of researchers have found that the area of perfusion deficit seen in cerebral blood flow (CBF)

Figure 16-8:
Dynamic uptake pattern of a Gd-based ECF-space agent in breast lesions. Enhancement of more than 90% in less than 90 s (pink area) after bolus injection occurs most likely in invasive ductal carcinomas only. Thus, such tumors can be identified in parametric images, where all pixels with enhancement > 90% at time < 90 s can be color-coded. The carcinoma is this picture appears in bright red.

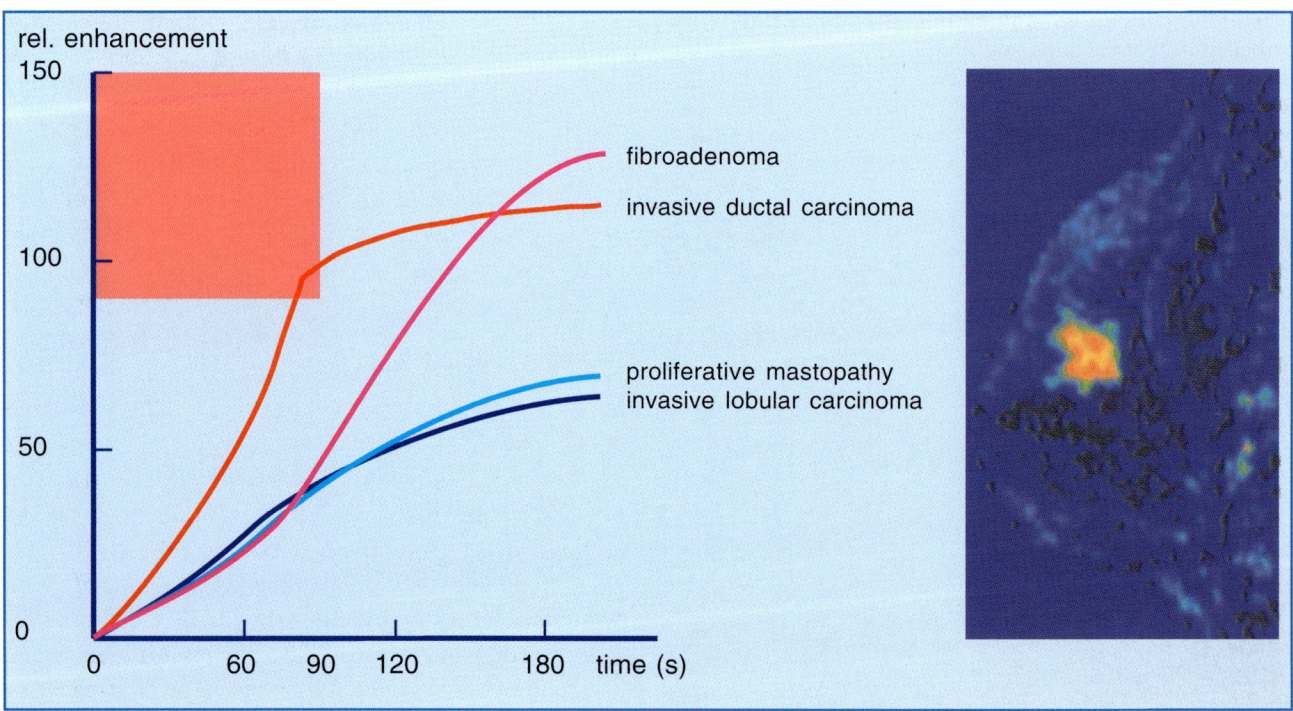

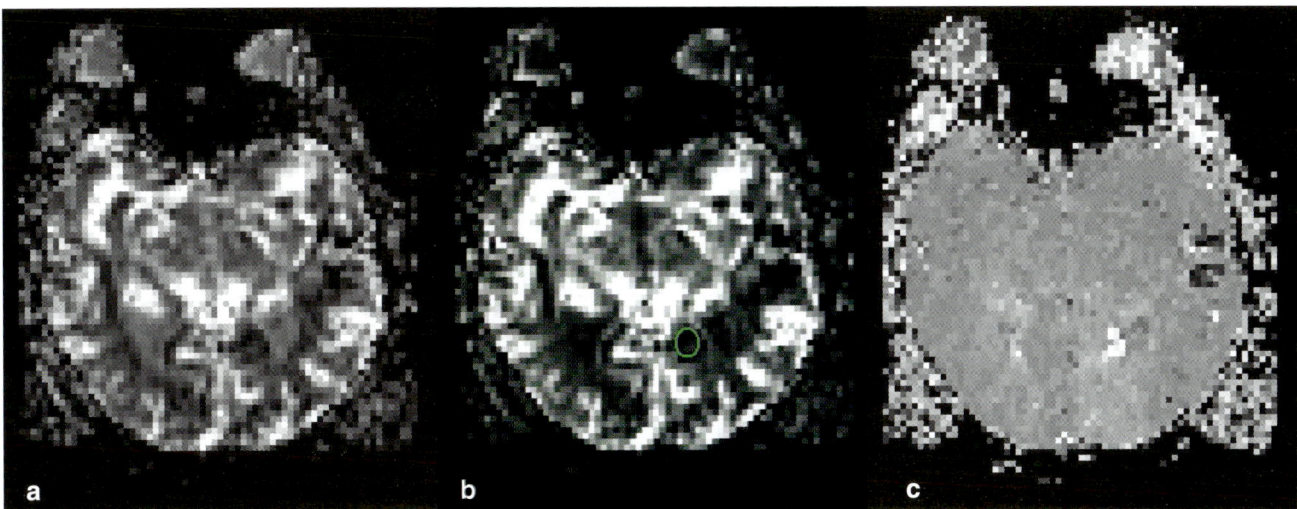

Figure 16-9:
Parametric images of the brain of a patient with recent stroke. (a) 'Area under the curve' image which is correlated to blood volume. The image was created by first converting the time-intensity curves into time-concentration curves by using mathematical processing. Non-perfused areas have a flat curve, thus the areas are small and therefore the non-perfused regions show as dark. In (b) a ROI has been drawn to indicate an ichemic region. (c) depicts a time-to-peak image of the same slice. The perfusion in the ischemic region is delayed; it shows up in light gray.
All images were produced with *Dynalize 1.0*.

Figure 16-10:
Based upon defined ROIs, parametric images of the heart can be obtained by combining anatomical and functional information. In this case, the heart is semiautomatically divided into ROIs which follow the supply area of the coronary arteries. The parametric image represents the cross-correlation coefficient (CCC) calculated one week after coronary infarction (dark area).

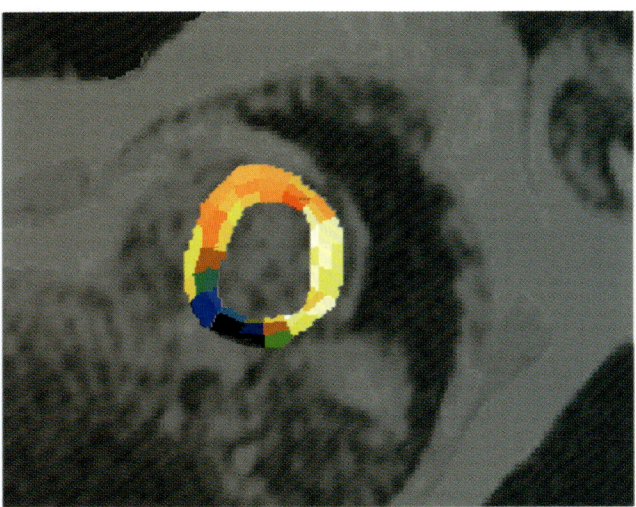

and mean transit time (MTT) maps may extend beyond the area of hyperintensity seen in diffusion-weighted imaging and that the size of the infarct finally seen on delayed T2-weighted images matches the area of perfusion defect.

Ultimately, the goal of perfusion imaging remains the visualization of the penumbra and thus the distinction between normal, salvageable and irreversibly damaged tissue. Tracer kinetics principles first employed in nuclear medicine can be applied to generate cerebral blood volume maps [2,19,20,24,29].

Regional cerebral blood volume (CBV) can be estimated by fitting first-pass transit curves to the pixel intensities of a series of images.

Heart Imaging. Screening for ischemic heart disease requires both high spatial and temporal resolution images with detection and quantification of abnormal wall motion, evaluation of cardiac metabolism, and measurement of regional myocardial perfusion [1,14].

In the heart, the much higher blood volume and the abundance of susceptibility artifacts in plain imaging commend assessment of perfusion by T1-weighted dynamic imaging during the first pass of an appropriately low dose of a paramagnetic contrast agent.

The main goal of myocardial perfusion imaging is the detection and delineation of hypoperfusion due to non-occlusive coronary artery stenosis.

Assessment of myocardial blood flow is difficult because a large fraction of extracellular contrast agents will extravasate into myocardial tissue during the first pass, making myocardial signal intensity dependent on both extraction fraction and flow. However, some research groups have succeeded in obtaining excellent delineation of hypoperfused areas using first-pass dynamic imaging with ECF-space contrast agents under stress, usually pharmacological stress (Figure 16-10)[9].

Other Applications. There are numerous other applications of dynamic imaging, including imaging of the liver, the kidneys, muscles and joints, the urinary bladder, and the prostate.

Image-processing is useless if applied randomly without a well-defined aim. Many approaches to explain results of dynamic imaging and image-processing are based on hypotheses which are still to be proved, and much research in this field is empirical and heuristic.

References

1. Atkinson DJ, Burnstein D, Edelman RR. First-pass cardiac perfusion: evaluation with ultrafast MR imaging. Radiology 1990; 174: 757.

2. Belliveau JW, Rosen BR, Kantor HL, et al. Functional cerebral imaging by susceptibility-contrast NMR. Magn Res Med 1990; 14: 538

3. Boetes C, Barentsz JO, Mus RD, et al. MR characterization of suspicious breast lesions with a gadolinium enhanced Turbo FLASH subtraction technique. Radiology 1994; 193: 777.

4. Flickinger FW, Allison JD, Sherry RM, Wright JC. Differentiation of benign from malignant breast masses by time-intensity evaluation of contrast enhanced MRI. Magn Res Imag 1993; 11: 617.

5. Gehrig G, Kikinis R, Kuoni W, von Schulthess GK, Kübler O. Semiautomated ROI analysis in dynamic MR studies. Part I: image analysis tools for automatic correction of organ displacements. J Comput Assist Tomogr 1991; 15: 725.

6. Gonzalez RC, Wintz P. Digital image processing. 2nd ed. Reading (U.S.A.): Addison-Wesley 1987.

7. Gribbestad IS, Nilsen G, Fjøsne HE, Kvinnsland S, Haugen OA, Rinck PA. Comparative signal intensity measurements in dynamic gadolinium-enhanced MR mammography. J Magn Reson Imaging 1994; 4: 447.

8. Heywang-Köbrunner SH, Beck R. Centrast-enhanced MRI of the breast. 2nd ed. Berlin, Heidelberg, New York: Springer 1995.

9. Higgins CB, Sakuma H. Heart disease: functional evaluation with MR imaging. Radiology 1996; 199: 307.

10. Jones RA, Haraldseth O, Müller TB, et al. K-space substitution: a novel dynamic imaging technique. Magn Res Med 1993; 29: 830-834.

11. Kaiser WA. Dynamic magnetic resonance breast imaging using a double breast coil: an important step towards routine examinations of the breast. Frontiers in European Radiology 1990; 7: 39.

12. Kvistad KA, Rydland J, Vainio J, et al. Breast lesions: evaluation with dynamic contrast-enhanced T1-weighted MR imaging and with T2*-weighted first-pass perfusion MR imaging. Radiology 2000; 216: 545-553.

13. Larsson HBW, Stubgaard M, Søndergaard L, Henriksen O. In vivo quantification of the unidirectional influx constant for Gd-DTPA diffusion across the myocardial capillaries with MR imaging. J Magn Reson Imaging 1994; 4: 433.

14. Lombardi M, Rovai D, Kværness J, et al. Myocardial perfusion in ischemic heart disease: the use of contrast agents in echocardiography and in magnetic resonance imaging. In: Rinck PA, Muller RN (eds.) New developments in contrast agent research. Proceedings of the 4th Special Topic Seminar of the European Magnetic Resonance Forum, 1994. Mons, Belgium: EMRF 1995. 137.

15. Maintz JB, Viergever MA. A survey of medical image registration. Med Image Anal 1998; 2: 1-36.

16. Nakagawa S-Z, Lin D, Bereczki G, et al. Blood volumes, hematocrits, and transit-times in parenchymal microvascular system of the rat brain. In: LeBihan D (ed.). Diffusion and perfusion magnetic resonance imaging. Applications to functional MRI. New York: Raven Press 1995. 193.

17. Orrison WW, Lewine JD, Sanders JA, Hartshorne MF (eds.). Functional brain imaging. St. Louis (U.S.A.): Mosby Year Book 1995.

18. Rinck PA, Myhr G. The gadolinium chelates: clinical applications. In: Dawson P, Cosgrove D, Allison D (eds.): Textbook of contrast media. Oxford: Isis Medical Media 1999, 333-353.

19. Rosen BR, Belliveau JW, Buchbinder BR, et al. Contrast agents and cerebral hemodynamics. Magn Res Med 1991; 19: 285-292.

20. Rosen BR, Belliveau JW, Chien D. Perfusion imaging by nuclear magnetic resonance. Magn Reson Q 1989; 5: 263.

21. Sebastiani G, Godtliebsen F, Jones RA, Haraldseth O, Müller B, Rinck PA. Analysis of dynamic magnetic resonance images. IEEE Transactions on Medical Imaging 1996; 15: 268.

22. Thompson HK, Starmer CF, Whalen RE, McIntosh HD. Indicator transit time considered as a gamma variate. Circ Res 1964; 14: 502.

23. Tofts PS, Berkowitz B, Schnall MD. Quantitative analysis of dynamic Gd-DTPA enhancement in breast tumors using a permeability model. Magn Res Med 1995; 33: 564.

24. Tofts PS, Brix G, Buckley DL et al. Estimating kinetic parameters from dynamic contrast-enhanced T1-weighted MRI of a diffusible tracer: standardized quantities and symbols. J Magn Reson Imaging 1999; 10: 223-232.

25. Tofts PS, Kermode AG. Measurement of the blood-brain barrier permeability and leakage space using dynamic MR imaging. 1. Fundamental concepts. Magn Res Med 1991; 17: 357.

26. Torheim G, Lombardi M, Rinck PA. An independent software system for the analysis of dynamic MR images. Acta Radiol 1997; 38: 165-170.

27. Torheim G, Rinck PA. Dynamic contrast-enhanced magnetic resonance imaging and image processing. In: Thomsen HS, Muller RN, Mattrey RF (eds.): Trends in contrast media. Berlin: Springer 1999, 285-295.

28. von Schulthess GK, Kuoni W, Gehrig G, Wüthrich R, Duewell S, Krestin G. Semiautomated ROI analysis in dynamic MR studies. Part II: Application to renal function examination. J Comput Assist Tomogr 1991; 15: 733.

29. Østergaard L, Smith DF, Vestergaard-Poulsen P, et al. Absolute cerebral blood flow and blood volume measured by magnetic resonance imaging bolus tracking: comparison with positron emission tomography values. J Cereb Blood Flow Metab 1998; 18: 425-432.

The Field-Strength War

Like almost everything in this world, MR machines come in different sizes: extra-small, small, medium, large, and extra-large. The technical terms in MR lingo for these sizes are ultralow, low, medium, high, and ultrahigh field machines. These terms refer to the magnetic field strength of the respective machine. The field strength is measured in Tesla (T), a unit that replaced the former unit of Gauss (G) some years ago, although Gauss is still used sometimes (10,000 G = 1 T).

Ultralow-field machines operate at a field strength below 0.1 T, low field between 0.1 and 0.5 T, medium field between 0.5 and 1 T, high field between 1 and 2 T, and ultrahigh field machines above 2 T.

In clinical surroundings, the national radiological protection boards allow machines as high as 2.0-2.5 T. Everything above this limit is considered potentially hazardous and thus should only admitted to research facilities — particularly if fast gradient-switching is used..

In describing MR machines, natural scientists prefer to talk about frequencies instead of field strengths. This is because different nuclei in the periodic system possess different resonance frequencies. At 1 T, for instance, protons resonate at 42.58 MHz, whereas at the same field strength, phosphorus nuclei resonate at 17.23 MHz. For clinical imaging purposes in medicine, this is of no importance because only proton MR imaging is used.

Strolling down the aisles of the world's biggest commercial exhibition of medical imaging equipment at the annual meeting of the Radiological Society of North America, one could find small machines operating at 0.06 T and huge machines operating at 4.0 T. Their magnets are different: below approximately 0.3 T, the magnets are permanent and resistive or electromagnetic, but above this field the magnets are superconductive. All these magnet types have their pros and cons.

Why does one find small ultralow field MR imagers and high field machines operating at fields 100 times stronger ? Why are there not only low or high field machines ?

The field-strength question has divided the MR community since the early 1980s. At that time, all MR machines operated at low fields, and many of the prototypes had strengths of approximately 0.15 T. Researchers did not believe that imaging at higher field would be possible because higher radio frequencies would not be able to penetrate the human body. Like many other predictions in MR imaging, this prediction was wrong.

MR images at that time were crude, blurry, and generally worse than CT images. Scientists working for the R&D divisions of companies producing MR equipment were asked:

»How do you get better image quality ?« They had a simple answer: »Increase field strength.«

From analytical applications of MR, it was known that the signal-to-noise ratio increases when you increase the field strength. The better your signal-to-noise, the better your image will be. Higher fields also require higher gradient strength to reduce the chemical-shift artifacts created by these fields. In turn, this led to better spatial resolution. So some manufacturers, driven by their research and marketing people, moved to high-field superconductive magnet systems. These systems were (and in some instances still are) huge, dinosaurlike machines. They were expensive, difficult to produce, and costly to maintain, but image quality suddenly became outstanding.

Another argument supported the development of high field machines; only these machines are able to produce *in vivo* MR spectra for phosphorus or proton spectroscopy. At this time, one of the aims in the development of MR in medicine was to combine imaging and spectroscopy to acquire morphological and metabolic information about the human body. The higher the field, the more detailed spectroscopic information will be.

However, *in vivo* spectroscopy did not take off, whereas the popularity of MR imaging exploded. Dedicated imaging machines became the rule, combined imaging and spectroscopy the exception.

Even for imaging, it became an ideology to plead for high fields. There is no rational scientific reason for this development; image quality and spatial resolution of low and medium field machines became as good as, and in some instances even better than, that of high or ultrahigh field equipment. Additional research revealed that the most important factor in medical imaging, tissue contrast, at least for certain diagnostic questions in the central nervous system, is best at medium fields and, in some instances, even decreases with higher fields [1, 2].

There was still no rational approach to the problem. At a 1983 magnetic resonance conference in San Francisco, a debate on field strength that had started on the platform was continued in the corridor of the conference center. The discussion nearly ended in a fist fight between the proponent of the high field ideology, whose company had put all its efforts into 1.5 T machines, and the proponent of low fields, whose company advocated MRI systems at 0.35 T.

The front lines in this war were mighty and the trenches deep. You were either part of one camp or the other. All large companies jumped on the high field side and promoted high fields with all the ammunition their marketing departments could provide. In some countries, millions of dollars of taxpayers' money were channelled into subsidies for the development of high field systems.

However, one morning in the early 1990s MR customers woke up and found that a gap was emerging. One company had decided to enter the mid-field market, another followed suit, and a third decided to compromise by offering an MR machine operating at a field strength in between the others.

The reasons for these steps were never publicly discussed, but people had realized that the signal-to-noise increase expected from the results in analytical NMR did not occur in the same way in whole-body MR imaging.

In whole-body MR imaging, signal-to-noise increased to a certain extent, and then the human body created additional noise that led to a flattening of the signal-to-noise curve at high fields. In addition, nobody had foreseen the new problems faced by users at higher fields, among them being the worsening of motion artifacts. Cost and hazards also increased with higher fields. At the same time, low and medium field machines became smaller, the quality of their diagnostic output better, and interventional MR became feasible.

Of course, the market for high field equipment is nearly unbroken because they are good diagnostic machines. They also have some advantages over low field equipment; for instance, ultrafast imaging, where scan time is reduced at the expense of signal-to-noise ratio, is generally more effective at higher fields. This facilitates another fashionable research area: functional imaging of the brain.

However, the new generation of buyers, the smaller hospitals and private practices, will prefer cost-efficient MR systems that they can use for most of the daily routine examinations. Bigger hospitals, and in particular those interested in spectroscopy and research in functional imaging, will go for high field systems, but for them also the second and third system will be medium or low field.

This has been realized by the marketing department of the biggest US-American manufacturer. Its marketing people had pushed for high field (1.5 Tesla) in the 1980s. Fifteen years later, it postulated that its new mid-field equipment (0.7 Tesla) is also high field. Definitions always seem to be in the eye of the beholder.

Derek Shaw worked for Varian, later for Oxford Instruments, and since 1983 for General Electric Medical Systems. He is one of the leading MR scientists in Europe. In 1996, he wrote the following statement in a book chapter:

»The early period of MRI ... was dominated by the 'field-strength war'. What was the best field strength for MRI ? These battles were essentially commercial, science being used to justify the company's competitive position ...

»Our pawn in the field strength battle was in vivo spectroscopy... As it became apparent that there was not going to be sufficient specificity available via T1 and T2 determinations, MRS ... was seen as a potential alternative ... MRS needed the highest field possible ...

»This need, along with the higher signal-to-noise ratios achievable at higher field strengths ... led, despite their extra costs, to the use of 1.5 T magnets ... Without this push to high field, MRI systems might be quite different today, probably lower down on the cost/performance scale.«[3]

However, the trend was towards high field. If all of this had been known or taken into consideration 15-20 years ago, more patients would have had access to MR imaging, and medical MR equipment might have been less expensive than it is today.

This is reflected in the sales of MR machines worldwide and in the sales revenue of MR equipment according to field strength. High field makes higher profit, which is a recurrent theme not only in medical technology.

References

1.   Hoult DI, Chen C-N, Sank VJ. The field dependence of MRI II. arguments concerning optimal field strength. Magn Reson Med 1986; 3: 730-746.

2.   Rinck PA, Fischer HW, Vander Elst L, Van Haverbeke Y, Muller RN. Field-cycling relaxometry: medical applications. Radiology 1988; 168: 843-849.

3.   Shaw D. From 5-mm tubes to man. The objects studied by NMR continue to grow. In: Grant DM, Harris RK. Encyclopedia of nuclear magnetic resonance. Volume 1, Historical perspective. Chichester: John Wiley and Sons. 1996, 623-624.

Chapter Seventeen

Common Artifacts in MR Imaging

Introduction

Shortly after the introduction of MR as a new imaging modality, it was hailed as a technique which did not suffer from the common beam-hardening artifacts which destroy the images in x-ray computed tomography. However, it was soon realized that an unfortunate side effect of the complex nature of MR imaging was a whole new set of artifacts.

Artifacts in MR images can take the form of variations in signal intensities or mispositioning of signals (Donald Duck artifact caused by dentures — Figure 17-1 and 17-2a). Such artifacts can mimic pathology to such an extent that examinations have to be redone or other diagnostic modalities have to be used to exclude pathology. Figures 17-2 and 17-3 show two examples of such artifacts.

Usually artifacts can be easily recognized when their causes are known. However, cases have been described where artifacts led to surgical intervention because pathology was falsely described[5].

Artifacts can be categorized into four main groups:
- magnetic field perturbations;
- RF artifacts and gradient-related artifacts;
- motion and flow artifacts;
- signal processing and mapping artifacts.

We will discuss each of these categories.

Figure 17-1:
If the sagittal MR brain scans you get on your screen remind you of these figures, there is something wrong.

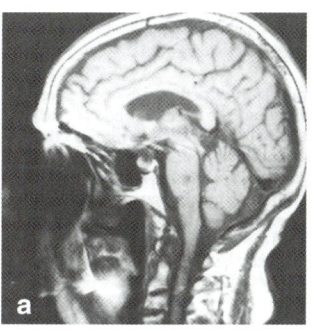 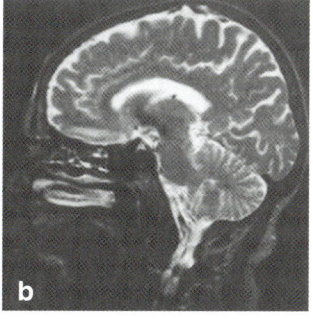

Figure 17-2:
Sagittal midline images through a head. (a) Intermediately weighted image, and (b) T2-weighted image (right). On the left image, a low signal intensity area is seen in the pons, suggesting a lesion. This lesion is not visible on the T2-weighted image. It is caused by image distortion created by ferromagnetic implants.

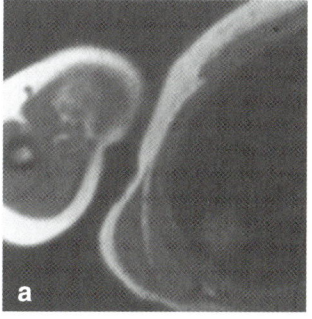 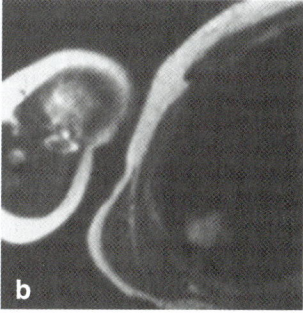

Figure 17-3:
Hemangioma of the right arm. (a) Transverse intermediately, and (b) T2-weighted depict an ill-defined high signal intensity lesion in the right lung. Follow-up studies on another day and the use of CT did not show such a lesion. The cause of the artifact remained unclear.

Magnetic Field Perturbations

The main magnetic field (or more precisely, the lines of magnetic flux) can be distorted by a number of factors outside the MR imaging suite, for instance by large stationary or moving ferrometallic objects such as elevators or passing vehicles.

The field has to be protected by shimming or shielding, which is generally performed properly by the manufacturer of the MR equipment during installation. Therefore, field inhomogeneities from the outside are at present seldom responsible for image artifacts.

Local Inhomogeneity Artifacts

Any *local* internal distortion of the magnetic field cannot be corrected by shimming. The most common causes of such local distortions are the presence of ferromagnetic foreign bodies and susceptibility effects. Ferromagnetic objects usually cause an area of total signal loss around the object and distort the signal intensity at the edge of this region (Figure 17-4 and 13-17).

For this reason, all external metallic objects such as jewelry and watches should be removed from the patient, and a change of clothing is advised to avoid problems with metallic zippers, etc.

Any implanted ferromagnetic material obviously has to be tolerated, but gradient-echo scans should be avoided in such cases since they are affected to a greater extent than spin echoes. Owing to their conductive properties, some nonferrous metal implants can disturb the magnetic field by low-level eddy currents.

Artifacts can also be caused by the ferromagnetic pigments used in eye and other makeup (e.g., mascara) and tattoos. This can result in a significantly reduced image quality, particularly in the case of studies of the orbit.

Susceptibility Artifacts

The susceptibility of a tissue tells us how easily it can be magnetized. The susceptibility values for most tissues fall within a fairly narrow range.

However, the presence of ferromagnetic material (e.g., localized concentrations of hemoglobin after a hemorrhage or high concentration of ferromagnetic contrast agents) or tissue-air interfaces lead to local variations in the susceptibility which result in a reduction of the quality of the local field (Figure 17-5).

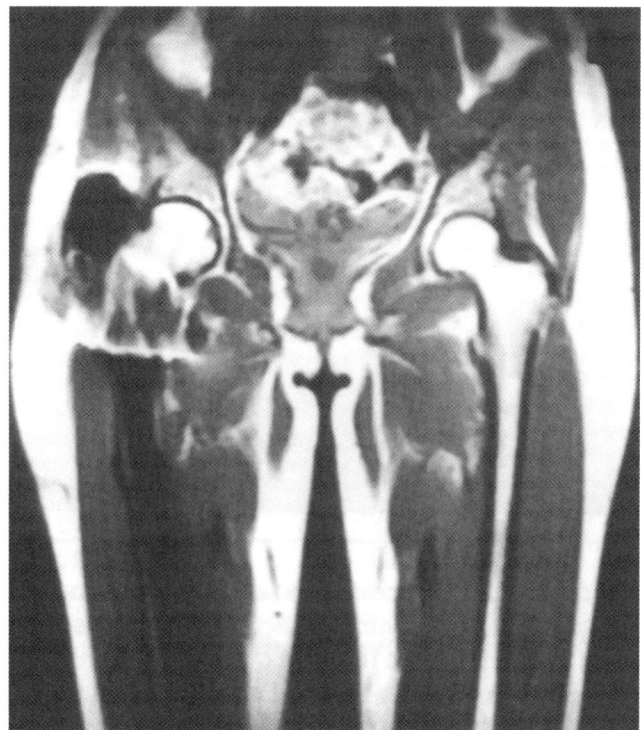

Figure 17-4:
Artifacts resulting from the presence of ferromagnetic material. Signal loss and signal distortion associated with a ferromagnetic prosthesis in the right leg.

Figure 17-5:
Susceptibility artifacts created by ingested ferromagnetic particles used as an oral contrast agent. Because the concentration of the particles is too high, the local magnetic field is disturbed and image artifacts are created.

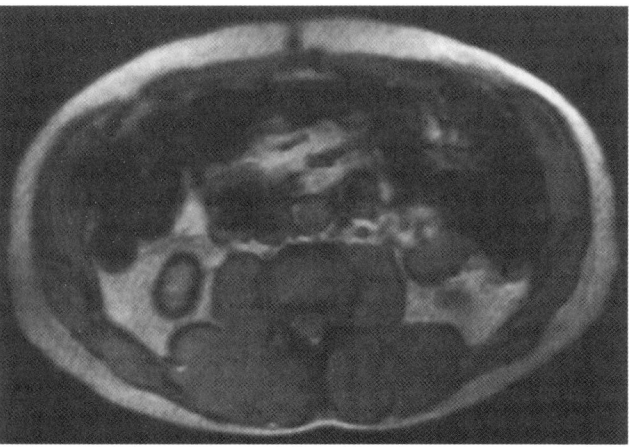

The form of the susceptibility artifact depends on the local conditions, and both increases and decreases in signal intensity are possible. Tissue-air interfaces which give rise to such artifacts can be found in the lungs, around the sinuses and in the nasopharynx. The lack of signal from the lungs is caused by the air in the lungs and the susceptibility artifacts produced by the interfaces between air and lung tissue.

The effect of susceptibility artifacts is reduced by using spin-echo sequences rather than gradient-echo sequences. The presence of local field variations can be determined by using either a phase image or a special pulse sequence which exploits the interaction between two echoes to produce a local field map.

In contrast-enhanced MR angiography contrast agents can induce susceptibility artifacts. If these artifacts are severe, they can be reduced by acquiring the full k-space; partial (asymmetric) echo sampling (and/or halfscan) should be reduced or avoided.

Figure 17-6:
(a) The ideal pulse shape is the square pulse which precisely defines the slice and homogeneously excites all spins within this slice. (b) However, the actual pulse profile tends to be Gaussian. These pulses can overlap and deliver RF energy to neighboring slices. (c) Increasing the interslice gap decreases the energy delivered to neighboring slices.

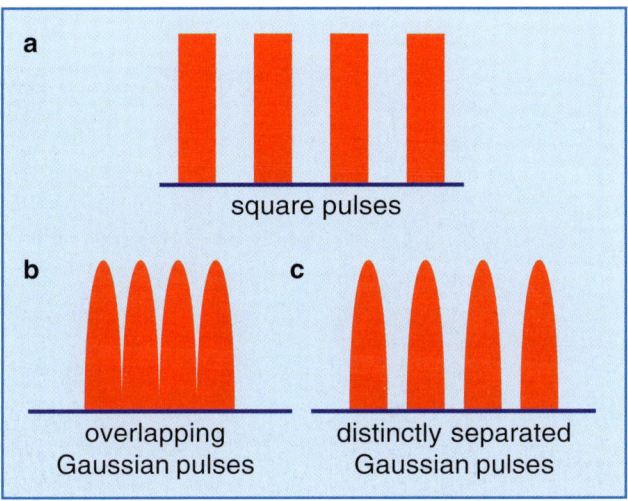

RF and Gradient Artifacts

Slice Profile Artifacts

When we discussed multiple slice imaging in Chapter 6, we mentioned the possible interference of the RF directed at one slice with the excitation of a neighboring slice. This undesired excitation is also known as *cross-excitation* or *crosstalk* (Figure 17-6).

The effects of cross-excitation are changes in image contrast. Countermeasures include an increase of interslice gaps or nonsequential excitation of anatomical slices.

The slice profile of an RF pulse can be distorted when repetition times not allowing full recovery of the signal are used [15]. For any given profile, there is always a transition zone where the value of the flip angle goes from zero up to the desired value. In the case of incomplete recovery of the magnetization, the strongest signal is produced at a flip angle which is less than 90°. The slice profile will therefore show maximum peaks on either side of a central dip.

This becomes particularly acute when very short TR values are used in conjunction with a large flip angle (e.g., spoiled FLASH sequence with good T1 contrast).

In such cases the desired contrast is not obtained since we will have a mixture of contrasts due to the variation in the flip angle across the slice. This can be overcome by using RF pulses which give very sharp transition zones or by using a 3D sequence.

Artifacts in Multiple SE Sequences

Multiple spin-echo or SE-based sequences are widely used in MR imaging. Problems can arise with such sequences due to the fact that the refocusing pulses will not be perfect 180° pulses across the whole slice, and in the transition zone (from 0° to 180°) at the edge of the slice, a whole range of flip angles will be present.

This means that the refocusing pulses will not only form the desired spin echoes, but also will generate other signals which, if not suppressed, can degrade the images obtained from the second echo onwards [6].

There are a number of solutions to this problem including the addition of spoiler gradients to the sequence (which increases the minimum echo times) or the use of phase-cycling to cancel out the unwanted components (which increases the scan time). It is also possible to move the artifacts to the edge of the image by using suitable phase schemes for the RF pulses. This generally reduces their effect on the region-of-interest.

Line Artifacts

A relatively common artifact is the presence of a high intensity line (sometimes looking like a 'zipper') at the center of the image, orientated in the phase-encoding direction. This is usually caused by RF leaking from the transmitter to the receiver. Since the leakage is at the resonance frequency, it will appear at the center of each projection. Slight variations in the amount of leakage in each projection cause the artifact to be smeared out across the field-of-view in the phase-encoding direction. This problem can be difficult to track down and eliminate completely, but can be removed by collecting two averages in conjunction with phase alternation of the excitation pulses.

Line artifacts in the phase-encoding direction away from the center of the image usually result from RF interference at a well-defined frequency. They usually are caused by polluting RF, e.g., commercial radio or television stations. The RF shielding supplied with commercial systems is usually adequate, but the door seals should be checked and cleaned periodically.

Central point artifacts are white or black dots in the center of the MR image. They are only seen on older MR equipment.

Motion and Flow Artifacts

Respiratory and Cardiac Motion

Motion artifacts are the most frequently observed artifacts in MR imaging and have severely hampered the use of MR imaging for abdominal studies (Figure 17-7). They require that ECG triggering be used for thoracic studies.

Motion can lead to blurring and ghost artifacts. Blurring of anatomical structures and interfaces results through averaging of moving structures (pseudo double exposure). This may obscure small lesions. Ghost images are partial copies of the parent image appearing at a different location; they are mostly caused by pulsatile flow.

Motion can be divided into two basic categories:
- motion occurring between acquisition of different lines in the scan;
- motion occurring between excitation and data acquisition.

The first category can be totally eliminated by physiological gating so that the data acquisition and motion are synchronous. This is widely used in cardiac studies and allows excellent, reproducible images of the heart cycle to be obtained.

Respiratory gating has also been tried, but the much lower frequency of the respiratory cycle results in very long scan times. A number of other techniques monitor the respiratory cycle and select the phase-encoding

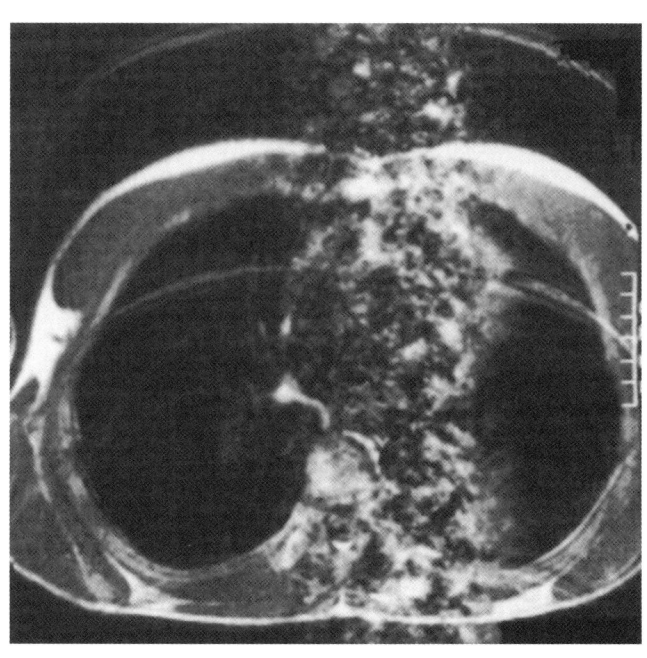

Figure 17-7 (right):
Ghost images resulting from respiratory and cardiac motion with both sets of artifacts being oriented along the phase-encoding gradient. The respiratory motion produces a number of distinct ghost images of the chest wall, while the cardiac motion results in the column of noise associated with the heart.

Figure 17-8 (overleaf left):
The misregistration of flowing spins leads to the depicting of blood outside the vessel lumen: flow artifact.

Figure 17-9 (overleaf right):
Gradient-echo image of a neck with the phase-encoding gradient oriented vertically. Flow artifacts are observed as a column associated with arterial vessels. Venous flow also produces artifacts, but at a much lower level since this kind of flow is less pulsatile.

steps in an order that minimizes the resulting artifact. The ROPE (respiratory-ordered phase-encoding) technique removes the periodicity of breathing upon k-space[1].

A more recent technique follows the normal sequence with a second acquisition which has no phase-encoding. This second echo, also called *navigator* echo, provides an indication of the amount of motion and can be used as the basis for a postprocessing procedure[3]. All these techniques suffer from the fundamental drawback that they assume bulk motion, i.e., everything is moving in the same direction at the same speed, which is not true in the abdomen. However, such techniques can improve image quality.

An alternative approach is to make the scan time short with respect to the respiratory cycle and hence limit the amount of motion. The ultimate example of this is echo-planar scanning where the image is formed from a single acquisition[11].

By using fast scanning techniques with acquisition times of a few seconds or less in conjunction with breath holding excellent images with few motion artifacts can be obtained.

Artifacts resulting from the second type of motion can be reduced by using motion-compensated gradients. The gradients in standard imaging sequences produce additional phase shifts for moving samples. Since the amount of motion will not be the same for each line of the scan (unless gating is used), a smearing of the signal in the phase-encoding direction will result. By using a modified form for the read and slice gradients, one can remove this additional phase shift and hence the artifact (MAST — motion artifact suppression technique)[10]. The drawback is that the minimum echo time will be somewhat longer for such sequences.

Flow

The origin of flow artifacts is very similar to that of motion artifacts, namely that the blood and CSF flow can be pulsatile. Thus, different flow velocities will be present in different lines of the scan. The read and slice gradients induce a phase shift for flowing material, resulting in a range of phase shifts being produced in the course of a scan[14]. The resulting artifact can take the form of a general smearing or a number of distinct artifacts in the phase-encoding direction (Figure 17-8).

The solutions are the same as for motion artifacts: the use of ECG gating to ensure that the same flow velocity is always observed and (or) the use of compensated gradients to null the phase shift from flowing material.

Flow artifacts are particularly severe in gradient-echo images since in a 2D scan, the flowing spins will not have experienced the preceding pulses and will therefore be fully recovered. This leads to a very strong signal from the flowing blood, and therefore to severe artifacts in the absence of gating and motion compensation (Figure 17-9).

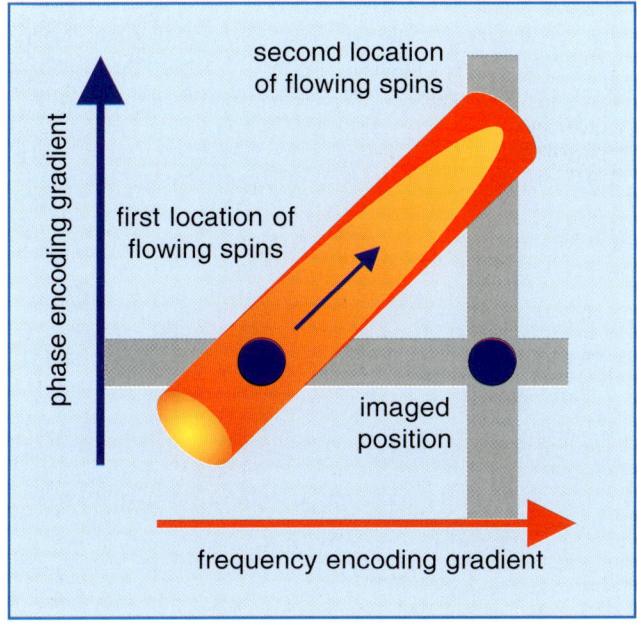

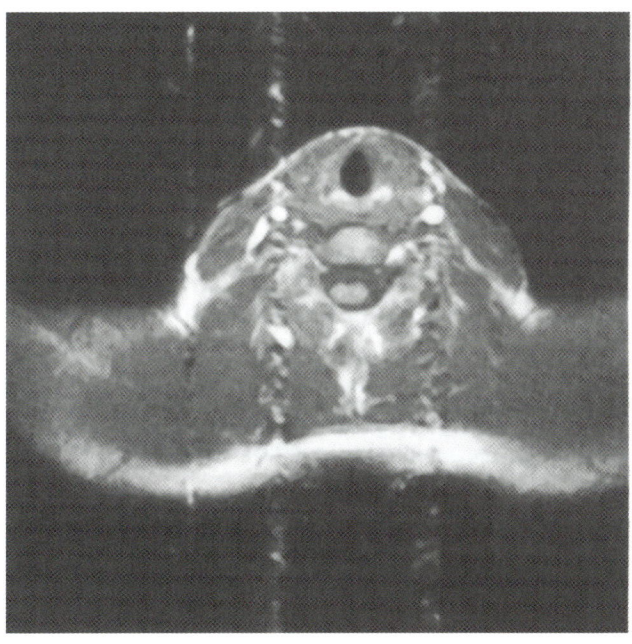

Where flow of blood or CSF degrades the diagnostic quality of the MR image, artifacts should be eliminated. This can be achieved by presaturation.

Here, additional RF pulses are applied which saturate spins outside the imaged region. Thus, blood flowing into this region is saturated, which causes a reduction in blood signal intensity and in flow artifact. Usually this is performed on both sides of the imaged slice because blood can enter the slice from either direction.

Figure 17-10 is a theoretical example of parallel presaturation.

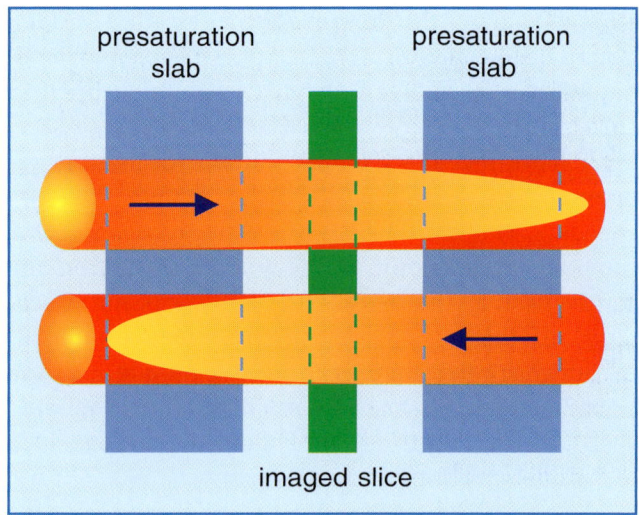

Figure 17-10:
Example of parallel presaturation. Presaturation excites spins outside the slice to be imaged. Thus, flowing blood arrives already saturated when the spins in the slice of interest are exposed to the excitation pulse. The blood signal is suppressed and does not give rise to artifacts. The arrows indicate the direction of flow.

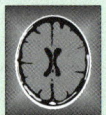 You can experience interactively image artifacts with **MR Image Expert**.

k-Space-Related Artifacts

Many artifacts in MR imaging are caused or enhanced by k-space. Some of them you can try out by playing with the k-space filters, as described in the chapter introducing k-space.

Some others are implemented directly in **MR Image Expert**.

Bit Errors

- Select *File/Open* or the *Brain* icon in the **MR Image Expert** menu bar. Double-click on the file **1.5 T: Normal brain, transversal slice**. Leave the parameters unchanged.

- Now select *Options/ Bit Error*, leave the selected choice (*ADC Bit Error: 1*) and click OK. Move the resulting k-space image to the right. The original image has changed into an image with stripe artifacts.
- Close the k-space image and repeat the experiment with different bit-error positions (1-128). The artifacts are always different.

RF Feed-Through

- Select *File/Open* or the *Brain* icon in the **MR Image Expert** menu bar. Double-click on the file **1.5 T: Normal brain, transversal slice**. Leave the parameters unchanged.

- Now select *Options/ RF Feed Through*, set RF Feed Through =5 and click OK. Move the resulting k-space image to the right. The original image has changed into an image with feed-through artifacts.
- Close the k-space image and repeat the experiment with different settings and compare the artifacts.

Undersampling / Aliasing

- Select *File/Open* or the *Brain* icon in the **MR Image Expert** menu bar. Double-click on the file **1.5 T: Normal brain, transversal slice**. Leave the parameters unchanged.

- Now select *Options/ Undersampling*. Move the resulting k-space image to the right. The original image has changed into an image with an aliasing (wraparound) artifact.
- Try this with different patient data files.

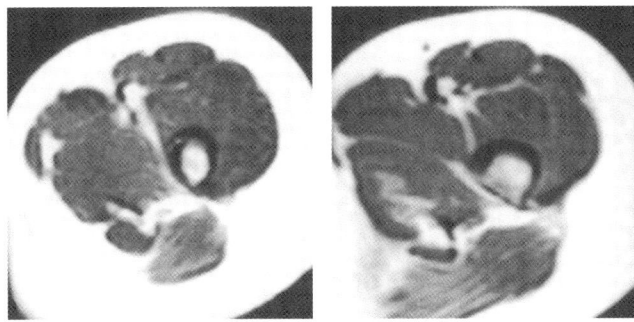

Figure 17-11:
High field strength (1.5 T) MR images showing chemical shift artifacts in the readout direction, which is oriented vertically in the image. The chemical-shift artifact is visible as a black rim between fat and muscle.

Signal-Processing and Signal-Mapping Artifacts

Chemical-Shift Artifacts

The chemical-shift artifact is caused by the difference in resonance frequency experienced by protons in different chemical environments. Protons contained in fat and water environments are separated by 3.5 ppm. Both the frequency and slice-encoding processes use frequency information. The signals from fat and water protons at the same position will result in different frequencies and hence a relative shift of one of the signal components (Figures 11-2 and 17-11). Since this is a frequency-dependent artifact, the effect will be most pronounced at high fields, with displacements of several pixels being visible in the readout direction. The artifact can be reduced by using stronger gradients, but this has the unfortunate side effect of decreasing the signal-to-noise ratio.

The problem can be overcome by suppressing one or other of the components prior to collecting each line of data. This can be done either by using presaturation techniques (which require good static field homogeneity) [4], or by using one of a number of add-subtract schemes (which increase the imaging time) [2,12].

As we have seen in Chapter 11, in gradient-echo studies, changes in the echo time lead to changes in the relative phases of the fat and water components of the signal. This can be used to change the contrast in the image or as the basis of a fat-suppression scheme [13].

Black Boundary Artifacts

Sometimes well-defined black contours following anatomical structures are seen. These artifacts are another class of chemical-shift artifacts. Pulse sequences prone to such artifacts are inversion-recovery and gradient-echo sequences. Figure 17-12 is an example of a GRE sequence of the abdomen.

The water and fat signals can be in-phase or out-of-phase. This was explained for the fat-suppression technique in Chapter 11. If this happens accidentally in volume elements with partial volume effects between water-rich and lipid-rich organs, the signal disappears and artifactual contours are seen.

To avoid these contours, in-phase echo times must be used (Figure 11-3). Alternatively, the examination should be performed with SE sequences whose 180° pulses refocus the phase shifts.

Figure 17-12:
Black boundary artifacts in the abdomen. Gradient-echo sequence with an echo time of 16 ms.

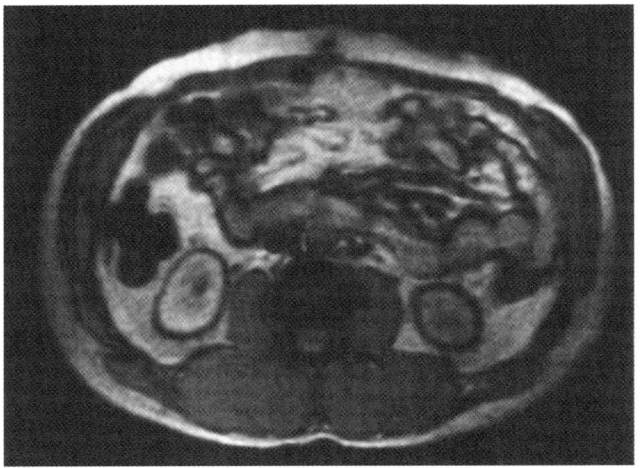

Truncation Artifacts

The truncation artifact is also known as ringing or Gibbs artifact. It appears as parallel striations, close to interfaces between tissues with different signal intensities, such as fat-muscle or CSF-spinal cord. Because these lines mimic regular structures, they can present interpretation problems if they are not recognized as artifacts.

Truncation artifacts are particularly severe when small image matrices are used and can be reduced by simply using a larger image matrix. Oversampling, while having no effect on the intensity of the truncation artifacts, does reduce their spacing which often results in the artifact becoming blurred and imperceptible.

Truncation artifacts are most commonly seen in the phase-encoding direction since increasing the matrix in this direction results in an undesirable increase in scan time. A combination of suitable scan orientation and increasing the data matrix in the frequency-encoding direction usually reduces the artifact to a tolerable level (Figure 17-13).

Truncation artifacts can also be reduced by applying a low pass filter; but as with all filtering the whole image, and not just the artifact, will be affected.

Aliasing

Aliasing causes data which lie outside the specified field-of-view to be wrapped back into the image. It can occur in both the phase- and frequency-encoding directions.

In the frequency-encoding direction the artifact results from the presence of signals with too high a frequency being mispositioned. According to the *Nyquist* theorem, frequencies must be sampled at least twice per cycle in order to reproduce them accurately.

Depending on the detection scheme used, the data can be folded back into the image on the same or on the opposite side. These high-frequency signals can be removed with a filter, but the response of such a filter will not exactly match the desired frequency range (the image bandwidth), leading to some artifact still being present or to a loss of signal at the edges of the image.

This problem can be overcome by doubling the amount of data collected (oversampling), either by doubling the sampling rate to the critical sampling frequency (*Nyquist* frequency) or by doubling the acquisition time. The latter scheme has the advantage of also improving the signal-to-noise ratio by $\sqrt{2}$.

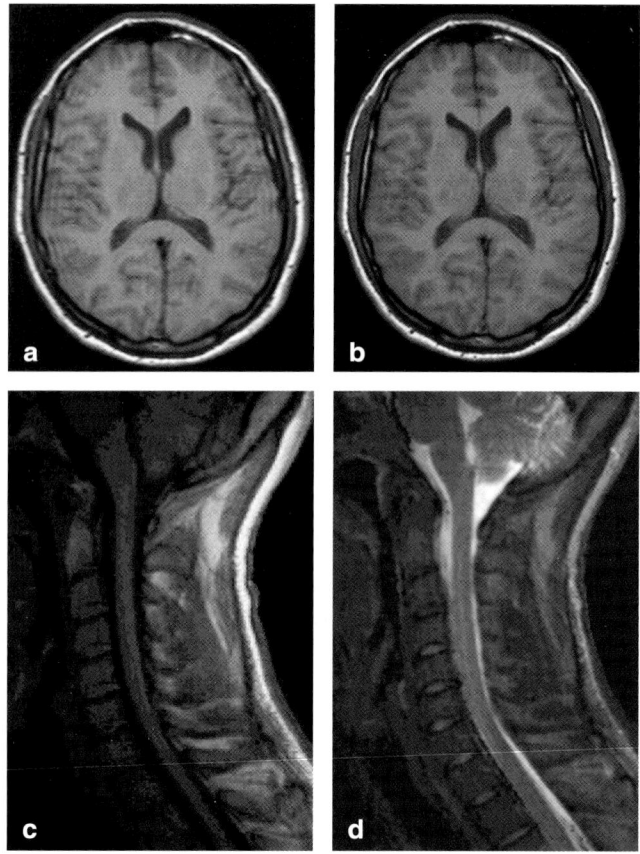

Figure 17-13:
Ringing (Gibbs, truncation) artifacts.
(a) 60% acquisition with ringing artifacts; (b) 80% acquisition, no artifacts visible.
(c) and (d) truncation artifact mimicking syringomyelia. (c) T1-weighted, (d) T2-weighted image.
As shown in (a) and (b), the reduction in such artifacts is achieved by increasing the matrix size from, e.g., 128×128 to 256×256, or by increasing the percentage of phase-encoding profiles.

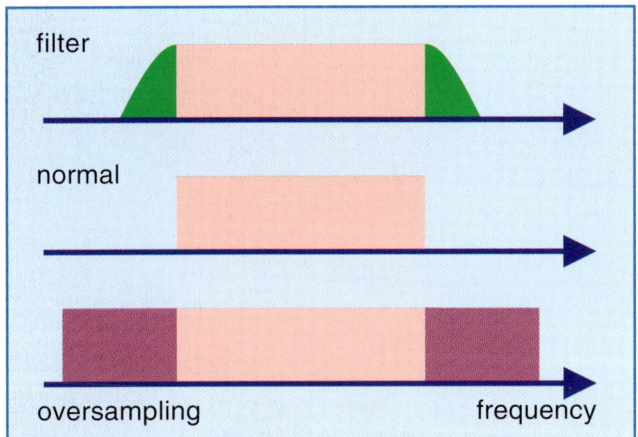

Figure 17-14:
Relation between the filter (top), the image frequencies for normal sampling (middle), and oversampling (bottom).

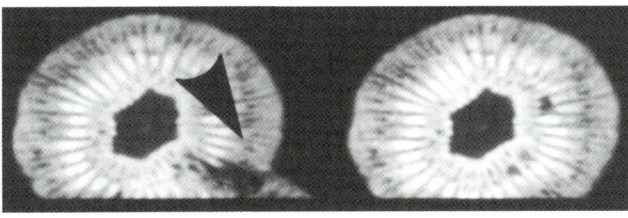

Figure 17-15:
Image of a kiwi fruit: (left) normal sampling, (right) result of oversampling. In both cases, the frequency gradient was orientated vertically. In the left-hand image aliasing of signal originating from outside the specified field-of-view results in the artifact visible at the bottom of the image (arrow).

We can then apply a filter which will remove all frequencies outside of the new image bandwidth but which will have no effect on the frequencies which correspond to the desired field-of-view (Figure 17-14). After the Fourier transformation the outer quadrants of the oversampled data are discarded, leaving us with the original field-of-view and no artifacts (Figure 17-15).

In the phase-encoding direction aliasing results in signal from outside the field-of-view being folded back into the image on the opposite side (Figure 17-16), because the two positions will produce identical phases.

Oversampling can also be applied to the phase-encoding direction, but will double the imaging time (since twice as many phase-encoding steps are required) and therefore is little used.

The usual practical solution is to orientate the phase-encoding direction in the image in such a way that no anatomical structures extend beyond the image boundaries in this direction. If this is not possible, a surface coil can be used to limit the volume from which the signal is obtained. This reduces or eliminates the problem. For 3D imaging, the backfolding can occur in both of the phase-encoded directions.

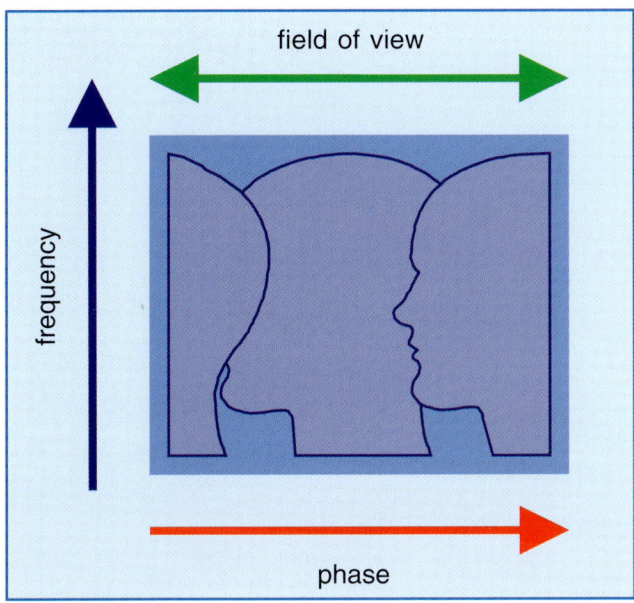
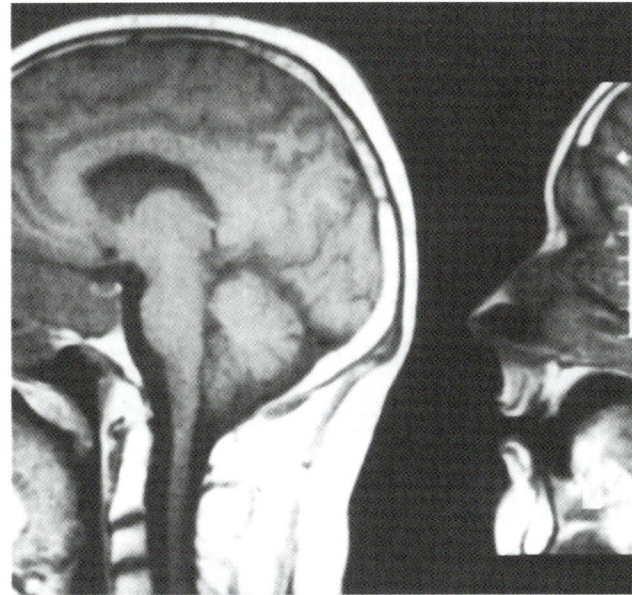

Figure 17-16:
Backfolding artifact resulting from the sample being larger than the field of view in the phase-encoding direction, which is orientated left-right in this example.

Quadrature Artifact

The magnetic resonance signal is detected using a receiver which has two channels, with the reference signal to the second channel being phase-shifted by exactly 90° with respect to the reference used for the first channel. Any maladjustment of this phase-shift results in a ghost image being observed, which is rotated about both the *x*- and *y*-axes with respect to the main image (Figure 17-17). This can be eliminated by adjusting the phase and gain of the receiver, which is most easily done by trying to minimize the quadrature peak in a transformed, off-resonance signal.

k-Space Artifacts

There are additional artifacts connected to k-space, e.g. from bad data points or spikes[9]. Some of them are discussed in the **MR Image Expert** Tutorial on page 206.

Table 17-1 summarizes the most common artifacts and their remedies. Artifacts caused by defective components, malfunctions of the imaging system, or artifacts connected to the equipment of specific manufacturers might be different from the common artifacts covered in this chapter[7].

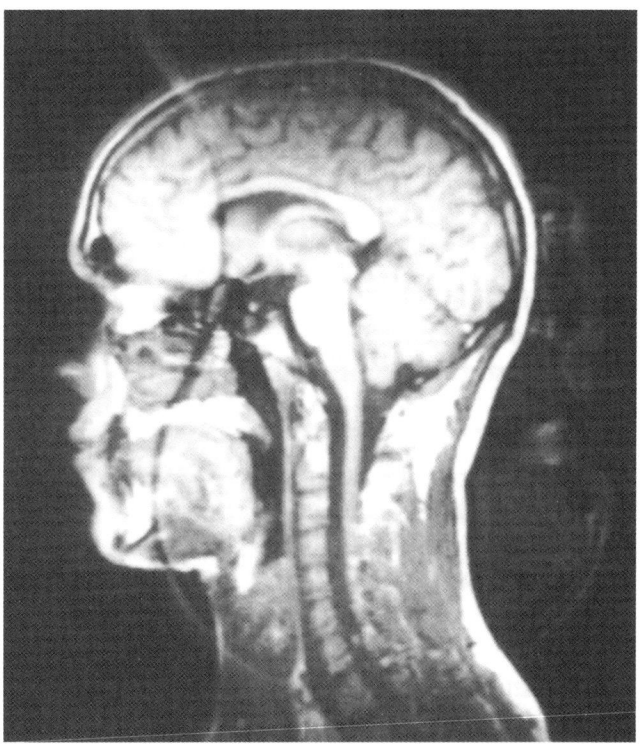

Figure 17-17:
Quadrature artifact resulting from an incorrectly adjusted receiver.

References

1.   Bailes DR, Gilderdale DJ, Bydder GM, et al. Respiratory ordered phase encoding (ROPE): method for reducing respiratory motion artifacts in MR imaging. J Comput Assist Tomogr 1985; 9: 835-838.
2.   Dixon WT. Simple proton spectroscopic imaging. Radiology 1984; 153: 189-194.
3.   Ehman RL, Femlee JP. Adaptive technique for high definition MR imaging of moving structures. Radiology 1989: 173: 255-263.
4.   Femlee JP, Ehman RL. Spatial presaturation: a method for suppressing flow artifacts and improving depiction of vascular anatomy in MR imaging. Radiology 1987; 164: 559-564.
5.   Food and Drug Administration of the United States of America. A primer on medical device interactions with magnetic resonance imaging systems. See: http://www.fda.gov/cdrh/ode/primerf6.html
6.   Graumann R, A. Oppelt A, Stetter E. Multiple spin-echo imaging with a 2-D Fourier method. Magn Reson Med 1986; 3: 707-721.
7.   Henkelman RM, Bronskill MJ. Artifacts in magnetic resonance Imaging. Rev Magn Reson Med 1987; 2: 126-256.
8.   Johnson BA and Kelly WM. Common MRI artifacts: an overview. MRI Decisions 1989; 3: 2, 17-23 and 3, 26-32.

9. Mezrich R. A perspective on k-space. Radiology 1995; 195: 297-315.

10. Pattany PM, Phillips JL, Chiu LC, et al. Motion artifact suppression technique (MAST) for MR imaging. J Comput Assist Tomogr 1987; 11: 369-377.

11. Rzedzian R, Pykett IL. Instant images of the human heart using a new whole body MR imaging system. Amer J Roentgenol 1986: 149: 245-250.

12. Szumowski J, Plewes DB. Fat suppression in the time domain in fast MR imaging. Magn Reson Med 1988; 8: 345-354.

13. Williams SCR, Horsfield MA, Hall LD. True water and fat MR imaging with use of multiple echo acquisition. Radiology 1989; 173: 249-253.

14. Van Dijk P. Direct cardiac NMR imaging of the heart wall and blood flow velocity. J Comput Assist Tomogr 1984; 8: 429-436.

15. Young IR, Khenia S et al. Clinical magnetic resonance susceptibility mapping of the brain. J Comput Assist Tomogr 1987; 11: 2-6.

Table 17-1:
Image artifacts and remedies. Modified from Henkelman and Johnson[7,8].

Artifact	Description	Causes	Solution

Aliasing	image wrap around
	   frequency-encoding
	   phase-encoding
	   slice selection
		undersampling of Fourier space	sample higher frequencies;
			band limit the data by filtration;
			limit excitation volume

Truncation	edge-ringing; syrinx-like stripe	discontinuity of the boundary of Fourier space	oversample filter; model the data
Stars/zippers	bands through center of image	RF feedthrough; residual FID or stimulated echo	phase cycle; displace to boundary with alternate phases; spoil stimulated echo with gradients
Errors in data	striped image; washed out image; line in phase direction	one bad data point; ADC over range; discrete RF noise	delete and interpolate bad data alternate receiver; shield room
Magnetic field perturbation	spatial distortion of the image or loss of signal	B_0 inhomogeneity; ferromagnetic implants, magnetic susceptibility, and particulate iron; chemical shift	shim the magnet; remove interfering object, if possible use a spin-echo; increase the read-out gradient
Non-ideal gradients	spatial distortion; asymmetric edge-ringing; banding in multiple echo	gradients non-linear; gradient amplifier saturates; interference between stimulated and spin-echo	need better gradient design; over compensate the amplifier; spoiler gradients
Ghosts	180° rotated about the origin; inversion in phase-encoding direction; displaced ghosts	quadrature imbalance; stimulated echo in multiple echo; digitalization errors on the DAC motion	correct; rephase all higher order echoes; keep phase gradients on integer amplitudes; many partial solutions
Flow	displaced image ghosts; anomalous intensities	incorrect phases due to motion; refreshed spins or loss of spins	gradient moment correction methods
Contours	dark boundaries between regions	inversion-recovery null point; chemical shift; motion shear	real reconstruction or change TI; use spin-echo or CSI; gradient moment rephasing

Chapter Eighteen

Safety of Patients and Personnel

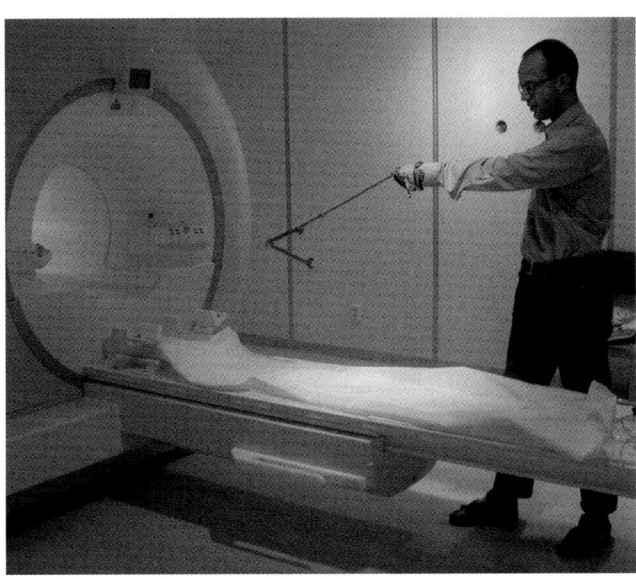

Figure 18-1:
Magnets can have fatal attractions. Be careful when you come close to MR equipment with ferromagnetic objects.

Introduction

Any new method in medicine, be it diagnostic or therapeutic, must be thoroughly checked for possible adverse side effects. More than 100 years ago, x-rays represented a major step forward, but then sobered radiologists and the public after the hazards of ionizing radiation were detected.

No ionizing radiation is involved in MR imaging. However, because of the known problems with x-rays and radioisotope examinations, magnetic resonance imaging and spectroscopy have been intensively examined for possible dangerous side effects.

During the last century, several hundred papers focusing on the effects or side effects of magnetic or radiofrequency fields have been published.

They range from anecdotal reports about therapeutical applications of magnetic fields as published by Zhang et al. [47] to reports on unwelcome side effects, such as Beischer's study [11].

This chapter cannot cover all potential sources of hazards. Several reviews of the literature have been published recently, e.g., by Shellock and Kanal [41], by Persson and Ståhlberg [31], and by Magin, Liburdy, and Persson [25].

Several of the side effects associated with MR are unique to this kind of medical diagnostic tool; others are similar to hazards of other diagnostic methods. Possible hazards can arise from or be connected to:
- static magnetic fields;
- varying magnetic fields (gradient fields);
- radiofrequency fields;
  and specifically:
- devices necessary to operate the imager (such as cooling gases) or to ensure the quality of life of the patients (such as intracorporal implants and extracorporal monitors);
- conducting loops such as electrical leads or accidental anatomical positions of the patient.

These hazards can affect patients, personnel, and other persons within the field of the magnet. They can be categorized as *acute* and *subacute*.

Acute Hazards

Acute hazards are created by the static magnetic field usually covering an ellipsoid region around the isocenter of the magnetic resonance imager (Figure 18-2). The range of this *fringe* or *stray field* depends on the field strength of the system, the type of magnet, and the kind of shielding used.

Ultralow- and low-field magnets possess a limited stray field of sometimes less than one meter radius from the isocenter. The stray field of large-bore, high-field systems may cover a radius of 15 or 20 meters, unless the magnet is heavily shielded (see also Table 3-3). Some examples of actual incidents created by MR machines are given in the insert on page 217.

External Objects and Devices

Projectiles. The most imminent danger for both patients and personnel in the magnetic field of an imaging system may result from ferromagnetic objects such as scalpels, scissors, pens, and even sand bags (not filled with sand but with iron shot) and gas dewars, which can be attracted by the magnet and thus behave like *projectiles*.

To prevent such accidents, the installation of a metal detector through which everybody has to pass before entering the MR suite has been recommended, but is rather cumbersome.

Every person working or entering the magnet room or adjacent rooms with a magnetic field has to be instructed about the dangers. This should include the intensive-care staff, and maintenance, service and cleaning personnel, as well as the crew at the local fire station.

The best protection against this danger is not to allow personnel other than those directly involved in patient examinations, i.e., the operator and the radiologist, into the magnet room. Constant education of everybody involved is also vital.

Monitors and respirators. The dependence on *physiological monitoring*, on *mechanical respiration,* and electric *infusion pumps* during MR examinations renders difficulties, and in certain instances does not allow such an examination.

However, with the development of appropriate monitoring and life-support equipment during the last few years, dependence is no longer a contraindication of MR imaging. Details on monitoring can be found in an article by Kanal and Shellock [20].

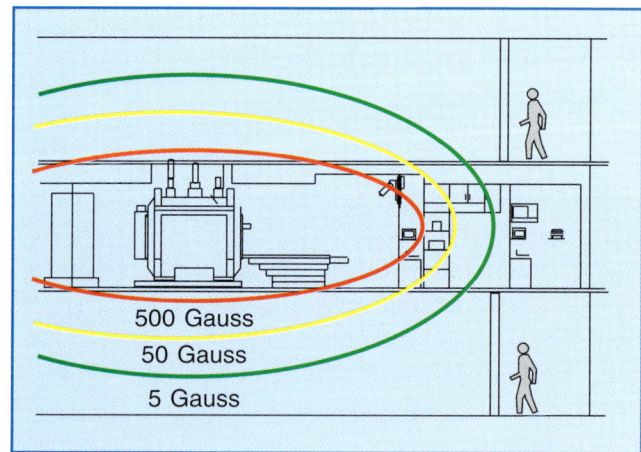

Figure 18-2:
The fringe field around the magnetic resonance system may stretch into adjacent rooms, floors, etc., and both influence electronic equipment and be a possible hazard to persons passing by.

The 5-Gauss safety line (green line) encircles the area that pacemaker carriers should not enter. This region can stretch beyond the magnet room itself. In this case, warning signs or similar notices should be displayed outside the magnet room, in neighboring rooms on the same floor, and on floors above and below. This danger has been reduced by shielded magnets.

Figure 18-3:
Danger and prohibition symbols and signs used in MR installations. (a) danger — strong magnetic field; (b) danger — high-frequency electromagnetic field; (c) active implants and metallic implants, such as pacemakers, prohibited; (d) loose ferromagnetic objects prohibited; (e) metal body implants prohibited; (f) magnetic media such as credit cards, diskettes, magnetic tapes prohibited; (g) mechanical watches, cameras and similar devices prohibited.

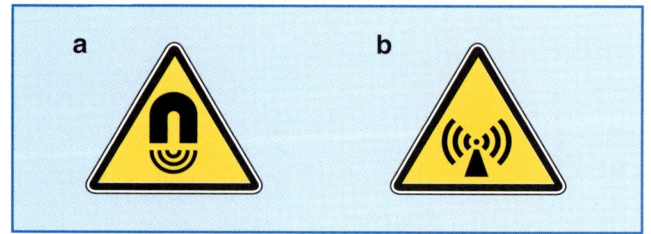

Contrast agents. Paramagnetic, superparamagnetic or ferromagnetic *magnetic resonance contrast agents* or other substances which have to be injected or applied in another way may present risks similar to those in any other invasive technique. The clinical experience of administering gadolinium-based or other agents intravenously to patients has shown that these agents are generally safe and well-tolerated. Only very few severe anaphylactoid reactions and cases of glottis edema have been reported. Still, all necessary precautions for intensive-care treatment have to be considered when injecting such contrast agents, particularly in patients with a history of allergy or drug reaction [37].

A precautionary 24-hour suspension of breast-feeding was generally recommended following the administration of gadolinium-containing contrast agents. However, it has been proposed that this suspension be reduced to 12 hours [18].

MR Equipment

Noise. The *noise* created by the switching of the gradients is an additional source of inconvenience for the patient and, occasionally, personnel. This noise is comparable to very heavy traffic. Noise levels increase with field strength.

Disposable earplugs for the patient are recommended in high-field systems. Noise-cancelling systems and special earphones are available, and active acoustic control systems are being developed [26].

Cooling gases. In superconductive magnet systems, helium and nitrogen are used as cooling gases. In the case of a quench, gases are released to the outside. Under normal circumstances, the gases should escape through a pipe system and not reach the magnet-room atmosphere. However, accidentally some gas could be released into the magnet room.

In this case, there are two potential dangers. Frostbite can be induced because the gases are extremely cold. Secondly, nitrogen is to be considered hazardous, in particular under pressure (whereas there is no danger of direct intoxication from helium). All personnel and patients must evacuate the area immediately and return only after proper ventilation of the magnet room. Oxygen monitors with an audible alarm, situated at an appropriate height within the magnet room are recommended safety devices [41].

Patient-Related Devices

Implants. A particular danger is presented by small metallic *surgical implants*. Hemostatic or other clips in the CNS can move in their position. Dislocation by magnetic attraction or torque presents a risk of hemorrhage. In other parts of the body, we consider this to be a minimal risk, because after the healing phase of six to eight weeks, fibrosis and encasement of the clip help to keep it in a stable position. The label *stainless steel* is not a guarantee for non-ferromagnetic steel.

Implants that involve magnets such as magnetic sphincters, stoma plugs, dental implants, etc., can be demagnetized by the MR imager. They should be removed prior to the examination.

An extensive overview of the behavior of implants is given by Shellock [40], including a list of several hundred devices which are not prone to dislocation. It also lists a selection of those metallic implants, materials, and foreign bodies that are potential risks for patients undergoing MR imaging examinations. An updated version can be found in Table 18-1.

Foreign bodies. Occult *ferromagnetic foreign bodies* incorporated in accidents are dangerous, in particular those close to the eyes. The patient's history may help to rule out such foreign bodies. Many patients, however, do not remember such accidents. In case of doubt, x-rays should be taken prior to MR imaging.

Ferromagnetic makeup and tattoos cannot only distort MR images, but also can be irritated and even be pulled into the eye by magnetic forces. Makeup should be removed before the examination, if possible.

Pacemakers. Research on the influence of magnetic and radiofrequency fields upon cardiac *pacemakers* reported that the RF radiation of the MR imager might disturb the function of demand pacemakers by closing the reed relay and switching to the asynchronous mode;

Aneurysm Clips

Drake (DR 14; DR 16; DR 24; 301 SS); Edward Weck, Triangle Park, NJ, U.S.A.
Downs multi-positional (17-7 PH)
Heifetz (17-7 PH); Edward Weck.
Housepian
Kapp (405 SS); Kapp curved (404 SS); Kapp straight (404 SS); V. Mueller.
Mayfield (301 SS; 304 SS); Codman, Randolph, MA, U.S.A.
McFadden (301 SS); Codman, Randolph, MA, U.S.A.
Pivot (17-7 PH); V. Mueller.
Scoville (EN58J); Downs Surgical, Decatur, GA, U.S.A.
Sundt-Kees (301 SS); Sundt Kess Multi-Angle (17-7 PH); Downs Surgical, Decatur, GA, U.S.A.
Vari-angle (17-7 PH); Vari-angle micro (17-7 PM SS); Vari-angle spring (17-7 PM SS); Codman.

Carotid-Artery Vascular Clamp

Poppen-Blaylock (SS); Codman, Randolph, MA, U.S.A.

Dental Devices and Materials

Palladium-clad magnet; Titanium-clad magnet; stainless-steel-clad magnet; Parkell Products, Farmingdale, N.Y., U.S.A. *

Heart Valves

Many of the commercially available heart-valve prostheses have been tested for ferromagnetism. The majority displayed measurable deflection forces. However, the deflection forces were relatively insignificant compared with the force exerted by the beating heart. Therefore, patients with these heart-valve prostheses may safely undergo MR imaging.

Intravascular Coils, Stents and Filters

Gianturco embolization coil; Gianturco bird nest IVC filter; Gianturco zig-zag stent; Cook, Bloomingdale, ID, U.S.A.
Gunther IVS filter; Cook, Bloomingdale, ID, U.S.A.
New retrievable IVC filter; Thomas Jefferson University, Philadelphia, PA, U.S.A.
Palmaz endovascular stent; Ethicon, Sommerville, N.J., U.S.A.

Note: Ferromagnetic coils, filters and stents typically become firmly incorporated into the vessel wall several weeks following placement, and therefore it is unlikely that they will become dislodged by magnetic forces after a suitable period of approximately 6 to 8 weeks has passed.

Ocular Implants

Retinal tack (SS-martensitic), Western European
Fatio eyelid sping/wire *

Otologic Implants

Cochlear implant; 3M/House
Cochlear implant; 3M/Vienna
Cochlear implant, Nucleus Mini 22-channel; Cochlear, Englewood, CO, U.S.A.
McGee piston stapes prosthesis (Platinum/17Cr-4Ni SS); Richards Medical, Memphis, TN, U.S.A.

Pellets, Bullets, Shrapnel, etc.

BB's, Daisy
BB's, Crosman
Bullet, 7.62; 0.39 mm (copper, steel); Norinco
Bullet, 0.380 inch (copper, nickel, lead); Geco
Bullet, 0.45 inch (steel, lead); North America Ordinance
Bullet, 9 mm (copper, lead); Norma

Note: The relative risk of performing MRI in patients with pellets, bullets, or shrapnel is related to whether or not they are positioned near a vital structure.

Penile Implants

Penile implant; OmniPhase, Dacomed Corp., Minneapolis, MN, U.S.A. *

Miscellaneous

Cerebral ventricular shunt tube, connector (type unknown)
Swan-Ganz Catheter, Thermodilution; American Edwards, Irvine, CA, U.S.A. (*There has been a report of a catheter melting in a patient undergoing MRI. Therefore, this catheter is considered contraindicated for MRI*).
Tissue expander with magnetic port; McGhan Medical, Santa Barbara, CA, U.S.A. (*see note under Dental Devices*)

* *Note: The potential for these metallic implants or devices to produce significant injury to the patient is minimal. However, performing MRI in a patient with one of these devices may be uncomfortable for the individual and/or may result in damage to the implant.*

Table 18-1:
Metallic implants, material, and foreign bodies with potential risks for patients undergoing MRI. If known, the manufacturer and his address are given. Prepared by FG Shellock and E Kanal. Modified from references [40-42]. SS = stainless steel.

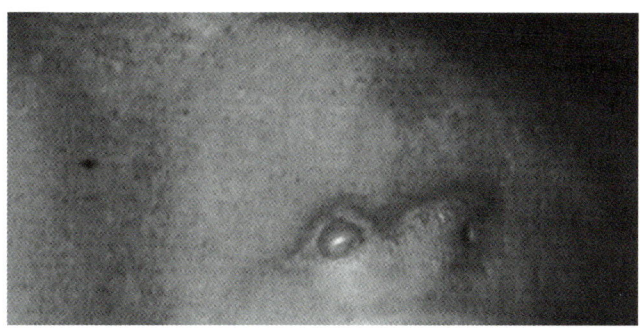

Figure 18-4:
Burn on the back of a patient who underwent MR imaging while lying on an ECG lead.

Some Examples of MR-Related Incidents

Static magnetic field

• A patient with an implanted cardiac pacemaker died during or shortly after an MR exam. The coroner determined that the death was due to the interruption of the pacemaker by the MR system.*

• A patient with an implanted intracranial aneurysm clip died as a result of an attempt to scan her. The clip reportedly shifted when exposed to the magnetic field. The staff apparently had obtained information indicating that the material in this clip could be scanned safely.

• Dislodgment of an iron filing in a patient's eye during MR imaging resulted in vision loss in that eye.

• A patient complained of double vision after an MR exam. The MR examination, as well as an x-ray, revealed the presence of metal near the patient's eye. The patient was sedated at the time of the exam and was not able to inform anyone of this condition.

• An i.v. pole was attracted to the magnet and struck a patient, cutting his arm. The patient required stapling of the cut.

• A pair of scissors was pulled out of a nurse's hand as she entered the magnet room. The scissors hit a patient, causing a cut on the patient's head.

• Two parts of a forklift weighing 80 pounds each were accelerated by the magnet, striking a technician and knocking him over 15 feet, resulting in serious injury.

Pulsed gradient magnetic or RF fields

• A patient received a 4 x 10 cm blistered burn to the left side of the back near the pelvis from an ECG-gating cable.

• A patient received blistered burns on the finger where a pulse oximeter was attached during MR scanning. A skin graft was required to treat the affected area.

• A patient with an implanted insulin infusion pump was placed in an MR scanner, resulting in movement of the device. The pump was removed from the patient and subsequently found to be nonfunctional.**

This event may also be attributed to the pulsed RF fields (*) or to the static magnetic field (**) [5].

varying magnetic fields may mimic cardiac activity. Magnetic attraction can provoke motion of the pacemaker in its pocket and thus move the conducting lead. Therefore, pacemaker patients or other persons bearing pacemakers should not be examined in, or come close to, an MRI or MRS system, although recently some exceptions have been described for new-generation pacemakers [43].

Pavlicek et al. reported a threshold for initiating the asynchronous mode of a pacemaker at 17 Gauss [30]. The national regulatory boards decided to limit the threshold for access to MRI areas to 5 Gauss. It seems advisable to mark this area by signs or lines on the floor (Figure 18-3).

It is of special interest for the observer of bureaucratic procedures that the 5-Gauss limit is ten times higher than the average earth magnetic field, but lower than the magnetic field in electric trains such as subways (up to 7 Gauss). The fields measured on the surface of the receiver of a telephone are 35 Gauss and of an audio headset 100 Gauss.

Similar considerations hold for pacemakers used for stimulation of the carotid sinus or intracorporal insulin pumps, for instance. Here, no adverse effects have been observed [38]. However, interference in electronic cochlear implants and ferromagnetic mechanical stapedial replacements has been reported [17].

Prosthetic heart valves are not considered to be dangerous in low fields. Patients should not undergo MR imaging in high fields if valve dehiscence is clinically suspected [44].

Wires, other metallic objects, and skin contact. Wire configurations such as *pacemaker lead wires*, *ECG and plethysmographic cables*, and *surface-coil connections* can act as antennae. Gradient and RF fields may induce current into these wires and thus cause fibrillations and burns (Figure 18-4). This presents a risk to the patient and must be eliminated prior to the examination.

This holds in a similar way for all *clothing containing metallic threads* or components, as well as all metallic objects such as *eye glasses, jewelry, hairpins, buttons, watches, bracelets, prostheses*, etc. All of these objects must be removed prior to the examination.

The patient's skin should not be in contact with the inner bore of the magnet. Large-radius wire loops should not be formed by leads or wires that are used in the magnet bore during imaging procedures.

If the patient's arms and legs are not completely covered with clothing, insulating material must be

Absolute Contraindications	Relative Contraindications
electronically, magnetically, and mechanically activated implants: cardiac pacemakers	*electronically, magnetically, and mechanically activated implants:* other pacemakers, e.g., for the carotid sinus; insulin pumps and nerve stimulators; lead wires or similar wires
ferromagnetic or electronically operated stapedial implants	non-ferromagnetic stapedial implants cochlear implants prosthetic heart valves (*in high fields, if dehiscence is suspected*)
hemostatic clips (CNS) metallic splinters in the orbit	hemostatic clips (body) makeup and tattoos congestive heart failure pregnancy (claustrophobia)

Table 18-2:
Contraindications for MR imaging and spectroscopy. Never forget that the magnetic memory of credit and similar cards, as well as magnetic devices such as tapes, will be erased by MR magnets. Leave home without them or leave them outside the magnet room.

placed between the legs and between legs and magnet. *Leg-to-leg* and *leg-to-arm skin contact* must be prevented in order to avoid the risk of burning due to the generation of high current loops if the legs or arms are allowed to touch.

IUDs. Most of the commonly used *intrauterine contraceptive devices (IUD)* do not move under the influence of the magnetic field, do not heat up during sequences usually applied for pelvic imaging, and do not produce major artifacts *in vitro* or *in vivo*. Thus, patients with either all plastic or copper IUDs can be safely imaged with magnetic resonance [27].

Joint and limb prostheses. Generally, such prostheses present no risk. However, they can introduce image artifacts. If possible, they should be removed prior to the MR examination.

Skin patches. Pharmaceutical products in transdermal skin patches may cause burns due to the absorption of RF energy. Such patches must be removed prior to MR examinations.

Other Considerations

Sedation. MR has become an important tool in pediatric imaging. Since some infants and children are unable to cooperate with the examiners, there is an increased demand for sedation.

Some infants sleep soundly through an MR examination, particularly if they have eaten; however, many infants and children up to eight years require sedation, even if they are accompanied by their parents into the scanner room. In most instances, teenagers can be treated like adults.

Details on sedation and procedures can be found in the literature [3, 4, 20].

Claustrophobia. This is a very real *psychological danger* for some patients. Claustrophobia and other psychological stress situations have been reported severe enough to interrupt the examination in about 1-4% of cases. In this respect, small and wide-bore MR imagers are advantageous because the percentage of claustrophobic incidents drops significantly. Explanation of the imaging procedure and the equipment prior to the examination helps to reduce claustrophobia significantly.

The possibility of the patient falling from the examination couch and hypotonic syndrome (due to heat, motionless horizontal lying for a certain time, and psychological agitation) are additional hazards.

Pregnancy. There is no evidence that MR can harm the fetus or embryo — MR imaging is used for fetography, particularly for imaging the brain (Figure 18-5). An epidemiological study by Kanal, et al. concluded that data collected from MR imaging technologists were negative with respect to any statistically significant elevations in the rates of spontaneous abortion, infertility, and premature delivery [21]. As a safety precaution, MR scanning should be avoided in the first

Operational Safety Rules

- Access to the area of high magnetic field (in the magnet room) should be limited to trained personnel or to screened patients and visitors who are accompanied by trained personnel.

- Entrance to the area should be controlled by a lockable door; the keys to the area should be issued only to trained personnel.

- All entrances to this area should be visible to the system operator.

- Appropriate warning signs must be posted (Figure 18-3).

Table 18-3:
Some important operational safety rules (modified from Price [32]).

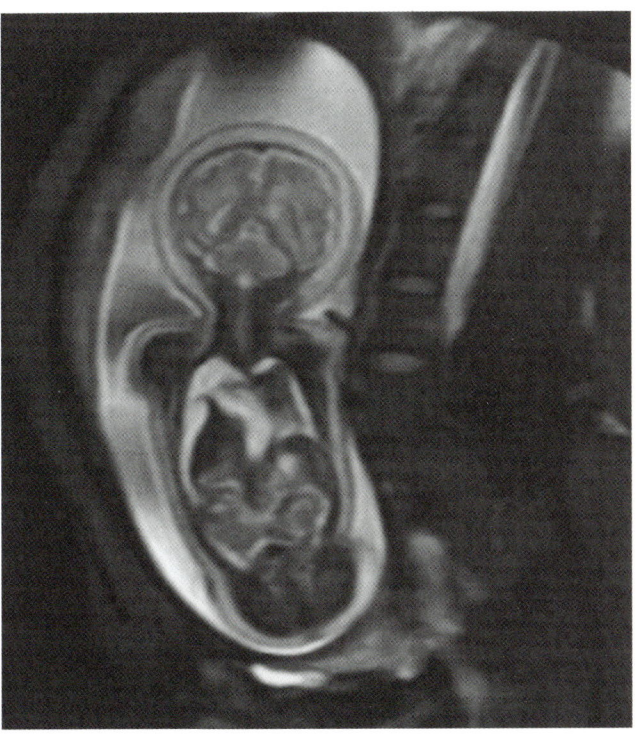

Figure 18-5:
MR fetography of a fetus in the 32nd week of pregnancy.

three months of pregnancy. MR imaging is indicated for use in pregnant women if other nonionizing forms of diagnostic imaging are inadequate, or if the examination provides important information which would otherwise require exposure to ionizing radiation such as x-ray or CT. Similar considerations hold for pregnant staff of a magnetic resonance department. Mainly for psychological reasons, it might be a wise precaution that pregnant staff members do not remain in the scan room during actual scanning; however, they are allowed to prepare and position the patient, administer contrast agents, and scan and film.

Legal Requirements

In the early 1980s, a number of national health and radiation protection boards first established recommendations concerning magnetic resonance imagers and spectroscopic units [15, 19, 33, 34]. All limits set by them were recommended levels, not mandatory ones.

Legal requirements in some European and Asian countries exist; some of them are without any scientific background, imposed by economic lobbies rather than learned societies. Others have no connection to reality or clinical routine [19, 29].

In the meantime, however, some manufacturers started using field strengths beyond 2.0 T, different pulse sequences, and gradient-switching procedures without any reported ill effects. Thus, these recommendations are partly outdated.

Adjustments made do not cover all possible medical applications of MR imaging, although the US-American Food and Drug Administration (FDA) extended the designation of 'nonsignificant risk' to MR systems with field strengths of up to 4.0 T in 1997 [5].

For legal reasons, the owner of MR equipment has to ensure that the equipment does fulfil the local requirements. In some countries, the regulations are more stringent than in others; in other countries, they are nonexistent. These requirements must be guaranteed by the manufacturer because the user in general is unable to check power output, gradient strength, or even field strength. This guarantee must cover authorized hardware and software updates after the initial installation. Specially designed computer programs usually supervise the power output of MR systems and will not allow or will interrupt any imaging or spectroscopy procedure exceeding those limits considered safe.

Relative and absolute contraindications for MRI and MRS investigations are summarized in Table 18-2, the most important operational safety rules in Table 18-3.

Subacute Hazards

The subacute risks of magnetic and RF fields have been intensively examined for a long time.

There are some publications associating an increase in the incidence of leukemia with the location of buildings close to high-current power lines with extremely low-frequency (ELF) electromagnetic radiation of 50-60 Hz [46], and industrial exposure to electric and magnetic fields [28]. However, a transposition of such effects to MRI or MRS seems unlikely.

According to the National Radiological Protection Board of the United Kingdom [35], the available experimental evidence weighs against electromagnetic fields acting directly to damage cellular DNA, implying that these fields may not be capable of initiating cancer in a manner that parallels that of ionizing radiation and many chemical agents. The results of some animal and cellular studies suggest the possibility that electromagnetic fields may act as co-carcinogens or tumor promotors, but taken overall, the data are inconclusive.

In the following paragraphs, we shall discuss some possible subacute hazards.

Static Magnetic Fields

In every MR examination, a large static magnetic field is applied. Field strengths for clinical equipment can vary between 0.2 and 2.0 T (2,000 and 30,000 Gauss); experimental imaging units have a field strength of up to 4.0 T, depending on the equipment used. In MRS, field strengths up to 12 T (120,000 Gauss) are currently used. No permanent hazardous effects of static magnetic fields upon human beings have yet been demonstrated [8]. However, there have been no long-term studies following persons who have been exposed to a static magnetic field.

Budinger calls the following five biophysical mechanisms into question whereby static magnetic fields might influence biological processes or an organism's behavior [14]:

Changes in enzyme kinetics. Up to 45 Tesla, no important effects on enzyme systems have been observed.

Orientation changes of macromolecules and living-cell subcellular components. The result of replicable experiments on the orientation effects of retinal rods in fields of 1 Tesla, the alignment of sickle cells at 0.35 T, and the orientation of certain bacteria and ani-

> **At present, exposure to MR examination procedures is considered safe for patients and personnel.**
>
> **There is no convincing evidence of any long-term or irreversible subacute effects of MR imaging or spectroscopy.**

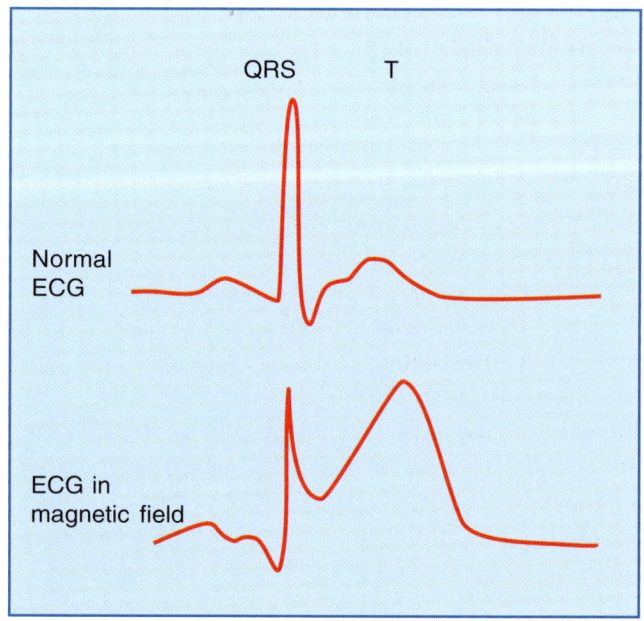

Figure 18-6:
Flowing blood can behave as a moving conductor in a magnetic field. The field can induce a voltage which will be highest during the part of the cardiac cycle with the fastest blood velocity. This coincides with the T-wave of the ECG and enhances the T-wave, potentially mimicking pathology.

mals might be explained by the physical torque rather than the sensing of the turning torque by nervous tissue.

Nerve conductivity. As early as 1893, the first results of experiments about a possible influence of static magnetic fields upon nerve tissue were obtained[6]. These and all later experiments showed negative results. There are apparently no effects on the conduction of impulses in the nerve fiber up to a field strength of 0.1 T generated by either changing the electrical resistance or the potential of the excitation[1,2]. Theoretical examinations argue that fields of 24 T are required to produce a 10% reduction of nerve impulse conduction velocity[23].

A preliminary study has indicated neurological effects in subjects exposed to a whole-body imager at 4.0 T[41]. Here, additional research is necessary.

Cardiac changes. A field-strength-dependent increase in the amplitude of the ECG in rats has been observed during exposure to homogeneous stationary magnetic fields. The minimum level at which augmentation could be observed was 0.3 T; at 2.0 T, the increase was by an average of 400%. The augmentation in T-wave amplitude occurred instantaneously and was immediately reversible after exposure to the magnetic field ceased (Figure 18-6). There have been no abnormalities in the ECG in the later follow-up[16]. The authors suggest that augmentation of the signal amplitude in the T-wave segment may result from a superimposed electrical potential. At field strengths of between 7 and 10 T, no *arrhythmia* could be proven[10].

According to the national radiation protection and health agencies, it is unlikely that cardiac fibrillation would occur as a result of induced flow potential in the major blood vessels or heart chambers at this level of field intensity.

No circulatory alterations coincide with the ECG changes. Therefore, no biological risks are believed to be associated with them.

Magnetohydrodynamic effects. A blood pressure increase of 28% is predicted theoretically for a field of 10 T. This is claimed to be caused by interaction of induced electrical potentials and currents within a solution, e.g. blood, and an electrical volume force causing a retardation in the direction opposite to the fluid flow. This decrease in flow velocity must be compensated for by an elevation in pressure. At 1.5 T, no significant changes are expected; at 6.0 T a 10% pressure change is expected[14,45].

In addition to Budinger's reflections, the following points are valid for discussion:

Genetic effects. There have been several reports that static magnetic fields may provoke genetic mutations, changes in growth rate and leukocyte count and other effects[36]. The results of these experiments could not be reproduced[39]. Inhibition of growth rate of Eschericha coli induced by low-frequency magnetic field could be shown. Nevertheless, some authors claim it be unlikely that mutagenic effects are introduced by fields lower than 1.0 T[24]. No reports have been published that persons exposed to magnetic fields, including personnel at MR departments, have a higher incidence of genetic damage to their children than found in the average population.

We believe, however, that this research needs further investigation and that *pregnancy* should be considered a relative contraindication for MRI and MRS. Taking into account that clinical MR imaging devices operate at field strengths of between 0.2 and 2.0 T, caution demands further experiments at higher field strengths.

Membrane transportation and blood sedimentation. Other potential hazards from static fields include, for instance, membrane transportation and blood sedimentation induced by the field. As Mansfield and Morris pointed out, static magnetic field gradients of 0.01 T/cm (100 G/cm) make no significant difference in the membrane transport processes. The influence of a static magnetic field upon erythrocytes is not sufficient to provoke sedimentation, as long as there is a normal blood circulation[24].

As discussed above, many results presented in publications about effects of static magnetic fields are contradictory and cannot be explained by biophysical or biochemical mechanisms. In some cases, the effects observed must be attributed to other causes which had not been considered by the researchers in the setup of the experimental protocol. Critical considerations of such experiments can be found in a number of reviews[14,31]. However, the data available are not comprehensive enough to assume MR imaging and spectroscopy are absolutely safe.

Varying Magnetic Fields

Varying magnetic fields are necessary for the localization of nuclei with magnetic properties within the sample.

A well described effect of varying magnetic fields is the so-called *magnetic phosphenes*[16], which were first observed some 90 years ago [7]. They are attributed to magnetic-field variations and may occur in a threshold field change of between 2 and 5 T/s. Phosphenes are stimulations of the optic nerve or the retina, producing a flashing sensation in the eyes. They seem not to cause any damage in the eye or the nerve.

Varying magnetic fields are also used to stimulate bone-healing in non-unions and pseudarthroses. The reasons why pulsed magnetic fields support bone-healing are not completely understood [9].

Rapid echo-planar imaging and high-performance gradient systems create fast-switching magnetic fields that can stimulate muscle and nerve tissues. The mean threshold levels for various stimulations are 3,600 T/s for the heart, 900 T/s for the respiratory system, and 60 T/s for the peripheral nerves. Guidelines in the United States limit switching rates at a factor of three below the mean threshold for peripheral nerve stimulation.

Radiofrequency Fields

Radiofrequency pulses are used in MR imaging for the excitation of the nuclei. RF fields may interact with both tissues and foreign bodies, such as metallic implants, in the patient. The main result of this type of interaction is heat.

The higher the frequency, the larger will be the amount of heat developed; and the more ionic the biochemical environment in the tissue, the more energy that will be deposited as heat [22,35]. This effect is well-known for homogeneous model systems, but the complex structure of various human tissues makes detailed theoretical calculations very difficult, if not impossible.

The specific absorption rate, SAR, helps to estimate RF heating effects. It increases with field strength, radiofrequency power and duty cycle, transmitter-coil type and body size. In high and ultrahigh fields, some of the multiple echo, multiple-slice pulse sequences may create a higher SAR than recommended by the agencies.

Hot spots may occur in the exposed tissue. At present, it seems unlikely that such hot spots in the body exist, but to avoid or at least minimize effects of such theoretical complications, the frequency and the power of the RF irradiation should be kept at the lowest possible level.

In several *in vitro* and *in vivo* experiments, no threatening increase in temperature could be shown [13,23]. Even in high magnetic fields, no local temperature increase

greater than 1°C occurred. The highest skin temperature increase described reached 2.1°C [41]. Eddy currents may heat up implants and thus may cause local heating. *In vitro* worst-case experiments performed with a large and very thin thermally insulated aluminium sheet at 1.5 T after 15 minutes of exposure showed a temperature rise of only 0.08°C.

According to the specific FDA criteria for SAR limits, the SAR must not be greater than:
- 4 W/kg averaged over the whole body for any 15-minute period;
- 3 W/kg averaged over the head for any 10-minute period; or
- 8 W/kg in any gram of tissue in the extremities for any period of 5 minutes.

Some European countries have issued SAR restriction too. No common denominator has been found.

References

Additional references and information are available at the following websites: http://kanal.arad.upmc.edu/mrsafety.html and at http://www.fda.gov

1. Abashin VM, Yevtushenko GI. Influence of a permanent magnetic field on biological systems. Biofizika 1975; 20: 276-280.
2. Abashin VM, Yevtushenko GI. A permanent magnetic field and the conduction of an impulse along the nerve. Biofizika 1975; 20: 281-285.
3. American Academy of Pediatrics, Committee on Drugs. Guidelines for monitoring and management of pediatric patients during and after sedation for diagnostic and therapeutic procedures. Pediatrics 1992; 89: 1110-1115.
4. American Society of Anesthesiologists House of Delegate Standards for basic intraoperative monitoring. American Society of Anesthesiologists' 1992 Directory of Members. Park Ridge, Il. 1992; 675-676.
5. Center for Devices and Radiological Health. A primer on medical device interactions with magnetic resonance imaging systems. U.S. Department of Health and Human Services. Food and Drug Administration. 1997. See: http://www.fda.gov/cdrh/ode/primerf6.html
6. D'Arsonval MA. Action physiologique des courants alternatifs à grande fréquence. Arch d Physiol 1893; 5: 401-406.
7. D'Arsonval MA. Dispositifis pour la mesure des courants alternatifs à toutes fréquences. C R Soc Biol (Paris) 1896; 3: 451.
8. Baker KA, DeVor D. Safety considerations with high field MRI. Radiol Technol 1996; 67: 25-52.
9. Bassett CAL, Pilla AA, Pawluk RJ. A non-operative salvage of surgically-resistant pseudarthroses and non-unions by pulsing electromagnetic fields. Clin Orthop 1977; 124: 128.
10. Battocletti JH, Salles-Cunha S, Halbach RE, Nelson J, Sances Jr A., Antonich FJ. Exposure of rhesus monkeys to 20,000 G steady magnetic field: effect on blood parameters. Med Phys 1981; 8: 115-118.
11. Beischer DE. Human tolerance to magnetic fields. Astronautics 1962; 7: 24-48.
12. Bottomley PA, Edelstein WA. Power deposition in whole-body NMR imaging. Med Phys 1981; 8: 510-512.
13. Budinger TF. Nuclear magnetic resonance (NMR) in vivo studies: known threshold for health effects. J Comput Assist Tomogr 1981; 5: 800-811.
14. Budinger TF. Safety of NMR *in vivo* imaging and spectroscopy. in: Budinger TF, Margulis AR: Medical magnetic resonance imaging and spectroscopy. A primer. Society of Magnetic Resonance in Medicine: Berkeley, CA, U.S.A. 1986; 215-231.
15. Bundesgesundheitsamt (der Bundesrepublik Deutschland) (German Federal Health Office). Empfehlungen zur Vermeidung gesundheitlicher Risiken, verursacht durch magnetische und hochfrequente elektromagnetische Felder bei der NMR-Tomographie und In-vivo-NMR-Spektroskopie. Bundesgesundheitsblatt. 1984; 27: 92-96.
16. Gaffey CT, Tenforde TS. Alterations in the rat electrocardiogram induced by stationary magnetic fields. Bioelectro-magnetics 1981; 2: 357-370.

17. Hepfner ST, Skelly MF. Radiofrequency interference in cochlear implants. N Engl J Med 1985; 313: 387.
18. Hylton NM. Suspension of breast-feeding following gadopentetate dimeglumine administration. Radiology 2000; 216: 325-326.
19. International Non-Ionizing Radiation Committee of the International Radiation Protection Association. IRPA/INIRC Guidelines. Protection of the patient undergoing a magnetic resonance examination. Health Physics 1991; 61: 923-928. *(Comment: guidelines of only limited value for manufacturers and users of MRI and MRS).*
20. Kanal E and Shellock FG. Patient monitoring during clinical MR imaging. Radiology 1992; 185: 623-629.
21. Kanal E, Shellock FG, Savitz DA, Gillen J. Survey of reproductive health among female MR operator (abstr.). Radiology 1991; 181 (P): 235.
22. Led JJ, Petersen SB. Heating effects in carbon-13 NMR spectroscopy on aqueous solution caused by proton noise decoupling at high frequencies. J Mag Res 1978; 32: 1-17.
23. Liboff RL. A biomagnetic hypothesis. Biophys J 1965; 5: 845.
24. Mansfield P, Morris PG. Biomagnetic effects. In: Mansfield P, Morris PG: NMR imaging in medicine. New York, London: Academic Press. 1981, 297-332.
25. Magin RL, Liburdy RP, Persson B. Biological effects and safety aspects of nuclear magnetic resonance imaging and spectroscopy. Annals of the New York Academy of Sciences; vol. 649, 31 March 1992.
26. Mansfield P, Haywood B. Principles of active acoustic control in gradient coil design. Magn Reson Materials Phys Biol Med 2000; 10: 147-151.
27. Mark AS, Hricak H. Intrauterine contraceptive devices: MR imaging. Radiology 1987; 162: 311-314.
28. Milham S. Mortality from leukemia in workers exposed to electrical and magnetic fields. N Engl J Med 1982; 307: 249.
29. Norris DG, Ordidge RJ. The regulation of MR examinations in Germany: a threat to scientific and technical progress for MR in Europe? MAGMA 2000; 10: 4-5.
30. Pavlicek W, Geisinger M, Castle L, Borkowski GP, Meaney TF, Bream BL, Gallagher JH. The effects of nuclear magnetic resonance of patients with cardiac pacemakers. Radiology 1983; 147: 149-153.
31. Persson BRR, Ståhlberg F. Health and safety of clinical NMR examinations. Boca Raton, FL, U.S.A.: CRC Publishers 1989.
32. Price RR. The AAPM/RSNA physics tutorial for residents. MR imaging safety considerations. RadioGraphics 1999; 19: 1641-1651.
33. Radiological Health, Bureau of (United States of America). Guidelines for evaluating electromagnetic risk for trials of clinical NMR systems. Department of Health and Human Services, Public Health Service, Food and Drug Adminstration. Rockville, 25 February 1982; and: US Food and Drug Administration: Magnetic resonance diagnostic device; panel recommendation and report on petitions for MR reclassification. Fed Reg 1988; 53: 7575-7579.
34. Radiological Protection Board, National (United Kingdom). Exposure to nuclear magnetic resonance clinical imaging. Radiography 1981; 47: 258-260.
35. Radiological Protection Board, National (United Kingdom). Electromagnetic fields and the risk of cancer. Documents of the NRPB 3, 1 Didcot, Oxon: NRPB. 1992.
36. Rinck PA. Risiken und Gefahren der NMR-Tomographie;

Vorschläge zum Schutz und zur Überwachung. Dtsch Med Wschr 1983; 108: 992-994.

37. Runge VM. Safety of approved MR contrast media for intravenous injection. J Magn Reson Med 2000; 12: 205-213.

38. Schroeder P-M, Miksicek A, Koralewski H-E, Römer T. Kernspintomographie bei Patienten mit Nervenschrittmachern (English abstract: MR in Patients with nerve pacemakers). Fortschr Röntgenstr 1987; 147: 198-199.

39. Schwartz JL, Crooks LE. NMR imaging produces no observable mutations or cytotoxicity in mammalian cells. AJR 1982; 139: 583-585.

40. Shellock FR and Curtis JS. MR imaging and biomedical implants, material, and devices: an updated review. Radiology 1991; 180: 541-550.

41. Shellock FG and Kanal E. Magnetic resonance: bioeffects, safety, and patient management. Philadelphia, PA, U.S.A.: Lippincott-Raven 1994.

42. Shellock FG and Kanal E. Bioeffects and safety. In: Stark DD and Bradley WG: Magnetic Resonance Imaging. St. Louis, MO, U.S.A: Mosby. 1999, Vol. 1: 291-306.

43. Sommer T, Lauck G, Schimpf R, et a.. MRT bei Patienten mit Herzschrittmachern: In-vitro- und In-vivo-Evaluierung bei 0,5 Tesla. Fortschr Röntgenstr 1998; 168: 36-43.

44. Soulen RL, Budinger TF, Higgins CB. Magnetic resonance imaging of prosthetic heart valves. Radiology 1985; 154: 705-707.

45. Tenforde TS, Gaffey CT, Moyer BR, Budinger TF. Cardiovascular alterations in Macaca monkeys exposed to stationary magnetic fields: experimental observations and theoretical analysis. Bioelectromagnetics 1983; 4: 1.

46. Wertheimer N, Leeper E. Electric wiring configurations and childhood cancer. A J Epidemiol 1979; 109: 273.

47. Zhang, Yong-shang, Heng-wang, Wu. Der Einfluß von magnetisiertem Wasser auf Harnstein. Eine experimentelle und klinische Studie. Acta Acad Med Wuhan 1984; 4: 31-37.

Acknowledgement

Figure 18-4 courtesy of C. Tempany; MR, The Newsmagazine of Magnetic Resonance. May/June 1993; 39.

MR Imaging from a European Perspective

From left to right:
Jean-Baptiste-Joseph Fourier, Felix Bloch, and Edward M. Purcell.

Looking back at the main protagonists involved in MR imaging is vital for an understanding of the development of the modality.

History differs according to who is telling the story and for which purpose. It also depends on who has written the 'documented history', particularly when you have not been directly exposed to the events. Although I have witnessed part of the history of MR imaging, my remarks will be anecdotal.

The topic is interesting, but rather sensitive. Therefore, I want to make this disclaimer at the beginning: I apologize to all those who have been instrumental to the development of MR imaging in Europe but are not mentioned — and those who are referred to but not properly quoted. In one short book chapter, it is impossible to do justice to all who have made substantial contributions.

Like any history, the history of MR imaging has no real beginning. There is no exact date when somebody claimed to have invented an imaging method based upon the phenomenon of nuclear magnetic resonance.

»Everything flows and nothing stays,« as Heraklitos pointed out.

One major contribution to the technique can be found in Napoleon's realm. Jean-Baptiste-Joseph Fourier served three years as the secretary of the Institut d'Egypte at the beginning of the nineteenth century, and later became prefect of the Isère département in France. However, the focus of his life was mathematics, and

without his Fourier transform we would not be able to create MR images.

In 1946, two scientists in the United States, independently of each other, described a physicochemical phenomenon which was based upon the magnetic properties of certain nuclei in the periodic system. This was named 'nuclear magnetic resonance', for short 'NMR'.

The two scientists, Felix Bloch and Edward M. Purcell, were awarded the Nobel Prize in Physics in 1952.

Purcell was born in Illinois in the United States of America. He worked at the Massachusetts Institute of Technology, MIT, and later joined the faculty of Harvard University.

Bloch was born in Zurich in 1905 and taught at the University of Leipzig until 1933; he then emigrated to the United States and was naturalized in 1939. He joined the faculty of Stanford University at Palo Alto in 1934 and became the first director of CERN in Geneva in 1962. In 1983 he died in Zurich.

Bloch was a protagonist for the interaction between Europe and the United States. NMR and MRI would not exist without this interaction.

At some stage, every European scientist wants to emigrate to the U.S.A. Some move transatlantic and some even stay for good. Others return. There is hardly any movement in the other direction. The historical reasons were different prior to and after the Second World War. Before the war, plain survival for many depended on emigration, or it was at least guided by political necessity. It was the attraction of the Statue of Liberty which made scientists move westward.

After the war, research facilities in the United States were more attractive than those in Europe because the

The Statue of Liberty: from the 1930s until the 1980s it promised freedom to many Europeans moving to the United States, including those involved in MR research. Today many of the R&D results originate in Europe again.

From left to right:
Otto Stern, Isidor I. Rabi, and Cornelis Jacobus Gorter.

Physica IX, no 6   Juni 1942

NEGATIVE RESULT OF AN ATTEMPT TO OBSERVE NUCLEAR MAGNETIC RESONANCE IN SOLIDS.

by C. J. GORTER and L. J. F. BROER

Zeeman-Laboratorium der Universiteit van Amsterdam
Communication No.266a from the Kamerlingh Onnes Laboratorium, Leiden

Zusammenfassung

Es wird das negative Resultat bekannt gegeben eines Versuches magnetische Kernmomente zu ermitteln durch Beobachtung von anomaler Dispersion im normalen Kurzwellengebiet. Der Grund des Fehlschlagens liegt wohl in der überaus kleinen Wechselwirkung zwischen den Kernmomenten und den Wärmewellen des Kristallgitters.

Résumé

On décrit une expérience ayant pour but d'observer des spins magnéti-

Beginning of Gorter's article about his failed experiments (1942).

Graphic designs of two magnetic-field-based diagnostic systems: left, from a Siemens patent (1967), and right, from R. Damadian's patent (1972). Both machines were one-dimensional and not conceived as imaging equipment.

academic system in the U.S.A. was more flexible than the university structures in Europe — and dollars were plentiful for research and for personal income.

Bloch and Purcell were not the only scientists working in the field. The 1920s had been roaring and inflationary, but also extremely fruitful in science. In 1924, Wolfgang Pauli suggested the possibility of an intrinsic nuclear spin. The year after, George Eugene Uhlenbeck and Samuel A. Goudsmit introduced the concept of the spinning electron. Two years later Pauli and Charles Galton Darwin developed a theoretical framework for grafting the concept of electron spin into the new quantum mechanics developed the year before by Edwin Schrödinger and Werner Heisenberg.

Pauli, Uhlenbeck, and Goudsmit went to the United States to work. The British stayed in Britain — at that time.

This development continued in the 1930s. After their initial pacemaking work, in 1933, Otto Stern and Walther Gerlach were able to measure the effect of the nuclear spin by deflection of a beam of hydrogen molecules. During the early 1930s, Isidor Isaac Rabi's laboratory at Columbia University in New York became a major center for related studies.

Rabi's research was successful, but only the visit by Cornelis Jacobus Gorter from the Netherlands in September 1937 finally showed how to measure the nuclear magnetic moment. Gorter had tried similar experiments and failed. Rabi accepted and realized Gorter's suggestions concerning his experiments, changed them, and was able to observe resonance experimentally. This led to the publication of 'A New Method of Measuring Nuclear Magnetic Moment' in 1938.

Gorter first used the term 'nuclear magnetic resonance' in a publication which appeared in the war-torn

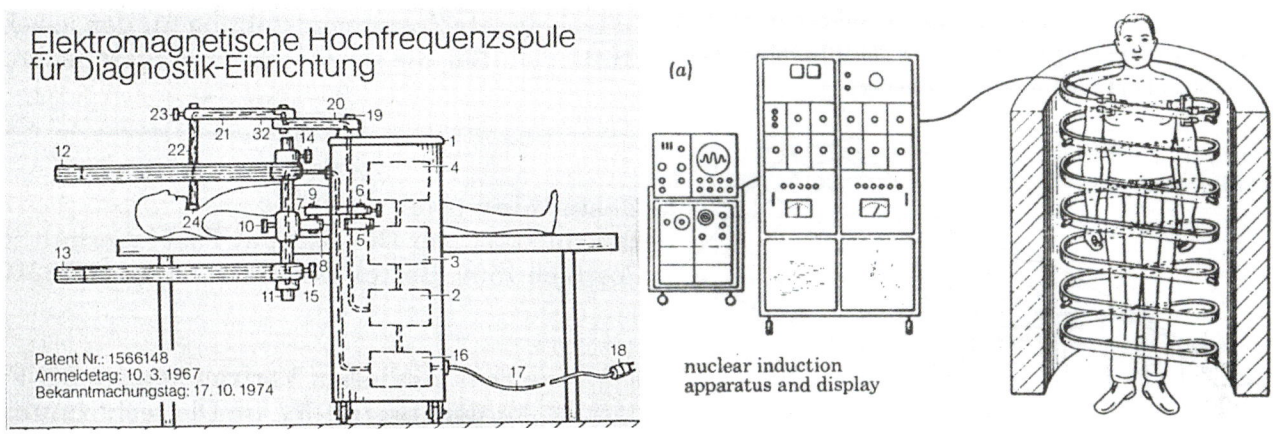

nuclear induction apparatus and display

From left to right:
Yevgeni K. Zavoisky, Paul C. Lauterbur, and Richard Ernst.

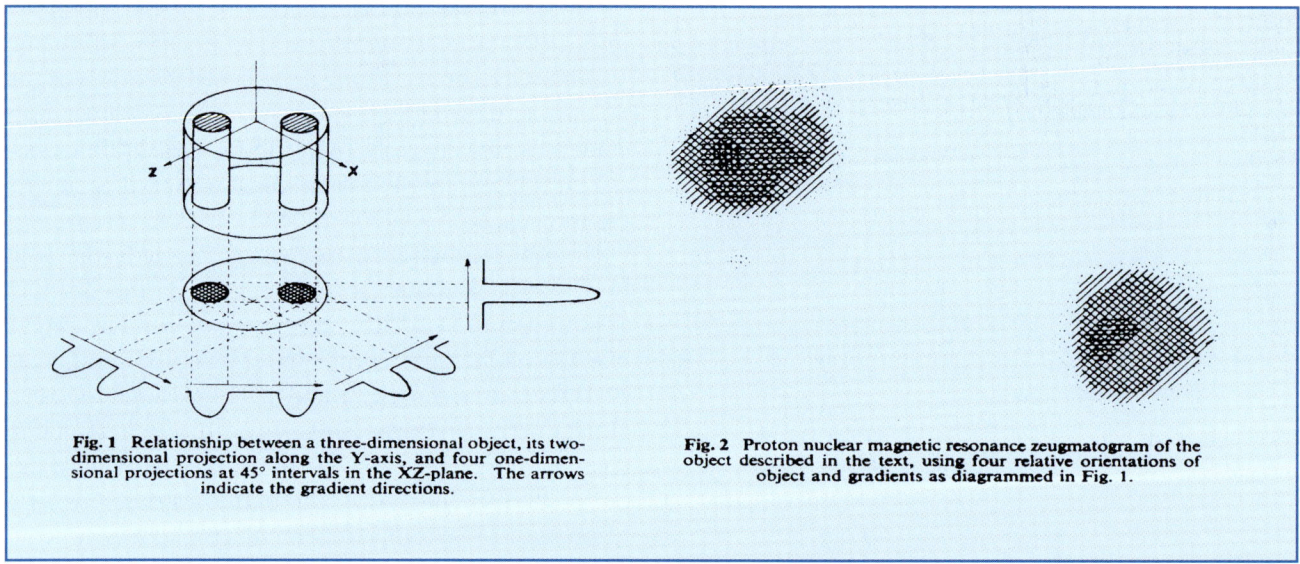

Fig. 1 Relationship between a three-dimensional object, its two-dimensional projection along the Y-axis, and four one-dimensional projections at 45° intervals in the XZ-plane. The arrows indicate the gradient directions.

Fig. 2 Proton nuclear magnetic resonance zeugmatogram of the object described in the text, using four relative orientations of object and gradients as diagrammed in Fig. 1.

Lauterbur's first published MR images acquired at the State University of New York at Stony Brook (from Nature 1973; 242: 190-191).

One of the first 2D-FT MR images from Richard Ernst ressearch group in Zurich (acquired by Anil Kumar in July 1974).

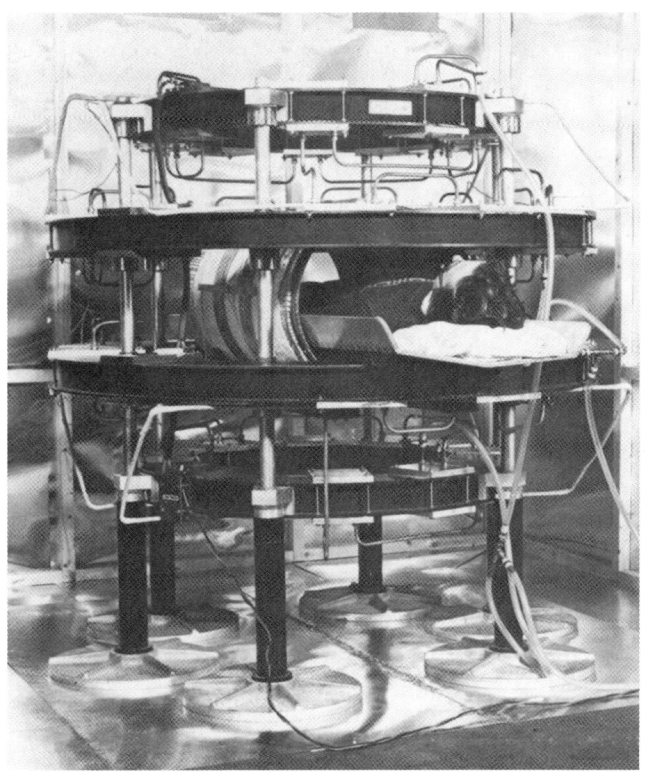

The first Aberdeen imager with Jim Hutchison inside (this picture was taken in 1979).

Netherlands in 1942, attributing the coining of the phrase to Rabi.

The Second World War had a major influence upon research — and its interruption. Germany, for instance, the leading country in science and medicine at the time, quit the race in the 1930s.

But there was another country in which major contributions to nuclear magnetic resonance were made. They originated in Kazan in Tatarstan, which was part of the Soviet Union at that time and is now an independent republic within Russia. Until recently, Russian contributions to NMR and radiology were frowned upon or not even discussed in the West.

Electron spin resonance was discovered at Kazan's university by Yevgeni K. Zavoisky towards the end of the war. Zavoisky had first attempted to detect NMR in 1941, but like Gorter he had failed.

The final breakthrough came with Bloch and Purcell in 1946.

During the next few decades NMR developed in a wide range of applications. Hardly any of them were medical, although *in vivo* NMR already had been performed since the early 1950s. In 1955, Erik Odeblad and Gunnar Lindström from Stockholm published their first NMR measurements of living cells and excised animal tissue. Odeblad continued working on tissues throughout the 1950s and 1960s.

In the late 1960s, Jim Hutchison at the University of Aberdeen in Scotland began working with magnetic resonance on *in vivo* electron spin resonance studies in mice. Others joined in this kind of research, among the better known being the research groups of Raymond Damadian at Downstate Medical Center in Brooklyn and Donald P. Hollis at Johns Hopkins University in Baltimore. Damadian's group measured T1 and T2 relaxation times of excised normal and cancerous rat tissue and stated that tumorous tissue had longer relaxation times than normal tissue. Hollis and his collaborators achieved similar results, but were more balanced and scientifically critical in their postulations and deductions. Initially, Damadian ignored the possibility of cross-sectional imaging and developed an analytical technique for *in vivo* measurement of tissue relaxation times, which he dubbed 'Fonar'.

Actual *in vivo* NMR spectroscopy took off in Oxford from 1974, with the group of Rex E. Richards and George K. Radda. Among others, David Hoult and David G. Gadian belonged to this group.

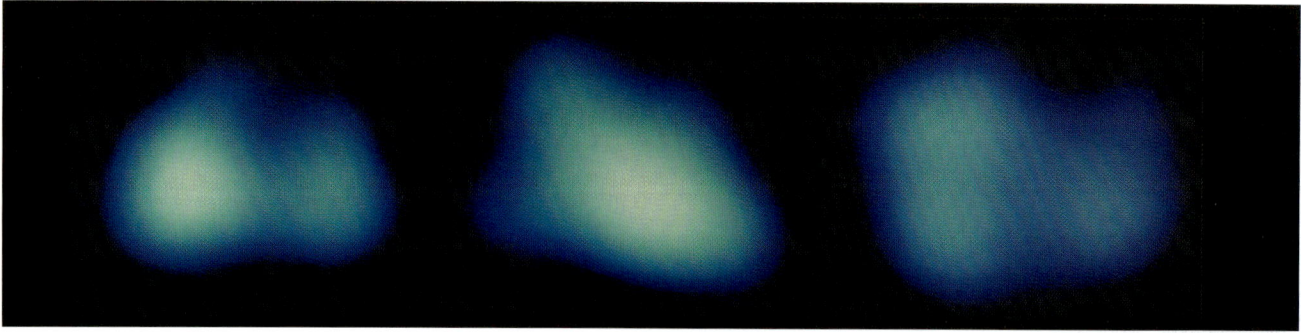

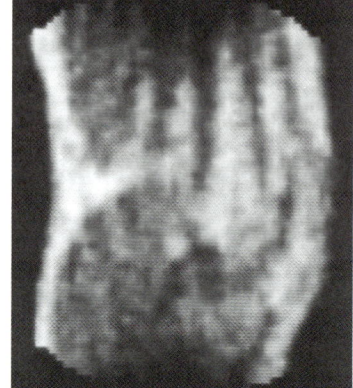
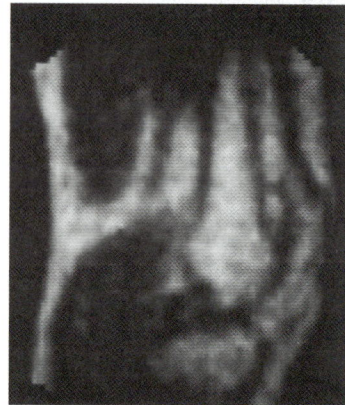

Top:
First *in vivo* fluorine ventilation images of a dog's lung acquired in 1982 by P.A. Rinck et al. (projection images; left: coronal, center: sagittal, right: transverse).
See also Rinck and Muller in Figure 6-1.

Right:
First magnetization transfer images of a knee acquired in 1982 by R.N. Muller et al. (left: before off-resonance radiation, right: with off-resonance radiation).

Spatial Encoding

All the experiments up to now had been one-dimensional and lacked spatial information. Nobody could determine exactly where the NMR signal originated within the sample. After MR imaging had first been described, several individuals and companies claimed that they had achieved imaging earlier, but their machines were not conceived of as imagers.

In roentgenology, the times of conventional imaging ended in September 1971, when the world's first CAT scanner was installed in England by EMI. In the same month, Paul Lauterbur of the State University of New York at Stony Brook had the idea of applying magnetic field gradients in all three dimensions and the CAT-scan back-projection technique to create NMR images. He published the first images of two tubes of water in March 1973 in the journal Nature. This was followed later in the year by the picture of a living animal, a clam, and in 1974 by the image of the thoracic cavity of a mouse.

Field gradients had been used before. Roger Gabillard from Lille in France, for instance, had imposed one-dimensional gradients on samples.

However, Lauterbur's idea revolutionized NMR because it opened the field to imaging. When he presented his approach to NMR imaging, named 'zeugmatography' at the International Society of Magnetic

Resonance (ISMAR) meeting in January 1974 in Bombay, Raymond Andrew, William Moore, and Waldo Hinshaw from the University of Nottingham, England, were in the audience and took note. As a result, Hinshaw and Moore developed their own approach to MR imaging with their 'sensitive point' method. An alternative imaging method was proposed in 1974 by Alan N. Garroway, Peter K. Grannell, and Peter Mansfield, also from the University of Nottingham. By 1975, Mansfield and Andrew A. Maudsley proposed a line technique which, in 1977, led to the first image of *in vivo* human anatomy, a cross section through a finger. In 1978, Mansfield presented his first image through the abdomen.

In 1977, Hinshaw, Paul Bottomley, Neil Holland, Moore, and Brian Worthington and collaborators succeeded with an image of the wrist. More human thoracic and abdominal images followed, and by 1978, Hugh Clow and Ian R. Young, working at EMI, reported the first transverse NMR image through a human head. Two years later, William Moore and colleagues presented the first coronal and sagittal images through a human head.

In the research group of John Mallard at the University of Aberdeen, Jim Hutchison, Bill Edelstein, Glynn Johnson, and Ted Redpath developed the spin warp technique. Margaret Foster, David Lurie, Francis Smith, and Ann Reid contributed to this effort. They

From left to right:
Jürgen Hennig, Axel Haase, Jens Frahm, and Peter Mansfield.

published a first image through the body of a mouse in 1974.

In April 1974, Lauterbur gave a talk at a conference in Raleigh, North Carolina (U.S.A.). This conference was attended by Richard Ernst from Zurich, who realized that instead of Lauterbur's back-projection one could use switched magnetic field gradients in the time domain. This led to the 1975 publication, 'NMR Fourier Zeugmatography' by Anil Kumar, Dieter Welti, and Richard Ernst, and to the method mostly used for MR imaging today.

The Great Transatlantic Brain-Drain

At this time, many of the researchers working in Britain went to the United States. It was a major brain-drain for British universities, but there was (and still is) little money in the British university system. Most of the researchers stayed abroad, whereas many of the Continental Europeans who worked in the U.S.A. in the late 1970s and early 1980s returned home.

Some of them had performed quite impressive research in the United States; among them was Robert

From left to right:
Graeme M. Bydder, Ian R. Young, Hanns-Joachim Weinmann.

N. Muller, who — in 1982 — described off-resonance imaging, a technique known today as 'magnetization-transfer' imaging (see also Figure 6-1).

The Return of Europe

In the 1980s, Continental Europe started to contribute intensively to MR imaging. Rapid imaging originated in European laboratories. Jürgen Hennig, together with A. Nauerth and Hartmut Friedburg, from the University of Freiburg introduced RARE (rapid acquisition with relaxation enhancement) imaging in 1986. This technique is probably better known under the commercial names of fast or turbo spin-echo.

At about the same time, FLASH (fast low angle shot) appeared, opening the way to similar gradient-echo sequences. This sequence was developed at Max-Planck-Institute, Göttingen, by Axel Haase, Jens Frahm, Dieter Matthaei, Wolfgang Hänicke, and Dietmar K. Merboldt.

FLASH was very rapidly adopted commercially. RARE was slower, and echo-planar imaging (EPI) — for technical reasons — took even more time. Echo-planar imaging had been proposed by Mansfield's group in 1977, and the first crude images were shown by Mansfield and Ian Pykett in the same year. Roger Ordidge presented the first movie in 1981.

Clinical Applications

At about this time, MR imaging started being clinically evaluated. One of the most admirable research groups worked at Hammersmith Hospital in London. The head of the group was Robert E. Steiner, but Ian R. Young and Graeme M. Bydder were the moving forces. Among others, Frank H. Doyle and Jacqueline M. Pennock supplemented this group.

Because MR imaging is at the crossroads between medicine and chemistry, physics, and computer science, groups with strong interdisciplinary relationships and cross-fertilization became scientifically extremely fruitful, which led to the 'odd couple' system, involving one physician and one scientist. At congresses, you would always see Graeme Bydder together with Ian Young, a seemingly ideal combination. There were (and are) other couples like them, but apparently this kind of relationship between radiologists and physicists does not fit into all European academic systems.

Early clinical imaging was extremely difficult, time-consuming, and often disappointing. Spin-echo imag-ing, for instance, was a bigger step than many imagine. Today it is taken for granted, and it has helped MR imaging immensely to become a routine technique.

Early MR images were mainly based upon proton-density differences, later upon differences in T1-weighting. By 1982-3, the Hammersmith and Wiesbaden groups pointed out that long heavily T2-weighted SE sequences were better at highlighting pathology. It took some years until this was generally accepted, mostly because many companies claimed that long TE was neither possible nor necessary.

Another European affair was the development of contrast agents. The possible concept had been described at universities in the United States by Maria Helena Mendonça-Dias and Paul C. Lauterbur, by Robert Brasch, and Gerald Wolf. However, most of the commercial development and scientific research took place in Europe. Schering submitted a patent application for Gd-DTPA dimeglumine in July 1981 in a project involving Hanns-Joachim Weinmann and Ulrich Speck. In 1984, Dennis H. Carr from the Hammersmith and Wolfgang Schörner from Berlin published the first images in men. Since the late 1980s, Magnevist has been commercially available, followed shortly afterwards by Dotarem from Guerbet in Paris.

MR Equipment

With the exception of the scientific instrument manufacturers, the hardware makers had no background in NMR. The most important scientific manufacturers were Varian in the U.S.A., JEOL in Japan, and Bruker-Spectrospin in Europe. Most scientific developments in MR imaging were done on Bruker machines. When you opened the cover of almost any commercial MR imager in the early and mid 1980s, you would find Bruker components.

The first hardware manufacturer to get involved in whole-body imaging was EMI in 1974. Later the company was taken over by Picker (today Marconi). Philips started research into MR imaging at the same time; P. Rob Locher, André Luiten, and Piet van Dijk were seen at many scientific meetings. Siemens got involved in 1977, Johnson & Johnson/Technicare in 1978/79, Instrumen-tarium at about the same time, and the others followed in the 1980s.

M&D Aberdeen was a company originating from the research group at Aberdeen University. It had one machine in Geneva, but it disappeared a long time ago, as have a number of other companies.

Another effort was the Finnish MR imaging machine. Raimo E. Sepponen, together with a number of other researchers, among them the surgeon Jorma T. Sipponen, aimed to develop a method and device for detection of internal hemorrhages. Their first clinical MR imaging model was installed at Helsinki University Central Hospital in June 1982 operating at a field strength of 0.17 T. The second unit operated at 0.02 T, and later units operating at 0.04 T, which at that time was politico-commercially a step in the wrong direction.

With one exception, all early magnets for MR imagers were produced by Oxford Magnets. Still today most magnets come from companies in the Oxford area.

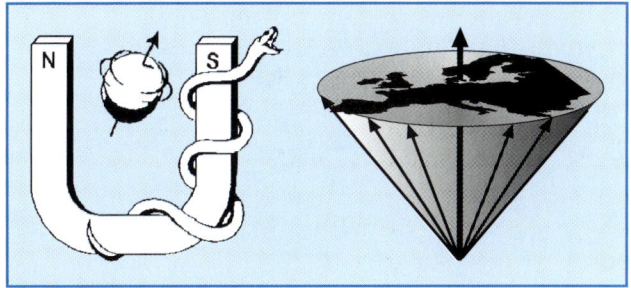

Logos of the European Workshop for Nuclear Magnetic Resonance in Medicine and the European Magnetic Resonance Forum Foundation.

Teaching, Training, Conferences

As Swiss-American scientist Felix Wehrli realized, there was an enormous need for user education in magnetic resonance imaging. The first European NMR imaging meeting was held in Nottingham in April 1976, followed by a second conference in Winston-Salem in North Carolina in the U.S.A. in 1981. Soon afterwards, the number of meetings exploded.

Another effort aimed at teaching users in Europe started also in the United States in the early 1980s: the European Workshop on Nuclear Magnetic Resonance in Medicine, now known as the EMRF Foundation. The first Annual Meeting of the European Workshop was held in Mons, Belgium, in 1983, followed by meetings all over Europe. Today, the EMRF Foundation specializes in smaller meetings and the presentation of sponsorships and grants. The major European MR meetings are organized by the European Society for Magnetic Resonance in Medicine and Biology which was founded in Geneva in 1983, the European Congress of Radiology, and national radiological, medical physics, and MR societies.

Reference

There are numerous original publications, too many to be listed here. An extensive overview of the history of NMR and, partly, MRI can be found in:

1.   Becker ED, Fisk CL, Khetrapal CL: The Development of NMR. In: Grant DM, Harris RK (eds.): Encyclopedia of Nuclear Magnetic Resonance. Volume 1. Chichester: John Wiley and Sons. 1996. 1-158.

Acknowledgements

The pictures were reprinted with the friendly permission of the owners and/or copyright holders: Raymond Andrew, EMRF Archives, the Nobel Foundation, and reference 1. For some images, no source could be determined.

MR Image Expert and Dynalize

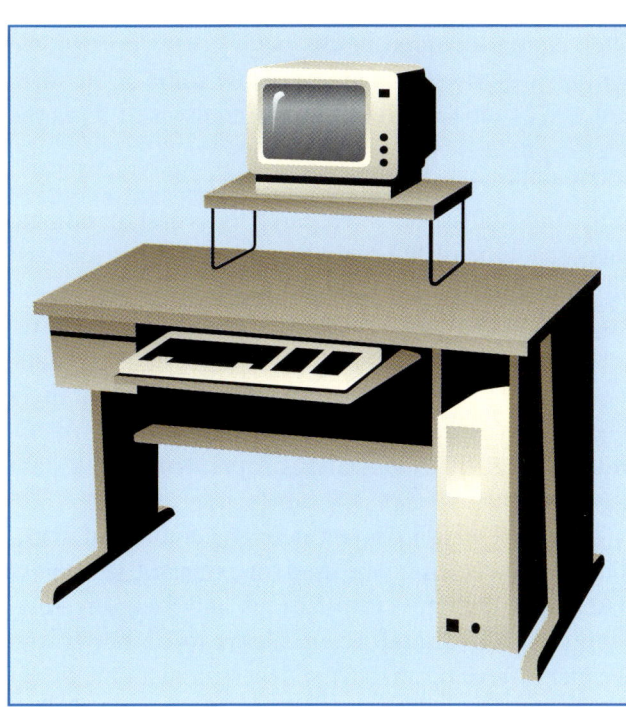

This book contains a CD-ROM with two software programs: *MR Image Expert®* and *Dynalize®*. An introduction to both programs is given here, and more details can be found in the accompanying text files.

MR Image Expert

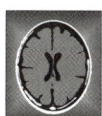

Magnetic resonance imaging is one of the intellectually most demanding and challenging medical technologies. Understanding the mechanisms which influence and change image contrast in MR imaging, in particular the relations between image contrast and pulse sequences and their parameters, is difficult and often requires much intuition and imagination.

Teaching tools, such as textbooks and slide presentations, are used to introduce the newcomer to the field, but the best teacher would be an MRI machine itself. However, learning by experience is time-consuming, expensive, and in cases involving human pathology, impossible.

Thus, the next best tool is an MR imaging simulator, a computer program which simulates an MRI machine without performing real examinations. Such a simulator should be as close to reality as possible.

MR Image Expert is a computer program which simulates the most important aspects of magnetic resonance imaging equipment. We have tried to make it as simple to use as possible.

By working through the exercises in the tutorials and playing with the simulator, you will get an understanding of how pulse-sequence parameters such as the repetition time *TR*, the echo time *TE*, the inversion time *TI*, and the flip angle *FA* influence the contrast in MR images. You will also become familiar with concepts like window settings and regions-of-interest (ROIs), signal averaging and its influence upon noise, as well as how contrast agents change the contrast of your image.

At the same time, the tutorials introduce the features of *MR Image Expert*. Going through the tutorials is the easiest and fastest way to get to know the software.

One thing you should never forget: simulations always remain simulations. There are too many variables in MR imaging which influence image contrast. This software has been developed as an educational tool, not as a diagnostic or research tool for pulse-sequence freaks.

The images displayed on the computer screen are of an extraordinary high quality, but they are not real MR images. The images are simulated by using a combination of different sophisticated techniques.

In many cases, the simulations are quite accurate. In other cases, there are minor differences between the simulated images and images created by your MR machine. For instance, it is very difficult, or even impossible, to simulate certain physiological parameters such as blood flow.

In addition, there is no standardization of MR imaging equipment; different manufacturers produce different machines applying similar techniques. Images made with the same parameters may look slightly different, depending on the machine and the magnetic field strength. This holds in particular for rapid imaging techniques, where manufacturers use different approaches without properly explaining how they acquire their images.

All image files you can use for simulations have originated from examinations of volunteers or patients on different MR machines. You will see that there are differences in image quality. Quality varies between different field strengths, although there is hardly any visible difference between 0.5 and 1.5 Tesla. Some of the raw data used stem from MR equipment which has been serviced and maintained up to a certain standard — which becomes visible if you compare these files with other cases.

Since various vendors of MR equipment use different terminologies and cannot agree upon a standard, we have introduced generic terminology. Synonyms of some identical or similar pulse sequences are given following each generic term, but because of the chaos in terminology, not all techniques were included. Actual image contrast can look different, depending on the specific pulse sequence implemented on your MR machine.

Pulse-sequence parameters, such as 'turbo factor', differ from company to company; for details and comparison of these parameters, contact the respective manufacturers' MR departments.

Dynalize

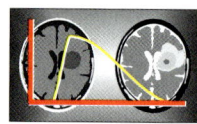

Dynalize is an independent PC-based software system for the analysis of dynamic contrast-enhanced imaging data. The version included in this book is a stripped-down demonstration version able to show some of the main features of the visualization and analysis of dynamic imaging data.

Dynamic imaging techniques are becoming increasingly important, both as research tools and for clinical purposes. One of the main requirements for successfully performing and assessing dynamic imaging examinations in clinical routine is fast, accurate, and reliable postprocessing and analysis of the acquired data. It is also of great importance to be able to compare results from different projects and different equipment. This requires that the analysis is performed in a standardized way.

Standard imaging equipment and connected workstations usually cannot be used for extended data analysis because patient examinations and image reading have priority. Therefore, it seems to be better and cheaper to develop an independent system for the evaluation and analysis of dynamic examinations which does not interfere with clinical routine. Such software should run on a personal computer.

Dynalize offers an integrated, standardized way of performing both image-processing and image analysis on all kinds of dynamic images, independent of the imaging equipment used to acquire the images. Data can be imported and processed through the ACR-NEMA-300/DICOM standard file formats.

A large number of existing modules provide a good starting-point for different kinds of dynamic analyses in radiology and cardiology. Region-of-interest (ROI) data and parametric values can easily be exported to, for instance, Microsoft Excel, Lotus 1-2-3, and other available software packages suitable for further analysis.

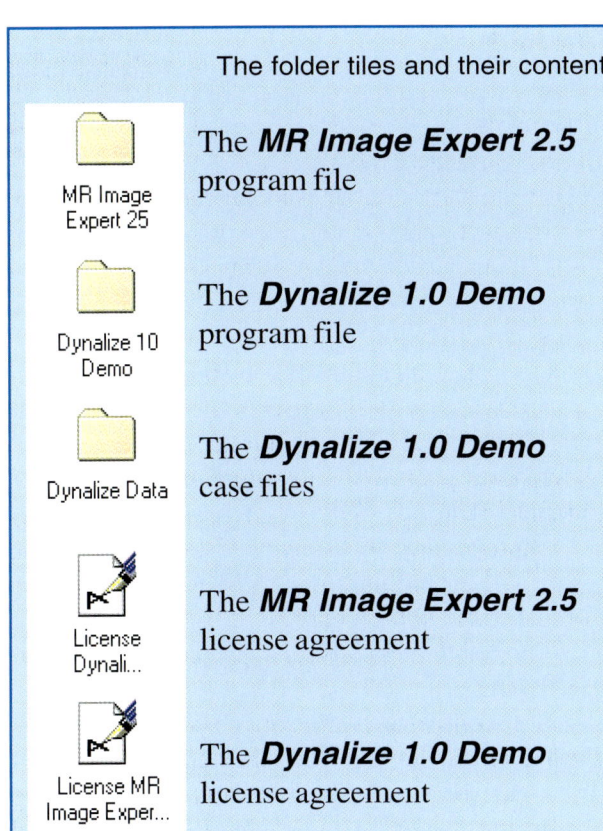

The folder tiles and their contents

The **MR Image Expert 2.5** program file

MR Image Expert 25

The **Dynalize 1.0 Demo** program file

Dynalize 10 Demo

The **Dynalize 1.0 Demo** case files

Dynalize Data

The **MR Image Expert 2.5** license agreement

License Dynali...

The **Dynalize 1.0 Demo** license agreement

License MR Image Exper...

MR Image Expert® *and Dynalize*®

Hardware and Software Requirements

- PC-compatible computer with Pentium CPU or higher;
- Graphics card with a minimum of 256 colors;
- Windows 95/NT/98;
- CD drive;
- At least 10 MB of empty space on the hard disk.

Installation

Dynalize 1.0 Demo
1. Insert the EMRF CD into the CD drive.
2. Open the **Dynalize 1.0 Demo** directory on the CD.
3. Start the program '*Setup.exe*' and follow the instructions.

MR Image Expert 2.5
1. Insert the EMRF CD into the CD drive.
2. Open the **MR Image Expert 2.5** directory on the CD.
3. Start the program '*Setup.exe*' and follow the instructions.

Notes

- Only the **Dynalize** software itself will be installed on your computer. All clinical cases will remain on the CD-ROM. To run **Dynalize** with clinical cases you always have to place the CD-ROM in the CD-ROM drive of your computer.
- **Dynalize 1.0 Demo** cannot follow shortcuts to directories; one has to access the directory directly.
- There is a problem with some versions of Windows 98 and **Dynalize 1.0 Demo**. When Dynalize is being closed, the program gives an error message. However, this should not affect execution of the program, since the error message is given after the program has ben shut down. This problem has only been observed with early versions of Windows 98.
- Versions for Macintosh computers are not available.

License Agreement

MR Image Expert ® • Version 2.5 · Release March 2001

TITLE. This licensed version of MR Image Expert® (= Software) is valid for a single user (= End-user). The License Agreement for MR Image Expert® is valid worldwide, including India and China. This license is not a sale of any of the rights of ownership of the original MR Image Expert® or any copy. In particular the EMRF Foundation (= Publisher) retains title and full rights of ownership of the original and any copy of the Software installed or copied or in use on all such physical media or on any other media such as but not restricted to diskettes or hard disks, CD-ROM or similar storage devices or operating memory.
The Software and the enclosed written materials are copyrighted and a trademark exists for the name 'MR Image Expert' and its logo.

For the avoidance of doubt the following are absolutely prohibited without the express prior written consent of EMRF Foundation:
• Unauthorized copying of the Software, including Software that has been modified, merged, or included with other software, or of the written materials.
• Reverse-engineering, disassembling, decompiling, or making any attempt to discover the source code of any of the Software.
• Sub-licensing, renting, leasing, or selling any portion of the Software or of the Software in its totality, with the exception of the use of the Software by the End-user. Any other transfer of the Software requires an additional written agreement between EMRF Foundation and the End-user.

The Software shall only be used for limited-access educational purposes. Use of the Software for public performance at conferences, seminars, workshops, etc. is not permitted. Special licenses can be acquired for the latter purposes.

COPYRIGHT. MR Image Expert®, Version 2.5 - Copyright © 2001 by EMRF Foundation, Minusio, Switzerland. All data and information on MR Image Expert® program are protected under all applicable copyright laws. No use of any data or information is permitted in any form or manner without the express written consent of EMRF Foundation.

LIMITED WARRANTY. EMRF Foundation warrants that the medium upon which the Software is provided by the Publisher to the End-user shall be free from defects in material and workmanship under normal use for a period of 90 days from the date of the End-user's receipt thereof.

DISCLAIMER AND LIMITATION OF LIABILITY. Medicine is an ever-changing science. Research and clinical experience are continually broadening our knowledge, in particular the knowledge of proper treatment and drug therapy. Insofar as this manual and the software program MR Image Expert® mention any dosage or application, or proposes certain imaging procedures, readers may rest assured that the editor, the coauthors, and the publisher have made every effort to ensure that such references are strictly in accordance with the state of knowledge at the time of production.
MR Image Expert® is an educational tool, not a diagnostic tool.
Some of the product names, patents, and registered designs referred to in this software and the accompanying manual are in fact registered trademarks or proprietary names even though specific reference to this fact is not always made. Therefore, the appearance of a name without designation as proprietary is not to be construed as a representation that it is in the public domain.
Except as expressly stated herein, the Software is provided "as is" without any warranty of any kind, express or implied, including, but not limited to warranties of performance or merchantability or fitness for a particular purpose. The user bears all risk relating to quality and performance of the Software. The performance of the Software varies with the various manufacturers' equipment with which it is used. Publisher does not warrant that the Software or the functions contained in the Software will meet the End-user's requirements, operate without interruption, or be error free. End-user's exclusive remedy for breach by EMRF Foundation of its limited warranty shall be replacement of any defective medium upon its return to Publisher within the warranty period, or, if Publisher is unable to provide a replacement that is free of defect, refund of the license fee paid by the End-user with respect to such medium.
In no event will Publisher be liable for any lost profits or other damages, including direct, indirect, incidental, special, consequential or any other type of damages, arising out of this Agreement or the use or inability to use the Software licensed hereunder, even if Publisher has been advised of the possibility of such damages.

License Agreement

Dynalize ® • Version 1.0 Demo · Release March 2001

TITLE. This licensed version of Dynalize® (= Software) is valid for a single user (= End-user). The License Agreement for Dynalize® is valid worldwide, including India and China. This license is not a sale of any of the rights of ownership of the original Dynalize® or any copy. In particular the EMRF Foundation (= Publisher) retains title and full rights of ownership of the original and any copy of the Software installed or copied or in use on all such physical media or on any other media such as but not restricted to diskettes or hard disks, CD-ROM or similar storage devices or operating memory.
The Software and the enclosed written materials are copyrighted and a trademark exists for the name 'Dynalize' and its logo.

For the avoidance of doubt the following are absolutely prohibited without the express prior written consent of EMRF Foundation:
• Unauthorized copying of the Software, including Software that has been modified, merged, or included with other software, or of the written materials.
• Reverse-engineering, disassembling, decompiling, or making any attempt to discover the source code of any of the Software.
• Sub-licensing, renting, leasing, or selling any portion of the Software or of the Software in its totality, with the exception of the use of the Software by the End-user. Any other transfer of the Software requires an additional written agreement between EMRF Foundation and the End-user.

The Software shall only be used for limited-access educational purposes. Use of the Software for public performance at conferences, seminars, workshops, etc. is not permitted. Special licenses can be acquired for the latter purposes.

COPYRIGHT. Dynalize®, Version 1.0 Demo - Copyright © 2001 by EMRF Foundation, Minusio, Switzerland. All data and information on Dynalize® program are protected under all applicable copyright laws. No use of any data or information is permitted in any form or manner without the express written consent of EMRF Foundation.

LIMITED WARRANTY. EMRF Foundation warrants that the medium upon which the Software is provided by the Publisher to the End-user shall be free from defects in material and workmanship under normal use for a period of 90 days from the date of the End-user's receipt thereof.

DISCLAIMER AND LIMITATION OF LIABILITY. Medicine is an ever-changing science. Research and clinical experience are continually broadening our knowledge, in particular the knowledge of proper treatment and drug therapy. Insofar as this manual and the software program Dynalize® mention any dosage or application, or proposes certain imaging procedures, readers may rest assured that the editor, the coauthors, and the publisher have made every effort to ensure that such references are strictly in accordance with the state of knowledge at the time of production.
Note that this version of Dynalize® is a research and educational tool, but not a diagnostic tool.
Some of the product names, patents, and registered designs referred to in this software and the accompanying manual are in fact registered trademarks or proprietary names even though specific reference to this fact is not always made. Therefore, the appearance of a name without designation as proprietary is not to be construed as a representation that it is in the public domain.
Except as expressly stated herein, the Software is provided "as is" without any warranty of any kind, express or implied, including, but not limited to warranties of performance or merchantability or fitness for a particular purpose. The user bears all risk relating to quality and performance of the Software. The performance of the Software varies with the various manufacturers' equipment with which it is used. Publisher does not warrant that the Software or the functions contained in the Software will meet the End-user's requirements, operate without interruption, or be error free. End-user's exclusive remedy for breach by EMRF Foundation of its limited warranty shall be replacement of any defective medium upon its return to Publisher within the warranty period, or, if Publisher is unable to provide a replacement that is free of defect, refund of the license fee paid by the End-user with respect to such medium.
In no event will Publisher be liable for any lost profits or other damages, including direct, indirect, incidental, special, consequential or any other type of damages, arising out of this Agreement or the use or inability to use the Software licensed hereunder, even if Publisher has been advised of the possibility of such damages.

Index